Heredity and Achievement

HEREDITY AND ACHIEVEMENT

A Book of Readings

Edited by
DANIEL N. ROBINSON
Amherst College

New York
Oxford University Press
London 1970 Toronto

Library of Congress Catalogue Number: 75-108737
Printed in the United States of America

To Margie and Robby, and to my wife, Francine.

Preface

The scientific literature, and the popular press as well, have shown a recent increase in concern over the issue of heredity and achievement. As in the past, the topic is often raised within the context of "racial" comparisons, and researchers in various social and political persuasions look for scientific information with which to defend their positions. Few areas of scientific inquiry enjoy as much public attention and even fewer are presented with as much passion and rhetoric. Thus, it is not surprising that the findings cited are often misinterpreted and that the subtleties implicit in research of this kind are ignored. The present anthology was assembled in the interest of assisting those who wish to gain a better understanding of the methods, results, and problems of behavior genetics.

The audience for whom this collection is intended is rather diverse; the material will augment courses ranging from beginning psychology and biology to education, anthropology, and sociology. Neither the editor's contributions nor the readings themselves constitute a textual basis for instruction in genetics or psychology. The introduction covers some of the more crucial concepts and frequently occurring terms in human heredity. Before each major section, some of the considerations of central importance to the material which follows are outlined. Each of the readings contains its own bibliography, and an assortment of suggested readings in each of the areas covered is included. Readers who wish to go deeper are referred to two very fine texts in the field: *Behavior Genetics,* by Fuller & Thompson (1960) and *Behavior Genetic Analysis,* by Hirsch (1967).

Since the present anthology was undertaken, two or three other collections have been announced which are advertised as being rather encyclopedic—a quality which the present collection does not enjoy. This relatively brief supplementary reader is designed to convey the history, nature, and applicability of behavior-genetic methods in the study of animal and human abilities.

I would like to thank the authors and the journals for their permission to reprint the works contained herein. In addition, I am most grateful for the cooperation and assistance provided by Mr. William Halpin of Oxford University Press. Finally, I especially acknowledge the tireless, devoted, and essential assistance of my wife, Francine.

D. N. R.

Amherst, Mass.
February 1970

Contents

Heredity and Achievement

Foundations

Over the centuries, philosophers have debated at length the question of "human nature." Notwithstanding the variety of philosophic and popular arguments, most of the controversy can be subsumed under two systems of thought; the *nativistic* and the *empiricistic*.

Nativistic philosophers subscribe to the view that the mental dispositions of man are implicit in his being man; i.e., they are *native*. Thus, it is argued in various moral philosophies that man is "basically" good or "basically" evil; that man is "fundamentally" rational or "fundamentally" animalistic. In the extreme, the nativist sees the environment as but a stage upon which all of man's inborn tendencies are acted. Thus, nativism is really an *instinct* psychology which accepts the environment as a condition for, rather than the cause of, human action and human achievement.

Empiricism is the philosophic foundation of that *behavioral* psychology which treats actions and achievements as the consequence of experience. In the extreme, the empiricist sees the organism as but a mirror which reflects the complicated outcomes of environmental influences.

Popular language is rife with expressions which take for granted the implications of nativism and empiricism. Clichés such as "it's in his blood," "you can't make a silk purse out of a sow's ear," and the "bad seed" reveal an outright acceptance of nativistic views of man. Similarly, "as the twig is bent, so grows the tree" and "sparing the rod spoils the child" are conceptions based upon the belief that experience is the principal force in the psychological formation of the individual.

It has been observed that radical nativism is a philosophy requiring no environment while radical empiricism is one requiring no organism![1] To be sure, such extreme positions are seldom confronted in philosophic literature. Rather, a variety of

[1] J. Fuller and W. Thompson, *Behavior Genetics* (New York, 1960).

eclectic approaches has been offered in which heredity and environment are accorded varying degrees of importance. While displaying commendable temperance, however, even the eclectic positions have contributed to the great confusion which suffuses many discussions of genetic and environmental factors. In its most naïve expression, eclecticism attempts to assign numbers—magnitudes—to the degree of genetic "versus" environmental control over a particular trait. It is not rare to hear the question "How much of personality is inherited and how much is learned?" or "What fraction of intelligence is determined genetically?" The naïveté of these questions indicates a failure to appreciate the nature of gene-environment interactions. It also suggests an improper understanding of gene action. While such misunderstandings are unfortunate in any field of science, they create special concern in genetic psychology because of the social implications of findings in this field. In an attempt to prepare the reader for the articles which comprise this book, it is necessary, therefore, to review those principles upon which modern behavior genetics is based.

Basic Principles

The first step in the creation of man is the union of male (sperm) and female (ovum) sex cells which are called *gametes.* This union or fertilization results in the formation of a *zygote* which possesses all of the genetic information present in the parental gametes. That is, the parent sex cells combine to create the *genome* of the offspring.

Within each gamete is a dense region called the *nucleus.* It contains the many substances necessary for the life and function of the cell. In addition, it houses thread-like microstructures, the *chromosomes.* Each gamete contains twenty-three chromosomes such that, with fertilization, the zygote comes to possess forty-six; twenty-three pairs of chromosomes, half from each parent sex cell. The zygote divides repeatedly by the process of *mitosis* and, with each successive division, the chromosomal strands "split" to maintain a constant number of chromosomes in the cells being formed. All of the cells in the body of the adult—with the exception of the gametes—will contain forty-six chromosomes. In the production of gametes, a different type of division occurs. It is called *meiosis* or *reduction division* because the gametes receive only half of the total gene pool. That is, of the forty-six parental chromosomes (twenty-three pairs), the gametes obtain only twenty-three. Any combination of twenty-three from the total pool of forty-six may be passed to the gamete. Thus, the number of possible combinations is large and it is not very probable that *only* initially paternal or maternal chromosomes will enter the gamete. More likely, some mixture of both will be present. If reduction division did not occur, each successive generation of gametes would contain twice the number of chromosomes present in the preceding generation. Instead, the number is kept constant by the meiotic "halving" of the number. The "half number" is called the *haploid* number, while the full complement of forty-six is called the *diploid* number.

Whether the zygote develops into a male or female depends upon the presence or absence of a particular chromosome. The male gamete possesses two chromosomes for sex—an X and a Y—and the female also carries two which are, however, both X's. If, in fertilization, the male's X unites with either of the female's X-X pair, the zygote will develop into a female (X-X). If the male's Y unites with the female's X (either one), the resulting X-Y pair will lead to a male. By "sex" of the offspring is meant the chromosomal makeup of the offspring's gametes, not the relative "masculinity" or "femininity" of the adult personality.

Each chromosome contains small entities known as *genes* which lie along the chromosomes as a "string of beads." The genic material is a complicated blend of organic molecules, the most important of which is deoxyribose nucleic acid—DNA.

The number of genes in human gametes is presently undetermined but is, no doubt, very large; perhaps 100,000. Normal development requires that all of the necessary genes are present. The full complement of genes is called the genome and the genetic composition of the individual is referred to as the *genotype.* It has been shown that certain adult characteristics depend upon certain genes for their expression. Moreover, the particular effects associated with genetic action derive from the position of the gene along the chromosome, i.e., the *gene locus*. Those characteristics of the individual which are strongly dependent upon gene action may result from the combined influences of a great many different genes. Some traits, however, appear to depend upon the presence or absence of but a single gene and these are called *Mendelian* traits or characters. An example of this is eye color.

Assume that a father has brown eyes and that both of his parents contained only genes for brown eyes. His wife is blue-eyed. It is known that the gene influencing the appearance of blue eyes is *recessive,* i.e., in the presence of the (dominant) gene for brown eyes, the gene for blue eyes is not expressed in the individual's eye color. Now, the father has received a "brown-eyes" gene (Br) from each of his parents so that his gene-pair for eye color is Br-Br. His wife, since she is blue-eyed and since this is a recessive gene, must contain the gene-pair Bl-Bl (a "blue" from each of her parents). The offspring of this marriage will possess the gene-pair Br-Bl and will all be brown-eyed with a recessive gene for blue eyes. Suppose that one of these offspring marries another brown-eyed person who also has a recessive "blue" gene. Their children may inherit any one of the following combinations:

Br-Br or *Bl-Br* or *Br-Bl* or *Bl-Bl*

Thus, *on the average,* one fourth of the children will be "pure" brown-eyed individuals. Half will carry a recessive gene for blue eyes but will be brown-eyed and the final fourth will be blue-eyed. It should be noted that 75 per cent of the children will be brown-eyed but only 25 per cent carry Br-Br gene pairs. Thus, the physical appearance of the eyes (the *phenotype*) does not always allow one to determine the underlying genetic composition (the *genotype*).

Just as some (Mendelian) traits are influenced by a single gene, there are certain genes which influence multiple traits. Such genes are referred to as *pleiotropic.* For example, a gene which affects the formation of a given enzyme can indirectly influence a whole host of physiological events which depend upon the presence of that enzyme. The mutation of a single gene can result in the failure to form an enzyme which controls the level of a certain chemical in the bloodstream. A consequence of increased concentrations of this chemical may be improper development of neural tissue. Such is the case in phenylketonuria, a condition resulting from a single gene mutation and producing mental deficiency.

How the genes are ultimately expressed in the varied aspects of the adult organism is only now being unraveled. The mechanisms of gene action are complex and even the highlights of present knowledge may confuse the beginner. Some very general features will be sufficient for the present purposes.

All of the myriad features of life are the result of cellular activities, activities which are controlled by the nuclear material of the cells. It appears to be the function of DNA to issue various "instructions" to cells, in response to which certain proteins are made, enzymes manufactured, growth and death processes initiated, etc.

The present model of DNA assumes a large, spirally coiled, double-stranded molecule with the strands linked by four bases; two *purines* (adenine and guanine) and two *pyrimidines* (thymine and cytosine). Alternating along the lengths of the two strands are molecules of sugars (deoxyribose) and phosphates. The arrangement of these molecules is invariant; only the sequences of purines and pyrimidines are free to alter. Thus, it is in the arrangements of these bases that geneticists look for the "language" of cellular processes. Different configurations of the purines and pyrimidines can lead to the production of different proteins, hormones, and enzymes, and, therefore, can determine the physiological events which are controlled by these substances.

The Norm of Reaction

Each genotype establishes a "norm" or "reaction range" for the physiological processes under its direction. Many modern geneticists accept a "one gene-one enzyme" theory of gene action which specifies the physiological norms of reaction in terms of the total number of enzymes a genotype will allow.

The norm of reaction may be viewed as the adaptive potential of the system. It embraces the total range of physiological responses available to the organism in any conceivable environment. The genotype of an individual may, for example, allow his cells to metabolize nutrients at any rate from n to n′. Given this genotype, the individual may be able to survive at any environmental temperature from —40° to +120° Fahrenheit. At colder or hotter temperatures his system (cells) will be unable to adapt. At such temperatures, the genotype's norm of reaction has been exceeded and survival is impossible.

Each genotype has a unique norm of reaction and, therefore, each individual is matched to some theoretically "ideal" environment whose properties are maximally compatible with that genotype. Obviously, the "ideal" is very seldom, if ever, confronted, and so individuals must accommodate a variety of environmental conditions, many of which are antagonistic to their genotypic biases. Under widely ranging environmental circumstances, physiological mechanisms are activated to maintain a state of equilibrium. That is, numerous physiological processes are devoted to compensating for the difference between the *real* environment and that environment which is "ideal" for a given genotype. These compensatory processes are referred to as *homeostatic,* i.e., equilibrium-preserving. Conditions may emerge which exceed the capabilities of the homeostatic mechanisms. When these conditions are widespread and of long duration, those species whose genotypes do not allow adaptation become extinct.

External conditions of life which are antagonistic to those life processes under genetic control have caused, over the past millions of years, the extinction of numerous plant and animal forms. The survival of any species requires of it the ability to produce genotypes adaptable to the pressures of the environment into which they are born. Changes in the environment must be accommodated either by the norm of reaction of a given genotype or by the emergence of a new genotype; a genotype which is different from all of those which could have resulted from the mating of any two different-sexed members of the species. Such new genotypes—those not contained in the combinatorial possibilities of the gene pool of the species

—are called *mutations*. They may result from the alteration of one gene or many genes.

From the foregoing, it can be seen that two processes alter the frequency with which particular genes occur in nature. First, there is what Darwin termed *Natural Selection:* the process by which certain environments favor the perpetuation of certain genotypes and the ultimate extinction of others. This process is evolutional. Then, there is the creational process of mutation whose genotypic product differs from all of the combinatorial possibilities of the existing gene pool. Together, selection and mutation function to allow the existence of life under changing conditions of the environment. Even in the absence of selection pressures and mutations, however, a genetic equilibrium would not necessarily be reached. That is, even if a species emerged within an unchanging environment "ideal" for its genotype and with no occurrence of mutants, other factors could result in biasing the frequency of certain genotypes. For example, the population could adopt inbreeding practices in preference to breeding among nonrelatives. Excessive inbreeding would produce genotypic homogeneity within the inbred lines. Such practices increase the likelihood of deleterious *recessive* genes emerging in the genotypes of the offspring. Should the recessives be lethal, inbreeding could lead to the elimination of that genotype, i.e., extinction.

Inbreeding is a specific instance of a general process referred to as *assortative mating*. Defined most rigorously, assortative mating is any mating pattern which is not random. Random mating allows the union of any two different-sexed genotypes in the population with the same probability as the union of any other two. That is, it is just as likely that incest will occur as it is that it will not. It is just as likely that rich and poor, black and white, old and young, will mate as it is that they will not. Under conditions of random mating and in the absence of selection and mutation, a genetic equilibrium would be reached. This equilibrium would prevent a disproportionate increase in the frequency of any given gene, i.e., the distribution of gene frequencies among members of the species would be constant. Because of selection, mutation, inbreeding, and assortative mating, such equilibria are theoretical possibilities but do not occur in nature. Rather—and in the case of man, this is especially so—breeding practices tend to produce dramatic departures from genetic equilibrium. Certain of the practices derive naturally from obstacles to random mating, e.g., geographic factors were insurmountable obstacles, until recently, to random mating. During the many centuries in which small tribes were geographically isolated from other human gene pools, mating was intratribal. This form of assortative mating resulted in the emergence of certain racial groups, i.e., populations whose gene frequencies differed from those in the gene pool of the larger population (species) from which they were drawn. Over many successive generations, greater and greater differences in gene frequencies will result if the sub-species (race, tribe, colony, community, etc.) remains isolated from the general breeding pool. That is, the local gene frequencies will drift further and further from the frequencies with which certain genes are appearing in the species as a whole. This process is referred to as *genetic drift* and is partially responsible for differences observed among the different "races" of man. One can see the consequences of genetic drift in a comparison of the frequency of certain blood types occurring in different sub-species, and this example is offered.

While there is still debate on the subject, current genetic theory favors the view that blood groups M, N, and MN result from three different forms (*alleles*) of a single gene at a fixed chromosomal locus. Under conditions of random mating, it would be expected that 25 per cent of the population would contain type M, 25 per cent type N, and 50 per cent, the mixed type, MN (there is no dominant allele for these blood types so that a paternal M and maternal N will lead to an MN zygote). As it happens, these expected percentages are very close to what is actually found in a representative sample of American whites; 29.16 per cent, 21.26 per cent and 49.58 per cent for M, N, and MN respectively. However, among the geographically isolated Eskimos of East Greenland, the observed frequencies are 83.48 per cent, 15.64 per cent and 0.88 per cent, while Australian Aborigines reveal frequencies of 3 per cent, 29.6 per cent and 67.4 per cent (Wiener, 1943). Thus, with isolation from the breeding pool, different subspecies can reflect marked genetic drift, i.e., marked deviations in gene frequencies relative to those found in the larger (species) breeding pool.

In light of the foregoing and for practical purposes, the concept of *race* is somewhat arbitrary. Anthropologists, for example, have varied in their estimates of the number of races of man from the classical three (Negroid, Caucasoid, and Mongoloid) to more than thirty. The anthropologist, of course, makes his distinctions primarily on the basis of phenotypes, e.g., skin color, bone structure, head size. Genetically, there are sub-species with broadly differing gene frequencies. Thus, Eskimos and Aborigines are genetically different in regard to the relative frequencies of genes associated with blood groups M, N, and MN. Notwithstanding this genotypic property of the two populations, however, there are *individual* Eskimos with blood groups found more among Aborigines than among Eskimos. Given a large assortment of traits, certain individuals will share many more of these than will others. One may select arbitrarily a particular constellation of traits deriving from a given distribution of gene frequencies and treat those individuals falling within that distribution as members of a race. Except in the trivial instance of monozygotes (identical twins), no two individuals ever have the same genotype. They may be assigned to the same racial category if their genotypes are drawn from the same pool. In the absence of prolonged assortative mating, genetic drift, and mutations, all individuals will obtain their genotypes from a common pool. The presence of racial distinctiveness indicates that these processes have attended the evolution of man. Despite their effects, however, there is considerable overlap among races, so that the assignment of a given individual to a distinct race may be highly arbitrary. Thus, while it is reasonable to look for differences between large populations of genetically differentiable races, it is naïve to search for "racial differences" between two individuals. The latter will always be genotypically distinct and any verification of this by some test or measurement is gratuitous. Because of this guaranteed difference in genotype—because each individual is biologically unique—no two individuals will interact identically with identical environments, even if the latter could be established. In fact, since the extent and nature of an individual's commerce with the environment are an expression of genotypic influences, there can be no environmental identity for two genetically dissimilar organisms. This fact assumes great importance in research concerned with "racial" differences in "mental" performance or "capacity," and the point must be expanded.

The races of man have evolved over a long period of time—perhaps 1.5 million years—and under widely varying environmental conditions. Through selection pressures, mutational alterations, and breeding practices, sub-species emerged which were (are) especially well suited to exploit the bounty and survive the challenges unique to their habitat. Success in such endeavors required great physical, temperamental, and intellectual resources. Emotionality had to be contained by the rational processes. The capacity for abstraction and the ability to predict the long-range consequences of action had to be developed to extents never before realized in phylogeny. Although selection pressures no doubt were different for the different emerging races, many of the fundamental demands of the environment were quite similar, e.g., the necessities of catching and killing food, securing safe quarters, fashioning tools for protection and convenience, evolving standards of behavior in the interest of the community. All extant races reveal highly sophisticated forms of these adaptive patterns. Earlier varieties of "man," less successful at hunting, invention, and socialization, did not survive. There is a theoretical formulation known as *Gause's Principle* which describes the basis of survival under the conditions confronted by early races. Gause's Principle asserts that, if two different genotypes are geographically coexistent and have identical ecological (environmental) needs, one of the genotypes will be eliminated; i.e., since the genotypic difference is bound to endow one group with greater adaptive efficiency—no matter how slight the edge—then the more efficient will survive and the less efficient will not. The present races of man were forged by the workings of this principle and represent, therefore, the survivors of extraordinary and chronic challenges. It would be highly unlikely that very significant differences would exist among races in regard to those characteristics which are vital to survival—for example, "intelligence." By the same token, since selection pressures did differ for different races, it is very likely that the (phenotypic) expressions of intellectual ability would differ and that cultural patterns of thought would be adopted by successive generations. In these instances, individuals lacking the ability to embrace the requirements of the culture were at a disadvantage. Over sufficient time and with the assistance of assortative mating within the culture, it is very possible that expressions of intellectual competence would become more or less homogeneous and compatible with cultural (subcultural) standards. The translocation of these racial genotypes to cultures calling for very different forms of intellectual expression could place the racial minority at something of a disadvantage. However, the disadvantage would not likely be of insuperable proportion. That is, the relocated race would contain genotypes whose norm of reaction surely *allowed* adaptation to the new requirements, even if it *preferred* some slightly different form of expression. It is not tenable to assume interracial "opposites" in intelligent behavior; rather, interracial preferences (*perhaps*). Even these may not be present to any measurable degree. As Fuller and Thompson have stated,

> The most diverse human cultures have common features related to the perpetuation of the species. It is difficult to conceive of a society in which intelligence, cooperation, and physical vigor would not have positive selective value. Hence it is likely that natural selection tends to oppose the establishment of major heritable behavior differences between races.[2]

Organization of the Readings

The readings which follow are in four major sections. Section I includes studies concerned with learning among *inbred* strains of animals. Of all the tools in behavior genetics, inbreeding allows the most accurate control and specification of genetic composition. Moreover, animal research permits care-

[2] Fuller and Thompson, *Behavior Genetics*, p. 324.

ful regulation of environmental conditions. Readings in the area of *learning* were selected because so much of what is considered to be the *psychological* features of organisms is rooted in the learning process. In this sense, the demonstration of strong genetic influence on learning would at once underscore equivalent genetic influence on the multitude of competences and characteristics which are products of the learning process.

Section II provides research and commentary representative of that literature devoted to genetic factors in personality formation. In this section, the reader is introduced to two other basic tools in behavior genetics research—the *familial concordance* method and the *twin* method. Each of these investigates similarities among individuals of different genetic relatedness. A trait is considered highly *heritable* when it is found in similar degrees among relatives, in (nearly) identical degrees among identical twins and in unrelated degrees among unrelated individuals—*all other conditions being equal.*

Section III offers measures of familial concordance, twin correlations, and interracial comparisons along a number of psychological dimensions, especially "intelligence."

In the last section, papers are presented which point to the need for behavior-genetic research, for an appreciation of individual differences, and for an understanding of the constant and vigorous interaction between genetic and environmental pressures.

Each section is preceded by a statement of the problem and by some of the methodological and conceptual difficulties which attend study in the area.

Suggested Readings

Dobzhansky, T. *Genetics and the Origin of Species.* Columbia University Press, New York (1959).

Oparin, A. I. *Genesis and Evolutionary Development of Life.* Academic Press, New York (1968).

Stern, C., and E. R. Sherwood (eds.). *The Origin of Genetics: A Mendel Source Book.* W. H. Freeman & Co., San Francisco (1966).

Watson, J. D. *The Double Helix.* Atheneum, New York (1968).

SECTION I

Heredity and Learning

No topic has received as much of the attention of experimental psychologists as the learning process. In part because of the impact of Darwin upon psychology and in part because of its obvious significance, learning has been viewed as man's chief mechanism for adaptation.

While everyone has some intuitive sense of what learning is, it has been difficult for scientists to arrive at a universally accepted definition. Since learning is measured by tests of retention, it isn't clear, for example, whether learning and memory are the same or different processes. Furthermore, learning is assessed in terms of behavior, so that it is difficult to distinguish learning from performance.

Representative of the rather general definitions of learning is: *the progressive modification of behavior as a consequence of experience.* Such definitions attempt to exclude instinct and maturation. Instincts, for example, are behavior patterns which emerge in the absence of any deliberate training (experience). Some behavior, such as the swimming behavior of fish, occurs after the maturation (growth) of necessary structures and, seemingly, without practice of any kind. The term *progressive* implies some kind of gradual, step-by-step alteration, and yet there are varieties of behavioral change which occur abruptly. Many types of problem-solving behavior are characterized by very sudden solutions. Apes, for example, will sit "puzzled" for some time, being unable to reach food placed outside their cages. Given two sticks, the apes will attempt abortively to reach the food with one of them. At some point the apes suddenly will *join* the two sticks, thereby creating one which is long enough to do the job. This form of behavior has been called *insight* and it would appear to be an exception to the view of learning as a progressive change in behavior.

A more specific and pragmatically satisfying definition of learning is that which considers it *a change in the probability of a response as a result of reward or punishment*. Such a definition is, of course, *ex post facto,* since the probability of a response is determined by its *observed* frequency of occurrence, i.e., performance. While reward and punishment have profound effects upon performance, it is not clear whether their application is necessary for learning. For example, rats may run loose in a maze whose terminal compartment contains no food reward. The rats may be fully fed before this exposure to the maze. Another group of hungry rats receive a small amount of food each time they reach the end of the maze. On successive days, it is observed that the rewarded rats improve progressively, i.e., they commit fewer and fewer errors day by day. The nonrewarded rats show no such improvement. However, if, after a week or so of such conditions, food is now introduced into the end-compartment, the previously unrewarded rats display a sudden improvement in peformance after but a single rewarded trial.[1] The implication of this finding is that, while performance did not improve during nonreward, learning was taking place. This phenomenon, known as *latent learning,* challenges the essentiality of reinforcement (reward and punishment) to learning. Even theorists who consider reinforcement indispensable may not agree on definitions of "reinforcement." In this respect, a certain circularity has been present in which learning is defined in terms of reinforcement effects and reinforcement in terms of effects upon learning.

[1] H. C. Blodgett. The effect of introduction of reward upon the maze performance of rats. *Univ. Cal. Pub. in Psychol.,* 1929, *4,* 113-34.

What all definitions have in common is the view that learning either is or leads to a behavioral change which is reliably related to a change in the environment (including the organism's *internal* environment). Even here there are logical problems. For example, surgical removal of one lung will reliably change the respiratory behavior of an organism. While this change in the "environment" produces a change in behavior, one would be reluctant to subsume the process under the rubric *learning*. To accommodate such difficulties, the definition should include the possibility for *systematic* alteration and *reversibility*. A working definition, then, is: *learning is a systematic and reversible alteration in behavior under conditions of environmental change.* One must eschew the temptation to attribute *adaptiveness* to the process, since organisms can and do learn maladaptive patterns of behavior.

It should be clear from the foregoing that any research concerned with genetic factors in learning will be limited by the complexities associated with the definition of learning. The earliest studies in the area were insufficiently sensitive to these complexities and, therefore, the interpretations of findings were successfully challenged by subsequent workers. The first paper in this section was among the earliest investigations of learning ability among different inbred strains of rats. E. C. Tolman was a distinguished psychologist whose lifelong contributions to the psychology of learning now occupy a prominent place in the history of that field. At the beginning of the twentieth century, the wide distribution of Mendel's genetic findings and theories combined with the Darwinian flavor of the "new" biology to excite interest in genetic control of animal behavior; a kind of *animal eugenics*. Papers along these lines appeared a decade before Tolman's, but his was the first to incorporate well-specified learning tasks and well-controlled environmental conditions. In crossing "maze bright" rats with one another and "maze dull" rats with one another, Tolman reports "very decided differentiation to appear in the F_1 generation,"[2] but he fails to find progressive improvements with continued inbreeding. Instead, "the differences tended to decrease." Another and even more influential pioneer in the field,

[2] Successive generations of common ancestry are referred to as F_1 (first filial), F_2 (second filial), etc.

Robert Tryon, began similar studies in 1925 and obtained even more mystifying results. Reliably, his F_1 and F_2 strains proved to be as variable in their performance as were their parents and their learning measures often overlapped those obtained by their parents. A naïve Mendelian prediction would run counter to such findings, i.e., successive generations (F_1, F_2, . . . F_n) should become "pure" (homozygous) for a given "trait" (e.g., maze learning), and later generations should, therefore, reflect progressively more similar (less variable) scores than their parents. The F_2, however, should be *more* variable than F_1 because of segregation. Mendel's law of segregation states that the offspring of a hybrid cross will reflect all of the combinatorial possibilities of the parent genotype. Thus, if parent-genotype Aa is crossed with Aa, the offspring will contain genotypes AA, Aa, aA and aa. Since 75 per cent of F_1 will reveal the dominant (A) character in their phenotype and only 25 per cent will reveal the pure recessive (aa), the F_1 will be *phenotypically* rather uniform, on the average. However, inbreeding F_1's on the basis of phenotype will result in the following crosses: AA $\times$ Aa; AA $\times$ aA; AA $\times$ AA; and Aa $\times$ aA. The F_2's will display a *recombination* of those genes segregated in F_1 and will, therefore, display *greater* variability. Neither Tolman nor Tryon obtained this result. Their "failure" was the result of failing to acknowledge the limitations of naïve Mendelian calculations in dealing with complex genetic systems. The rat contains twenty-one chromosome pairs and the number of possible gametes emerging in the F_1 of such an animal is 2^{21}, i.e., the number of combinations of maternal and paternal chromosome pairings. Thus, 2,097,152 gametes are possible in F_1. The F_2 genotypes possible from the crossing of any of these gametes is staggering. As Hirsch points out in the matter of the "Tryon effect," ". . . these experiments never had the slightest chance of making the measurements for which they were intended."[3] Many, many generations of inbred strains would be necessary to achieve the Mendelian predicted values of variability.

In addition to the foregoing, Tolman's study was limited

[3] J. Hirsch (ed.) *Behavior Genetic Analysis* (New York, 1967), p. 419.

by its tacit acceptance of "maze learning" as some singular "trait." It was soon observed, however, that "maze bright" rats were not necessarily "bright" at much else. It was not at all clear what had been inbred but it was clear that "general" intelligence had not. Learning depends upon many specific factors (sensitivity, attention, motivation, etc.), all of which in one way or another may be influenced genetically. Subsequent research has attempted to specify more unambiguously which behavioral variables are being manipulated by selective breeding. The papers by Brody, McClearn, Collins, and Bovet *et al.*, are representative of the best efforts along these lines.

The last paper in Section I, that written by Verplanck, draws attention once again to the nature-nurture pseudoquestion. This paper stands as a warning and a lesson to those interested in relations between heredity and learning.

Suggested Readings

Fuller, J., and W. Thompson. *Behavior Genetics.* John Wiley & Sons, Inc., New York (1960).

Hirsch, J. (ed.). *Behavior Genetic Analysis.* McGraw-Hill, Inc., New York (1967).

The Inheritance of Maze-learning Ability in Rats[1]

EDWARD CHACE TOLMAN

The problem of this investigation might appear to be a matter of concern primarily for the geneticist. None the less, it is also one of very great interest to the psychologist. For could we, as geneticists, discover the complete genetic mechanism of a character such as maze-learning ability—i.e., how many genes it involves, how these segregate, what their linkages are, etc.—we would necessarily, at the same time, be discovering what psychologically, or behavioristically, maze-learning ability may be said to be made up of, what component abilities it contains, whether these vary independently of one another, what their relations are to other measurable abilities, as, say, sensory discrimination, nervousness, etc. The answers to the genetic problem require the answers to the psychological, while at the same time the answers to the former point the way to those to the latter.

But as far as the present investigation is concerned it must be admitted that only the most elementary answers of either

[1] The present paper is in the nature of a first report upon a relatively extensive program which is being worked on in this laboratory. The original impetus which started the problem came from Prof. Warner Brown; and most of the actual labor has been and is being performed by others beside the writer. In particular, it was only through the very exceptional industry and persevering labor of Mr. Frederick J. Adams that most of the 300 odd maze records here reported were obtained. To Prof. Raymond Franzen, the writer owes great indebtedness for suggestions as to the statistical treatment of the results. And the actual computations involved were in large part performed and supervised by Miss Barbara Burks. The writer wishes also to acknowledge the receipt of 3 grants from the Research Board of this University amounting in all to $540.

From *Journal of Comparative Psychology*, 1924, *4*, 1-18. Reprinted by permission of the Williams and Wilkins Co.

sort have as yet been obtained. The preliminary problems of technique and method of attack have proved all important.

The problem, as originally formulated, was merely the very simple one of trying by selective breeding to establish two independent strains of rats—one "bright"[2] and the other "dull."[2] Eighty-two white rats (41 males and 41 females) plus a certain number who for one reason or another had to be discarded, obtained from the stock of the anatomy department of this university, constituted the initial or P generation. This initial group had wild blood in them due to one wild grey male that had been mated into the anatomy stock some six years previous to the beginning of this experiment. And among the near relatives of the animals chosen for this experiment were many grey, black, black-hooded and grey-hooded individuals. The animals chosen for this experiment, however, were all of them white. The age of these initial 82 animals, although not known exactly, averaged about ninety days. The animals were received in small lots of about 20 each and were tested as received in a maze made of 7½ inch unpainted redwood, covered with chicken wire, of the shape and dimensions shown in figure 1.

In all the tests, both of these initial 82 rats and of their descendants, the general method of procedure was the same and as follows:

Feeding. The animals were given at least five days of preliminary feeding in the food box. In some cases, however, this preliminary feeding lasted longer. The food used was bread and milk with sometimes sunflower seed mixed in. The duration of the feeding was approximately twelve minutes and took place in the food box.

The Two Mazes. When any given batch was being run, the males were run in one maze and the females in another; the two mazes were of the same shape and construction and were placed similarly as regards the light, which was artificial. In successive batches the males and females were interchanged as regards the two mazes.

Experimenter. In all the cases reported Mr. Frederick J. Adams acted as experimenter.

[2] In the sense of ability to learn a maze.

Scoring. The animals were in all cases run 1 trial a day for ten consecutive days and obtained on each day a score for number of errors and time in seconds. They also obtained a score for the number of perfect runs achieved in the total 10 trials. An error was counted as an entrance of a full body's length or more into a blind *while going forward,* or a retracing, of more than one section of the true path—irrespective of how much more of the true path or how many blinds it might involve. The time was measured in seconds from the closing of the door behind the rat, after he had left the starting box, to the closing of the food box behind him, after he had entered the food box. These doors were manipulated by strings.

General Plan of the Investigation. Of the original (P) 82 rats, 9 supposedly bright males and 9 supposedly bright females were chosen to start the bright families; and 9 sup-

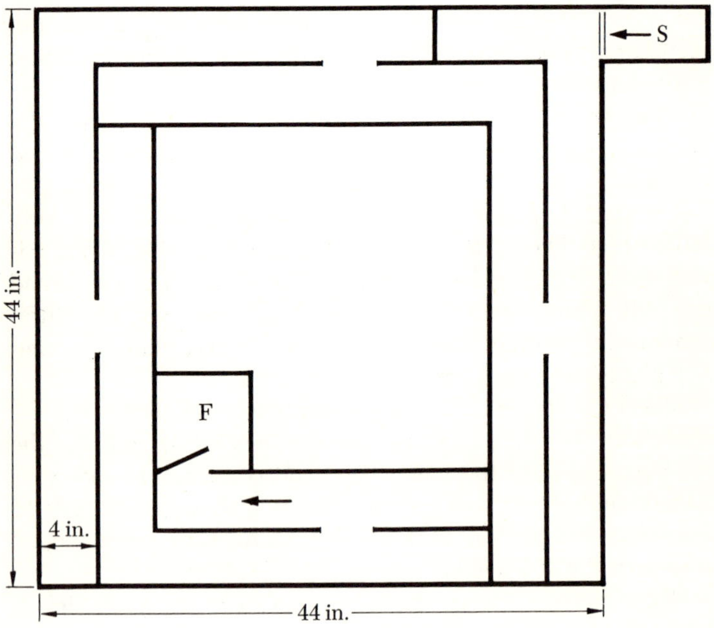

Fig. 1

posedly dull males and 9 supposedly dull females to start the dull families. The original 82 rats were run in successsive batches of about 20 each and the two brightest males and the two brightest females and the two dullest males and the two dullest females were chosen in each batch. Some batches proved as a whole, however, brighter or duller than others, so that this method of having to select within a batch did not result in always picking the absolutely speaking brightest and dullest out of the total 82. The selection was based upon a rough pooling of the results as to errors, time, and number of perfect runs. In presenting our results, however, the data as to errors, time, and perfect runs will be given separately.

The family relationships among these first 82 rats were not known, so that in selecting bright males to be mated with bright females and dull males to be mated with dull females we may have been, in some cases, mating litter mates with each other. The F_1 generation was constituted by the litters from these original bright and dull matings. The F_2 generation was obtained by further selective breeding within the bright litters and dull litters of the F_1 generation, and in each case a brother was mated with a sister.

Statistical Treatment of Results from Original Group

In working up our data, we have considered our original 82 rats as an unselected group which could be used as a standard in determining the best statistical treatment for the remaining data. As Hunter and his students (4, 8, 9, 12) have shown, there is considerable question as to the possibility of measuring reliable individual differences in rats at all. And for that reason a discovery of the best statistical method of treating results was important.

Our first point of attack was to see what indications of reliability could be shown by correlating the error scores for different runs against one another. Table 1 shows the intercorrelations (all the coefficients presented in this report are Pearson Product-Moments coefficients) for errors between successive trials 1 to 10 for all 82 rats.

The result is certainly discouraging. Apparently no two trials correlate with any degree of reliability. This is similar to the results reported by Hunter for time (9), only, if anything, even worse. In general, however, it may perhaps be said that the correlations near the diagonal, that is, between closely adjacent runs, tend to be higher than those at a distance from the diagonal, i.e., than those between far-distant runs. Hunter found a similar result, and his figures indicated it with even more certainty.

TABLE 1

Errors

n = 82

	2	3	4	5	6	7	8	9	10
1	0.145	–0.065	0.033	0.033	0.024	0.057	0.167	0.094	0.025
2		0.068	0.154	0.311	0.194	–0.002	0.019	–0.020	–0.122
3			0.116	0.109	–0.017	–0.129	–0.111	–0.099	–0.013
4				0.358	0.097	0.111	0.227	0.056	–0.068
5					0.123	0.209	0.172	–0.006	0.172
6						0.181	0.098	0.088	0.079
7							0.264	0.132	0.326
8								0.020	–0.020
9									0.305

Our next attempt was to correlate the *combined* error scores for a group of early runs against a group of late runs. Since the correlations between individual runs, and particularly those between early and late runs, was very low, we could hardly expect those between *combined* scores for the first part and the last part of the learning to be any higher. And such, indeed, proved to be the case (table 2).

Although no correlations (or even negative correlations) were thus obtained between the first and second halves of learning, it occurred to us that, if the results found by Hunter and somewhat suggested by our data in table 1 are correct—if, that is, the adjacent parts of the learning are nevertheless somewhat correlated—that better results might be obtained by combining alternate runs and then correlating odd vs. even. We therefore tried this out with the results shown in table 3. In certain cases, to save time, we obtained correlations for the 41 males, rather than for all 82, males plus females.

These correlations while unsatisfactorily low are at least better than those of the preceding tables. Furthermore, it will be noted that the highest correlation of all, 0.379, was obtained when runs 3 to 10, only, were included. In other words, it would appear that runs 1 and 2 were, as might be expected, very much subject to chance so that when included they tend to bring down the correlations. *We have, therefore, from this point on computed all our results from the data of runs 3 to 10 only.*

TABLE 2

Errors

n = 82

Runs	Coefficients of Correlation
2, 3, 4, vs. 5, 6, 7	−0.108
2, 3, 4, 5, vs. 6, 7, 8, 9	0.002
2, 3, 4, vs. 5, 6, 7, 8, 9	−0.229

TABLE 3

Runs (odd vs. even)	n = 41 (Males) r	n = 82 r
1, 3, 5, 7, 9, vs. 2, 4, 6, 8, 10		0.276
2, 4, 6, 8, vs. 3, 5, 7, 9	0.344	0.296
3, 5, 7, 9, vs. 4, 6, 8, 10	0.179	0.379
3, 5, 7, vs. 4, 6, 8		0.324

Further consideration of the whole situation suggested several possible treatments which might raise this correlation. Three different schemes were tried out. And, although we did not adopt any one of them in the actual working up of the further data of this report, we feel that they are of sufficient theoretical and methodological interest to be presented and to be kept in mind in the planning of future work.

The first one of these methods consisted in weighting the scores of the individual runs in terms of their respective standard deviations. Since the absolute numbers of errors are very much larger in the earlier runs than in the later, the adding of the crude error scores together means that the final result will depend much more upon the earlier runs than upon the later. Run 3, for example, will have a very undue weight in the total odd score, and so also will run 4 in the total even score. In order to get around this difficulty, therefore, we tried adding not the crude scores together, but the scores divided by the standard deviations of the respective runs. Thus the errors in trial 3 were divided by the standard deviation for errors in trial 3, and those in trial 4 by the standard deviation in trial 4 and so on. This caused the score in each run to have an equivalent effect upon the combined score.

The resultant correlation in the case of 41 males is shown in table 4.

TABLE 4

Weighted error score

41 males

Runs	r
3, 5, 7, 9 vs. 4, 6, 8, 10	0.317

Comparing this with the correlation of only 0.179 obtained for the 41 males for the same runs (see table 3) when crude score was used, it appears that weighting does tend to raise the correlation, which means that a weighted error for all 8 runs would be a more reliable measure of the rats' error tendency than an unweighted score. The extra labor involved, however, is very great, and for the present we did not feel it to be warranted.

A second method of increasing reliability which suggested itself arose out of the observation that for the most part it seemed to be purely an occasional vagary in a few individual rats that brought the correlations down. The idea occurred to us, therefore, that in planning a future investigation it might prove desirable to discard all rats who in some one individual trial tended to vary excessively from the mean for that trial. A future investigation might in short be limited to *consistent* rats only, i.e., rats for whom the maze was really a reliable measure. In order to see if this would work out, to see, that is, if we could thus divide our rats into consistent and inconsistent, we tried discarding from the original group all rats whose error score in any single trial *varied two standard deviations* or more from the mean for that trial. This resulted in our discarding 12 males and 8 females. These discarded animals, it is to be noted, were not ones whose *total* scores were necessarily widely variant from the average; but merely rats who in some single trial or trials (usually it was only one trial) tended to be variant from their own average. The resultant correlations are shown below, not only for odd vs. even, but also for 3, 4, 5, 6, vs. 7, 8, 9, 10 (see table 5).

TABLE 5

Runs	29 Males	62 Males Plus Females
3, 5, 7, 9 vs. 4, 6, 8, 10	0.449	0.420
3, 4, 5, 6 vs. 7, 8, 9, 10	0.384	0.261

It is evident that the correlations go up somewhat for both odd vs. even (compare with table 3) and also for first half vs. second half (compare with table 2). The increase in correlation in the case of the odd vs. even is not very great in the case of the total (82). We conclude, therefore, that the method would have no very great advantages to recommend it for future work.

Our third method of trying to get higher consistencies depended not upon eliminating any rat's score, but rather upon trying to iron out the meaningless excess variation in that score. When a rat made an unusually great number of errors in any given trial (due perhaps to excitement) or an unusually small number (due perhaps to some lucky chance) such extremes did not seem to us true measures of his real ability. Instead, however, of discarding him altogether, we decided to try a method of scoring which would *per se* discount those extremes—a method of scoring which we could apply without fault of favor to all rats. This method consisted in scoring simply in terms of greater than, less than, or equal to the mean for each given trial. In other words, his score for errors in any given trial was merely 1, 2, or 3 according as he made less than, equal to, or more than the average number of errors for that trial. The results for the 41 males and for all 82, trials 3, 5, 7, 9, vs. 4, 6, 8, 10, are shown in table 6.

TABLE 6

Corrected error score

Runs	41 Males	82 Males Plus Females
3, 5, 7, 9 vs. 4, 6, 8, 10	0.537	0.509

These figures are the biggest of any we have yet obtained. Applying the formula[3] for determining the probable correlation on this basis of a score obtained from a whole maze learning (all 8 runs) against another whole maze learning, we would have as the probable correlations of the corrected scores for all 8 runs combined $r = 0.699$ for 41 males, and $r = 0.675$ for the 82 males plus females. These reliabilities are high enough to be almost satisfactory. One practical difficulty, however, in the application of this method to *our data as actually obtained* presented itself. The method would have been all right, if the data had been obtained so as to lend itself readily to working up into terms of correlation coefficients. For, in such a case, the parents could have been scored in terms of their means and the children in terms of theirs. But in the present instance the experiment was so laid out that the chief re-

[3] Usually known as Brown's formula. $\frac{2r}{1+r}$. See Kelley, "Statistical Method" p. 206.

sults were to be found not in terms of correlations but in terms of comparisons of group averages (i.e., between "brights" on on the one hand and "dulls" on the other). Such being the case, there was no common mean in terms of greater than, less than and equal to which it was reasonable to score these contrasted groups. As a matter of final procedure, therefore, we decided to reserve this corrected score as a possible method—for the future, and for the present to employ merely the crude unweighted scores.[4]

We may turn now to a treatment of time scores. We did not attempt to try out all the different methods we used with errors—since the labor necessary seemed too great to be worth it. We tried merely the three methods of crude score, score weighted in terms of the standard deviations for the separate trials and score corrected in terms of above, below, or equal to the mean, for runs 3, 5, 7, 9, vs. 4, 6, 8, 10 and for the 41 males only. The results are shown in table 7.

TABLE 7

Time

41 males

Runs 3, 5, 7, 9 vs. 4, 6, 8, 10

Crude score	0.588
Weighted score	0.528
Corrected score	0.523

It will be noted that here the crude score is just as good, if not better than, either of the others. Why this should be so we are not certain. A possible guess would be that the time scores are less given to extremes than errors because a rat tends to compensate. If, in other words, he is making an excessive number of errors, he tends to hurry up, and vice versa,

[4] In a minor investigation (3) which can be considered as more or less of an appendix to this report it was found as a matter of fact that although the corrected score method gave *apparently* much higher reliabilities within each individual maze, they did not raise the correlations between mazes.

if he is making few errors, he tends to take it more slowly. A careful study of time and error records side by side might throw light on this matter and must be undertaken in the future.

Before turning to our final genetic results, there is one more statistical treatment of the data of the original group to be presented. It consists in the intercorrelation between the 3 criteria, errors, time, and number of perfect runs.[5] The results are shown in table 8.

TABLE 8

$n = 82$

r_{ET}	0.515
r_{EP}	−0.624
r_{TP}	0.005

E = errors 3 to 10 combined (crude score).
T = time 3 to 10 combined (crude score).
P = number of perfect runs out of the total 10.

They come out as expected, save perhaps the zero correlation between time and perfect runs. To better understand this, however, we merely need to turn to the partial correlations, which are shown in table 9.

TABLE 9

$r_{ET.P}$	0.663
$r_{EP.T}$	−0.731
$r_{TP.E}$	0.487

The outstanding feature of these partials is that the partial correlation between time and perfect runs goes up to a respectable *positive* figure. In other words, when the effect of errors on time has been rendered constant, it appears that the slower (more cautious?) rat makes the greater number of perfect runs, whereas the rapid (slap-dash?) rat continues to make a few final errors (due to too much haste). This is a

[5] No significant internal correlation for the number of perfect runs suggested itself.

most interesting and important result and among other things it suggests a certain scepticism as regards any real significance in the time score. For it would seem to mean that time is a measure of two quite independent factors. On the one hand, it goes up in a purely mechanical fashion with errors—the more errors, the more time they take—but on the other hand, *per se,* it seems to be negatively correlated with errors, or at any rate with *final* errors. It would seem, therefore, that the usual practice (5, 7) of favoring time as the one best single measurement is perhaps an unfortunate one.[6]

So much for the preliminary statistical treatment of the results for the P generation. Let us turn now to the genetic problem, i.e., to the effects actually resulting from selective breeding.

Results of Selective Breeding

As before stated, we selected out of the original 82 rats of the P generation 9 bright males to be mated with 9 bright females, and 9 dull males to be mated with 9 dull females. The offspring of these matings constitute the F_1 generation, and fall into two distinct groups, "bright" and "dull," accordingly, as they are the offspring of the original bright matings or of the original dull matings. The F_2 generation was the product of brother and sister matings in the F_1 generation, a "bright" brother mated with a "bright" sister in the case of the "brights" and a "dull" brother with a "dull" sister in the case of the "dulls."

Not all the original 18 families (9 bright and 9 dull) continued till the F_2. This was due either to death or infertility.[7]

Two sorts of results have in general been computed and will be presented: one, group-means as between "brights" and

[6] This suspicion of the time score was borne out in the results we obtained between the two mazes (3). There, although the internal correlations within each maze by itself were high for time, the correlations between the two mazes were nil for time, although high for errors and perfect runs.

[7] Breeding has actually been carried on through 5 or 6 generations but with too small numbers to make the results altogether significant.

"dulls," and two, correlations between parents and children in the cases of P_1 vs. F_1; and F_1 vs. F_2.

Tables 10, 11 and 12 present the former data, and table 14 the latter. The results for errors and time and perfect runs have been tabulated separately.

TABLE 10

Errors 3 to 10

	Original Unselected Group	"Brights"	"Dulls"	Selected To Be Bright Parents	Selected To Be Dull Parents
	(n = 82)			(n = 18)	(n = 18)
P m	12.1			7.2	17.4
σ	15.6			5.1	8.6
		(n = 50)	(n = 59)	(n = 16)	(n = 13)
F_1 m		6.2	15.7	4.9	27.2
σ		3.5	13.9	3.0	22.7
		(n = 49)	(n = 39)		
F_2 m		10.5	16.4		
σ		4.9	7.7		

An examination of tables 10, 11 and 12 indicates that selection had the effect that the bright parents had in general very decidedly brighter children than had the dull parents. It appears, however, that the further selection performed in the F_1 generation did not make the effect more pronounced in the F_2 generation. In fact, in all three tables the differences between the "brights" and the "dulls" is less in the F_2 generation than in the F_1. And in each case it appears that this was due not so much to the coming up of the "dulls" as to the sinking of the "brights." How to explain this we are not at present sure. It seems possible that it may have been due to the strict inbreeding which was practiced in obtaining the F_2 generation. Our original stock was decidedly heterogeneous and inbreeding may well have uncovered some recessive taint as say, nervousness. This possibility seems to be borne out by such

TABLE 11*

Time 3 to 10

	Original Unselected Group	"Brights"	"Dulls"	Selected To Be Bright Parents	Selected To Be Dull Parents
	(n = 82)			(n = 18)	(n = 18)
P m	19.1			13.5	29.6
σ	11.9			5.4	25.2
		(n = 50)	(n = 59)	(n = 16)	(n = 13)
F_1 m		17.6	26.7	15.2	35. 6
σ		13.9	24.1	6.5	36. 2
		(n = 49)	(n = 39)		
F_2 m		17.5	19.5		
σ		6.1	9.9		

* The figures given in this table are in terms of 10-second units.

TABLE 12

Perfect runs—10 runs

	Original Unselected Group	"Brights"	"Dulls"	Selected To Be Bright Parents	Selected To Be Dull Parents
	(n = 82)			(n = 18)	(n = 18)
P m	3.6			5.7	1.6
σ	1.5			1.1	1.1
		(n = 50)	(n = 59)	(n = 16)	(n = 13)
F_1 m		5.0	2.5	5.5	1.4
σ		1.5	1.7	1.3	1.4
		(n = 49)	(n = 39)		
F_2 m		3.9	2.2		
σ		1.7	1.8		

behavior tests as have been tried on Dr. King's inbred rats (2, 11, 13, 14). It is also possible that it was due to, or in some way affected by, the respective ages at which the animals were tested. The intention throughout was to test all animals at three months of age; but in the F_2 generation they tended to heap up on us with the result that some of them were tested late. This brought it about that they averaged somewhat older and this may have been part of the cause for the apparent sinking in ability of the brights (see table 13). To draw any

TABLE 13

Ages (days)

P (not known exactly but approximately ninety days)

	BRIGHT			DULL		
	n	m	σ	n	m	σ
F_1	52	96.6	14.3	55	88.9	13.4
F_2	49	108.6	16.0	39	124.5	17.5

certain conclusion upon this point, however, much further information would have to be obtained (e.g., the correlation and recession lines for age and maze-ability). Or, thirdly, it may have been due to something about our general living conditions which tended to cause our strain to degenerate, although at present we can think of no such cause. Our diet was practically the same as that of the original anatomy rats, i.e., garbage, save during the actual tests and for a couple of months during the summers. A future, better controlled investigation must settle between these possibilities.[8]

We turn now to table 14 which presents the data for correlations of parents with children: P_1 vs. F_1 and F_1 vs. F_2. The two sets of variants in all these correlations consist of average figures for the parents on the one side, and average figures for their litters on the other.

[8] The evidence obtained from our later generations, not here reported, from a group of brain weights which we measured, suggest that the matter was due to something more fundamental than mere age of testing, i.e., that it was either a matter of inbreeding or of nutrition.

TABLE 14

	Errors 3 to 10	Time 3 to 10	Perfect Runs
P vs. F_1 ($n = 18$)	0.458	0.297	0.858
F_1 vs. F_2 ($n = 16$)	0.320	−0.267	0.779

In evaluating these figures it must be remembered that they result from arbitrarily selected groups and not from purely normal chance distributions. Of the 18 P vs. F_1 families, 9 were due to arbitrarily selecting the brightest individuals for mating and 9 from arbitrarily selecting the dullest individuals. Similarly of the 16 F_1 vs. F_2 families 8 were very bright and 8 very dull. In other words, we have only the two extremes of each distribution; and this would tend to raise the correlation from what it would be for the whole distributions. However that may be, the figures taken at their face value are of considerable interest in that they pretty clearly substantiate our previous indications that time is an uncertain and ambiguous quantity. The time correlations are nil.[9]

And the second point of interest is that comparing the figures for errors and for perfect runs, the correlations for the latter are *much* the higher. One explanation that suggests itself is that since crude scores rather than corrected ones were used for computing the error figures, the error correlations were brought down beyond what might be conceived as their true or ideal values. The perfect run scores, on the other hand, were probably less affected by chance upsets, and hence more truly measured the real ability.

A second possibility, however, is that perfect runs actually measure something which is really more truly and directly inherited than is that which is measured by errors. That is, our choice lies either between (*a*) assuming that the perfect run correlations between parents and children are higher than the error correlations, simply because the perfect run scores are in themselves a more reliable measure of one and the same

[9] Zero correlations have since been found for time, for one set of animals in two successive mazes (3).

thing that the error scores are a measure of, or (*b*) assuming that the perfect run scores really measure something different from the error scores and that this something different is more nearly inherited than is that which is measured by errors.

As between these two possibilities we shall not at present attempt to decide and it may well be that both factors play a part.

Before concluding, we may give the intercorrelations between time, errors, and perfect runs for the F_1 and F_2 generations. We already have them for the P generation (tables 8 and 9). For purposes of comparison we will group them together for all three generations in table 15.

TABLE 15

	$n^P = 82$	$n^{F_1} = 109$	$n^{F_2} = 88$
r_{ET}	0.515	0.579	0.543
r_{EP}	−0.624	−0.671	−0.557
r_{TP}	0.005	−0.412	−0.179
$r_{ET.P}$	0.663	0.448	0.540
$r_{EP.T}$	−0.731	−0.582	−0.557
$r_{TP.E}$	0.487	0.038	0.179

All the three generations present practically the same picture. The only exception is that in the F_1 generation the partial correlation between time and perfect runs when the effect of errors is eliminated is zero, not positive. With this particular group of rats, the effect of "cautiousness" seems not to have been very important.

Finally, in conclusion, we may point out that our results based upon more numbers substantiate those of Bagg (1). But just how complicated a genetic trait maze-learning ability is, how many genes are involved, etc.—these are all problems for the future.[10]

[10] At the present moment we are putting our energies upon the construction of some self-recording mazes five or six of which can be run simultaneously. This will very much lessen the labor involved and allow the carrying out of a much more extensive and better controlled investigation.

Our main findings may be summarized as follows:

1. Practically zero correlations were obtained between numbers of errors made by the same rats in successive trials.

2. A correlation of 0.379 was, however, obtained between errors in trials 3, 5, 7, 9 vs. errors in trials 4, 6, 8, 10.

3. The highest internal correlation for errors was obtained from a "corrected score" in which the animal was scored in each trial simply in terms of whether he made more, less, or equal to the average number of errors for that trial. This correlation, trials 3, 5, 7, 9 vs. trials 4, 6, 8, 10 was 0.509 for 82 rats.

4. The intercorrelations between trials 3, 5, 7, 9 and 4, 6, 8, 10 for time was 0.588, and no higher for a "corrected time score."

5. From the intercorrelations of time, errors, and perfect runs with each other, it appears that time is an ambiguous measure that increases with errors, but at the same time increases with perfect runs.

6. Selective breeding in the P generation caused a very decided differentiation to appear in the F_1 generation between the descendants of "bright" parents and the descendants of "dull" parents. This appeared in all three scores—errors, time, and perfect runs. Further selective breeding caused no further differentiation in the F_2 generation. In fact, the differences tended to decrease. Whether this was due to inbreeding or to some extraneous factor we have not as yet determined.

7. The correlations between parents averaged and children averaged (P vs. F_1 and F_1 vs. F_2) were relatively high for perfect runs, low for errors, and very low, even negative, for time.

References

(1) Bagg, Halsey J.: Individual differences and family resemblances in animal behavior. A study of habit formation in various strains of mice. Archives of Psychology.

(2) Basset, Gardner C.: Habit formation in a strain of albino rats of less than normal weight. Behav. Mon., 1914, ii, no. 4.

(3) Davis, F. C., and Tolman, E. C.: A note on the correlations between two mazes. Jour. Comp. Psychol. (Forthcoming.)

(4) Heron, William T.: The reliability of the inclined plane problem box as a measure of learning ability in the white rat. Comp. Psychol. Mon., 1922, i, no. 1, pt. I.
(5) Hibbert, Helen B.: Time versus distance in learning. Jour. Animal Behav., 1914, iv, 60-69.
(6) Hibbert, Helen B.: The effect of age on habit formation in the albino rat. Behav. Mon., 1915, ii, no. 6.
(7) Hicks, Vinnie C.: The relative values of the different curves of learning. Jour. Animal Behav., 1911, i, 138-156.
(8) Hunter, Walter S.: Habit interference in the white rat and in human subjects. Jour. Comp. Psychol., 1922, ii, 29-59.
(9) Hunter, Walter S.: Correlation studies with the maze with rats and humans. Comp. Psychol. Mon., 1922, i, no. 1, pt. II.
(10) Kelly, Truman L. Statistical Method. Macmillan, 1923.
(11) King, Helen D. Studies on inbreeding. II. The effects of inbreeding on the fertility and on the constitutional vigor of the albino rat. Jour. Exper. Zool., 1918, xxvi, 335-278.
(12) Maupin, Oakland: The reliability of the maze as a measure of learning ability in the white rat. (In preparation.)
(13) Utsurikawa, N. Temperamental differences between outbred and inbred strains of the albino rat. Jour. Animal Behav., 1917, vii.
(14) Yerkes, Ada W.: Comparison of the behavior of stock and inbred albino rats. Jour. Animal Behav., 1916, vi, 267-296.

A Note on the Genetic Basis of Spontaneous Activity in the Albino Rat

ELIZABETH G. BRODY

In August 1942, the author (1) reported the results of an experiment in selective breeding for activity and inactivity in the albino rat which supplemented the initial study by E. A. Rundquist (2), and in addition presented data from cross matings between the two strains. It will be recalled that the inactive strain which had been selectively bred for twenty-five

From *Journal of Comparative and Physiological Psychology*, 1950, *43*, 281-88. Reprinted by permission of the author and the American Psychological Association.

generations could not be continued past this point due to the infertility of the females. This paper will present results obtained from selectively breeding for a new strain of inactive rats, as well as further data on the continuation of the active strain.

The new strain of inactives had been selectively bred for ten generations and the active strain had been continued through the F35 generation when the war intervened. Since it was impossible to obtain sufficient help to make the selective matings and to run the animals in the activity cages, the strains were maintained by permitting unrestricted matings within a strain without regard for parental activity. From the fall of 1942, no selective breeding was conducted. In the fall of 1946, the author made matings from within the active strain and within the inactive strain and ran these offspring in the activity cages to determine the effect of four years of non-selection upon activity within each strain.[1]

In all, the activity records of three non-selected generations of active and inactive rats were obtained. To avoid human selection of any kind in obtaining these three generations large wooden cages were constructed in which all of the animals within a strain were placed, the females being removed to separate cages as soon as pregnancy was detected.

The new strain of inactive animals was started by selecting the most inactive rats from among the second group of cross-bred animals (1). Most of these parents were of the F_1 inactive back cross and F_2 generation, and the only criterion for selection was that they be phenotypically inactive.

In table 1 are presented the means and standard deviations of the revolving drum records of these ten generations. It will be noted that there is no clear-cut tendency for the means to decrease during the ten generations of selective breeding; however, the males are as inactive as the original inactive strain. On the other hand, while the activity of the females also

[1] The experimental conditions under which these animals were tested for activity were identical in all respects with those employed prior to the war. With the exception of the mating methods the animals were raised and cared for under exactly the same conditions as described in the author's previous article.

TABLE 1

The mean activity and variability per thousand revolutions during a fifteen day period of a new strain of inactive rats during ten generations of selective breeding

Generation	Males			Females		
	N	Mean	S.D.	*N*	Mean	S.D.
F_1	33	10.7	15.7	41	56.0	60.4
F_2	36	10.4	18.7	40	57.5	65.1
F_3	48	15.2	25.2	51	48.9	54.5
F_4 (1st)[1]	31	11.4	14.4	19	26.6	26.4
F_4 (2nd)	27	7.6	10.2	23	38.5	54.6
F_5 (1st)[1]	46	5.3	6.6	54	42.0	41.2
F_5 (2nd)	16	14.7	18.2	13	40.4	44.1
F_6	39	5.9	6.3	37	56.9	50.8
F_7	35	5.3	7.5	34	49.7	49.4
F_8	18	13.6	18.1	13	25.0	40.4
F_9	22	3.9	3.8	30	31.7	38.4
F_{10}	38	5.6	17.4	36	55.6	73.4

[1] Two sets of F_4 and F_5 matings were made. The first, in each case, represents offspring from matings made prior to running the animals in the activity cages and the second, offspring from matings made after running. Therefore, only the second sample was selectively bred.

shows no consistent downward trend, the means and standard deviations are considerably greater than those of the original inactive strain. This is especially true when a comparison is made with generations F_{18} to F_{25} (1). In obtaining the first inactive strain, it will be recalled that matings were not made within a strain during the first four generations, but were based on phenotypes alone. The F_5 generation of the original strain is thus roughly comparable to the F_1 of this new, inactive strain. Taking this into consideration, the activity of females from the new inactive strain approximates closely that of the original strain during the comparable generations. The male activity in the original strain decreased somewhat during the first two generations of mating with the inactive strain, i.e., to the F_7 generation; but from that point on, it remained relatively constant, fluctuating from generation to generation to about the same extent as this new strain but with no further downward trend.

The finding that selective breeding had little or no effect on the inactivity of the males after the initial separation of the strains whereas activity of the inactive females did decrease suggested that a single gene for inactivity might be sufficient to produce inactive males whereas two were rquired to produce inactive females. Results from cross matings of the two strains also appeared to indicate that the differences between the two strains could be accounted for by a single gene which was dominant for inactivity in the males and recessive in the females.[2] However, this alone would not wholly account for the effects of selection in the inactive strain since none of the inactive-strain males from the F_5 generation on in the original strain exceeded 140,000 revolutions and none exceeded 110,000 revolutions in the new strain. One would certainly expect to obtain a few very active males so long as there was any heterozygosity within the strain. Furthermore, one would expect a more consistent decrease in the activity of the female offspring from one generation to the next if inactive females were inactive by virtue of homozygosity of the genes involved, even though the heterozygous and homozygous males couldn't be differentiated on the basis of the phenotype.

Thus, some genes having a modifying effect on activity in addition to the basic gene pair would appear to be involved. Evidence was presented on the relation of activity to litter size which suggested that the decrease in heterozygosity resulting from inbreeding did have an effect on activity as well as on litter size. To determine whether any relationship existed between fluctuations in litter size and fluctuations in activity from one generation to the next in the new inactive strain as in the original strain, rank order correlation coefficients were calculated. The correlation between the mean activity of the fe-

[2] Single genes that produce readily distinguishable differences in quantitative traits have been identified by a number of other investigators, e.g., (1) Overbeek, I. van "The growth hormone and the dwarf type of growth in corn." *Proc. nat. Acad. Sci.*, 1935, 21, 292-299, and (2) Castle, W. E. "Size genes in mice." *Proc. nat. Acad. Sci.*, 1942, 28, 69-72.

males and the mean number of females per litter was + .72, and for male activity and number of males per litter + .33. These correlations are fairly close to the product moment correlations calculated from the smoothed data of the original strain which were + .81 and + .46 respectively. Evidently activity, particularly among the inactive strain females, is related in some way to litter size. Although the exact nature of the inter-relationship has not been established, there is certainly evidence that the relationship is genetic and not due to the dependency of both activity and litter size on environmental fluctuation. This is clear from the type of relationship found for the original strain and also is borne out by the comparison of rank order correlations for the new strain; these correlations are + .83 for the mean number of males and females per litter and — .16 for males and female activity. That the genes which are involved in the relationship between activity and litter size are not the same as those which differentiate the inactive from the active strain was demonstrated in the original study which showed that differences in litter size between the two strains during the F_{16} to F_{25} generations were not significant.

In table 2 are presented the means and standard deviations of the active strain from the F_{30} to F_{35} generation. There is considerable fluctuation from one generation to the next, but the amount of fluctuation is similar to that found prior to the F_{30} generation. The differences between most of these means are significant, as were those for the first 29 generations, indi-

TABLE 2

The mean activity and variability of the active strain rats from the F_{30} to F_{35} generations

Generation	Males			Females		
	N	Mean	S.D.	*N*	Mean	S.D.
F_{30}	32	154.7	43.4	24	188.8	50.2
F_{31}	28	122.5	50.8	26	168.8	47.1
F_{32}	44	142.5	58.0	52	237.1	31.7
F_{33}	44	211.7	69.8	42	231.7	55.1
F_{34}	20	266.0	34.9	29	286.0	68.4
F_{35}	31	203.1	68.6	34	272.9	58.9

cating that the different generations cannot be assumed to be random samples from the same population. It was concluded that the differences from one generation to the next within the active strain were probably due to environmental influences since selective breeding did not result in any consistent change in either the amount of variation or the mean activity.

Further evidence that genetic differences are not the principal source of this variation has been obtained from several analyses of variance calculations. The litters within a single generation do not differ from each other more than would be expected on the basis of random sampling from the same population. One would expect that if genetic factors were operating to determine differences between generations, they would certainly operate to cause significant differences within a single generation. This does not mean that no differences in activity within the active strain are genetically determined, but only that environmental variation and, no doubt, errors of measurement are of such a magnitude that the genetic differences cannot be statistically identified. It does indicate, however, that the genetic differences that remain within the active strain are not due to the same gene pair which differentiates the active and inactive strains which we have here isolated.

What happens to the active strain when selective breeding is not practiced for a period of four years? The answer to this can be found in table 3. It will be noted that the mean activity of these three generations of males is lower than during any of the preceding six generations, and that of the females is lower for the F_{III} generation than for the F_{30} to F_{35} genera-

TABLE 3

The mean activity and variability of the active strain rats after four years of non-selection

Generation	Males			Females		
	N	Mean	S.D.	*N*	Mean	S.D.
F_I	19	113.7	86.5	29	174.3	66.2
F_{II}	43	83.6	64.5	51	190.9	92.1
F_{III}	25	87.8	43.0	24	134.2	76.9

tions. However, we already noted the variation from one generation to the next within the active strain; so we should compare these results with all the active strain data. Taking all generations into consideration, the mean activity of these last generations is somewhat low, but not lower than in several generations in which selection was involved (1). The mean of the F_{29} males, for instance, was only 73.4 thousand revolutions, and of the F_{21} females, 125.1 thousand. Thus, these animals are not less active than the animals of some of the generations prior to the discontinuance of selective breeding. Permitting the animals to mate at random within the active strain for a number of years has certainly not resulted in an elimination of the strain, nor has it reduced activity to a level characteristic of either the old or new inactive strain.

TABLE 4

The mean activity and variability of the new inactive strain after four years of non-selection

Generation	Males			Females		
	N	Mean	S.D.	*N*	Mean	S.D.
F_I	31	79.8	76.8	18	84.4	94.4
F_{II}	49	43.0	63.0	46	165.2	108.1
F_{III}	25	29.8	34.7	24	109.2	78.7

Table 4, however, makes it apparent that the effect of non-selection upon the inactive strain is radically different from that on the active strain. The means and standard deviations indicate that the activity of both the males and females is considerably greater than was true for this strain during selective breeding. Though there is much variation among the three generations the results are quite similar to those of the F_2 generations of cross-bred animals. Since activity increases when selection is not practiced, it is apparent that heterozygosity must have been present within the inactive strain.

The question, of course, is how could the inactive-strain males appear to be pure for inactivity? When selection is relaxed the activity of both the males and females is increased which makes it necessary to assume that heterozygosity was present all along. One possibility would be that the females which were sufficiently inactive to be selected for breeding were homozygous and, therefore, all the male offspring would have at least one dominant gene. The males selected for breeding could be either heterozygous or homozygous so two types of female offspring would continue to be produced. This, as was suggested, should tend to reduce the number of recessive genes rather rapidly unless the homozygous males were not as likely to produce offspring as the heterozygous males. Support of this hypothesis is suggested by the relationship between litter size and activity. Also, there were data presented in the previous article which suggested that some of the homozygous males may be eliminated prior to birth. If this is true, one would certainly expect increased activity when selection is discontinued. Thus, with time the inactive strain should attain an equilibrium which would approximate the activity of the F_2 generation obtained from cross-breeding. Conversely, if the active strain is pure for the non-inhibiting gene, relaxation of selection should have little or no effect on the activity of the strain. The hypothesis that modifiers are probably involved in determining differences within a strain would account for some decrease in activity among the active strain when selection was no longer practiced.

In addition to the preceding data a test of the consistency of activity over a period of several months was conducted on the first generation of the post non-selection animals. These animals were first run in the activity cages at the age of approximately three months and after running they were placed in mating cages. About three and a half months after the first running when the animals were approximately six and a half months old, they were placed in the activity cages again. In table 5 are presented the means and standard deviations of the two samples of activity for the active and inactive strain males and females as well as the correlation coefficients.

It will be noted that the correlation between the two samples of activity for the active females indicates a tendency for the animals to maintain the same relative position although there is obviously some fluctuation in activity as meas-

TABLE 5

Mean activity, variability and coefficients of correlation between two samples of active- and inactive-strain activity

	N	First Run		Second Run		Correlation
		Mean	S. D.	Mean	S. D.	
Active Strain						
Females	29	174.3	66.2	196.0	71.6	.673
Males	19	113.7	86.5	111.8	56.7	.868
Inactive Strain						
Females	18	84.4	94.4	97.2	74.0	.791
Males	31	79.8	76.7	31.8	47.0	.555

ured after a three and a half month interval. That the males are considerably more consistent is shown by the correlation in the high eighties. Though the females do differ to some extent in activity, with a mean activity 21.7 thousand revolutions greater in the second running, it should be pointed out that the difference is not statistically significant. This is important because it will be recalled that the differences among different generations were statistically significant. Therefore, errors of measurement and/or environmental conditions operating while the animals are in the cages are probably not significant factors in determining the differences among generations in activity. The correlation among inactive females is somewhat higher than among active females, but considerably lower among the inactive males than among the active males. The actual size of the latter correlation does not adequately represent the relationship, however, which was markedly curvilinear. Those males that were initially inactive remained inactive, whereas a considerable number of those that were active became very inactive or moderately inactive.

SUMMARY AND CONCLUSIONS

After the inactive strain initially separated by Rundquist (2) had died out, the author attempted to isolate a new inactive strain using cross-bred animals as a parental group. The results obtained during ten generations of selective breeding approximate those of the comparable generations in the original strain. The new inactive-strain males were as inactive as those in the original group, but the females were more active than the last few generations of the first inactive strain.

During a four-year period no selective matings within the two strains were made and no measurements of activity were undertaken. The activity of the active-strain males and females following this period of non-selection tended to be somewhat lower than in the generations immediately preceding non-selection, but the means and standard deviations did not fluctuate outside the limits of the preceding thirty-five generations. The inactive-strain animals, however, were considerably more active than any of the preceding inactive-strain generations which indicates that heterozygosity was present within the strain prior to the cessation of selection. As in the original inactive strain, there is evidence that litter size and activity tend to fluctuate concomitantly.

The first generation of rats which followed the period of non-selection was run in the activity cages twice, once at the age of approximately three months and again at six and a half months. The difference in mean activity between these two periods was no greater than that to be expected from random samples of the same population; this is in contrast with the significant differences present among generations. This substantiates the hypothesis that environmental factors operating at the time the animals are run in the cages or errors of measurement do not account for the between-generation differences. The animals tend to maintain their relative positions with respect to each other in activity, though some of the correlations are relatively low.

It seems reasonably clear from our data that there is a major controlling gene pair which differentiates these active and inactive strains of rats but that this gene pair alone does not by any means account for all of the differences in activity. Even though one gene pair is sufficient to differentiate between the strains the problem remains unsolved as to how the genic control is effected. The difference in behavior which we find at least in part to be genetically determined undoubtedly are reflected by physiological differences as well. Some of the ambiguities and discrepancies that appear can be answered only by correlating genetic with physiological data in future studies.

Could selection in the new inactive strain have been continued to the point wherein the females were as inactive as those in the original strain without reducing the fertility and thus again causing loss of the strain?

As has been found in numerous other breeding experiments, a decrease in vigor, vitality, and fertility frequently accompanies inbreeding. This decrease may have no effect upon the character for which selection is being conducted but this may not be true in the case of activity. As a result of hybridization the phenotypically inactive females in the new strain of inactives readily produced offspring whereas in the last few generations of the original strain they did not. Does this difference necessarily follow because in the former case the females were homozygous for the inactive gene pair and in the latter not, or is it a concomitant effect of differences in many modifying genes which indirectly affect both activity and fertility as a part of general vitality and vigor? The use of physiological techniques for determining differences in fertility might well provide an answer to this question.

Physiological comparisons might also enable us to differentiate those inactive animals within the active strain which differ genetically from inactive-strain animals but are phenotypically like them in terms of behavior. A study of possible hormonal differences between active and inactive animals would be not only valuable but this and other physiological researches are necessary if we are to achieve any true understanding of the inheritance of activity.

References

1. Brody, Elizabeth G.: The genetic basis of spontaneous activity in the albino rat. *Comp. Psychol. Monogr.*, *17*, No. 5, 1942.
2. Rundquist, E. A.: The inheritance of spontaneous activity in rats. *J. comp. Psychol.*, 1933, *16*, 415-438.

The Genetics of Mouse Behavior in Novel Situations[1]

GERALD E. MCCLEARN

Thompson (1953) has made a significant contribution to the study of the genetics of behavior by systematically testing 14 inbred strains and 1 hybrid strain of *Mus musculus* for food drive, emotionality, and exploratory behavior. Food drive was measured by the amount of food consumed following deprivation, emotionality was measured in terms of the tendency to defecate in the food-drive situation, and exploratory behavior was determined by recording the number of squares traversed on a checkerboard-type floor which held barriers at intervals. Large interstrain differences were found on all tests, food drive and emotionality were shown to be negatively correlated, and no relationship was found between food drive and exploratory activity.

This paper presents the results of two further studies on the genetics of exploratory behavior of mice.

EXPERIMENT I

The principal purpose of the first study was to determine the situational generality of the behavior, i.e., the extent to which the same ordinal relationships among the various strains could be obtained in a variety of situations which presumably measure exploratory behavior. In addition, observations were made on the stability of strain rank orders on retest, and on sex differences within strains.

[1] This research was supported by National Science Foundation Grant G1783, and was conducted in the Psychology Laboratory of Allegheny College, Meadville, Pensylvania. I wish to express my gratitude to Robert Smith for his faithful assistance in the conduct of this research.

From *Journal of Comparative and Physiological Psychology*, 1959, 52, 62-67. Reprinted by permission of the author and the American Psychological Association.

Method

Subjects

Six strains of mice were chosen for study: C57Br/cd, C57-BL/10, LP, AKR, BALB/c, and A/Jax. The origin, degree of inbreeding, and tumor characteristics of these strains have been described (Committee on Standardized Nomenclature, 1952).

Ten females and five males of each strain were obtained from the Roscoe B. Jackson Memorial Laboratory at approximately 40 days of age. They were approximately 65 days old when the testing program commenced.

Colony Conditions

Sexes and strains were housed separately. As far as possible, five animals were housed per cage.

The temperature was controlled between 70° and 75° F., and a dehumidifier prevented rapid changes in humidity. An artificial light cycle was maintained, with lights on from 7 A.M. to 7 P.M.

Apparatus

Four apparatuses were used: Arena, Hole in Wall, Open Field, and Barrier.

Arena. This apparatus is a modification of that used by Thompson, differing principally in the extent of control of extra-apparatus stimuli. A 30-in. square floor is subdivided into 36, 5-in. squares. Barriers 3⅞ in. high are erected at staggered intervals on the square borders. This floor is enclosed by walls 36 in. high at the front and 48 in. high at the back. The sloped top contains a one-way mirror. The front wall slides upward in channels, providing access to the inside of the apparatus. Illumination is provided by four 10-w. frosted bulbs, one in each corner 34 in. from the floor. A starting box is mounted outside the apparatus, and a guillotine door permits the animal to enter the corner square from the starting box.

Hole in Wall. This apparatus consists of two chambers, approximately 3 in. by 4 in. by 3 in., separated by a partition. A guillotine door covers a hole ⅞ in. in diameter in the partition. When raised to expose the hole, the door activates a microswitch, which starts a timer. The starting chamber is covered by a transparent plastic cover that can be locked in place. The goal chamber is covered by an opaque wooden top and is relatively dark inside. The floor of the goal chamber is hinged, and when depressed by the weight of S, activates a second microswitch, which stops the timer.

Open Field. This apparatus is an elaboration of one used by Fredericson (1953). A gray unobstructed floor, 30 in. square, is provided to fit within the Arena apparatus, resting above the barriers. Lighting and observation conditions are, therefore, the same as for the Arena. A centered square 21.2 in. by 21.2 in. is outlined on the floor. The area enclosed by the square is thus equal to the total area of the margins surrounding the square. A small 5-in. square is also drawn in the center of the floor. An inverted funnel is placed on a corner of the outer square. A string through the top of the apparatus permits the raising of the funnel.

Barrier. In this apparatus, a 12-in.-square floor is subdivided into four 6-in. squares by intersecting barriers. From the center to a point 3 in. out, the barriers are 7 in. high, preventing S from climbing diagonally from one square to another. The remaining half of each barrier, next to the wall, is 3 in. high. This floor is placed in a box 12 in. by 12 in. by 36 in. high, which contains a one-way mirror in its top. Illumination is provided by four 7½-w. frosted bulbs located in the corners 34 in. from the floor. The front wall of the box is hinged to permit introduction and retrieving of the animal.

Except for the boundaries of the squares in the Arena and Open Field, all apparatuses are painted neutral gray.

Procedure

Arena. The mouse was placed in the starting box. The guillotine door was raised, and a timer was started. The number of square entries was recorded on a hand tally counter for a period of 5 min. A "square entry" was defined as the placing of both forepaws across a line. At the termination of the trial, the front wall was raised, and S was retrieved.

Hole in Wall. The animal was placed in the starting chamber, and the plastic cover was locked in place. Fifteen seconds later the guillotine door was lifted, exposing the hole and starting the timer. When S had crawled through the hole and placed its weight on the floor, the timer stopped. The time was recorded, and S was immediately removed from the apparatus. If S did not enter the goal compartment within 180 sec., the trial was terminated.

Open Field. The S was placed under the inverted funnel, which restrained it until the front door could be lowered. When the funnel was lifted, a timer was started. The time S spent within the large square was recorded cumulatively by means of a hand-operated switch which controlled a timer. The number of times S crossed the line with both forepaws was recorded on a hand tally counter, and the number of contacts with the small center square was counted. A "contact" required entry with both forepaws.

After 2 min. the trial was terminated and S retrieved.

Barrier. The mouse was placed in a back compartment, the front door was closed, and a timer started. The total number of barrier crossings was recorded. A "crossing" required that all four feet touch the floor on the other side of the barrier. Each trial lasted 2 min.

Testing Schedule

Two trials were given to each S on each apparatus. A six-day rest was given between Trial 1 and Trial 2 of each test, and a three-day rest between Trial 2 of any test and Trial 1 of the next test. Two *E*s conducted the testing, with a different *E* conducting the first and second trials for any given animal. It was not feasible to counterbalance the order of presentation of tests, so all animals were run in the same order: Arena, Hole in Wall, Open Field, and Barrier.

[Testing was conducted between 8 A.M. and 10 A.M. in a windowless room. A ventilating fan masked extraneous noises.

Commercial mouse food and tap water were available in the living cages at all times.

Results

Nonparametric statistics were extensively used in the analysis for several reasons. Most important is the uncertainty about equality of intervals of the scales used. It appears likely, for example, that the difference in exploratory behavior represented by the difference between zero and one barrier climbing is greater than that represented by the difference between three and four climbings. In addition, there were indeterminate scores in the Hole in Wall, and inspection revealed that other distributions were skew. All nonparametric tests were computed according to Siegel (1956).

Stability under Repeated Testing

Because the focus of interest in this study is on behavior in novel situations, it is not essential that the relationships be reproduced on Trial 2, which is no longer novel. However, other things being equal, it would be desirable to have tests on which phenotypic expression is not greatly modified subsequent to one exposure.

Table 1 shows Spearman rank correlation coefficients, which were computed from the ranked strain medians of Trial 1 and of Trial 2 for each test. A value of .829 is required at the .05 level. Only the measures of seconds within square and the

Table 1

Rank-Order Correlation Coefficients Between Strain Medians on Trial 1 and Trial 2

Measure	*Rho*
Arena	.94
Hole in Wall	.88
Open-Field crosses	.90
Open-Field seconds	.49
Open-Field contacts	.79
Barrier	.96

contacts with center square on the Open Field test fail to reach this level. Because these items are both related to the number of crosses into square, which shows a significant correlation, the latter is the only exploratory measure of the Open Field test which will be discussed.

Strain Differences

A summary of the Trial 1 results is presented in Table 2. The data have been rounded to the nearest whole numbers. Because of the necessity of committing some Ss to a breeding program, some observations were made on less than 15 Ss. For these instances, the number observed is shown in parentheses.

For Arena results, a Median test over all strains gave a $p < .0001$.

For the Hole in Wall, the A/Jax median was indeterminate, because only 40% of the animals had entered the goal chamber within the 180-sec. time limit. A Median test over all strains yielded a $p < .001$ for this test.

The results of the Barrier test are best described in terms of the percentage of animals which climbed a barrier one or more times. Median values for the strains with only one or two animals climbing are not very meaningful, but it is instructive to compare the C57BL/10 median of 7.0 with the next highest median, 0.8 for C57BR/cd. It appears that a C57BL/10 animal is not only more likely to be a "climber" but, compared with "climbers" from other strains, will climb more barriers within the 2-min. interval. Because of the large number of expected cell frequencies less than five, an over-all Median test was not possible for the Barrier test.

An over-all Median test on the Open Field results gave a $p < .001$.

A general consistency in order may be seen over all four measures, with C57BL/10, C57BR/cd, and LP generally obtaining higher exploratory scores than AKR, BALB/c, and A/Jax. To evaluate this consistency, the strain medians were ranked from most exploratory to least exploratory on each test, and a Kendall's coefficient of concordance was computed. The resulting W is .90, for which $p < .01$.

TABLE 2

First-Trial Results for All Strains on All Tests

Measure	Strain					
	C57BR/cd	C57BL/10	LP	AKR	BALB/c	A/Jax
Arena						
Mdn square entries	311	267	286	207	78	11
Range	199–365	203–313	143–397	158–284	3–238	1–41
Hole in Wall						
Mdn sec.	9	9	17	40	46	149+
Range	3–28	3–48	7–180+	12–110	20–180+	19–180+
Open Field						
Mdn no. of crosses	9	7	7	2 (10)	1	1 (9)
Range	3–19	6–14	0–18	0–12	0–12	0–5
Barrier						
Per cent climbing	58 (12)	80	6	13	0	0 (7)
Range (no of climbings)	0–6	0–17	0–8	0–1	0	0

In addition to the strain differences in median scores or percentages described above, substantial differences in variability are apparent in the ranges shown in Table 2. To determine if some strains were consistently more or less variable than others in respect to all measures, a coefficient of concordance was computed on the ranked ranges. The resulting *W* is .64, which is not significant.

Sex Differences

Sex differences were tested within each strain by use of the Mann-Whitney test, except for a few cases when Fisher's exact procedure was more appropriate. The results of all tests which gave a $p < .05$ are shown in Table 3.

TABLE 3

Sex Differences in Performance

Strain	Measure	♀	♂	*p*
BALB/c	Arena, Mdn line crossings	10	170	<.05
C57BR/cd	Hole in Wall, Mdn sec.	6.5	16.0	<.02
C57BR/cd	Barrier, per cent climbing	88[a]	0[b]	<.05

[a] $N = 8$.
[b] $N = 4$.

EXPERIMENT II

The determination of differences in behavior of a number of inbred strains which have been selected for some other, nonbehavioral, characteristic yields the following information: (*a*) the strains are known to differ genotypically, and (*b*) the strains differ in behavior. When rearing and testing have been performed under standardized conditions, the reasonable interpretation may be made that genotypic differences are in some way related to the phenotypic differences. The nature of the relationship cannot, however, be specified in any detail, because there is no systematic way of ordering the genotypic values of the strains. It is possible, however, by mating animals of different strains, to manipulate genotype, and the relationship between behavioral phenotype and the genotypic values of the derived generations may be examined.

The purpose of Experiment II was to apply the tests used in Experiment I to the F_1 hybrids of two of the most extremely different inbred strains, C57BL/10 and A/Jax, as an initial step in the description of the genotype-behavior relationship.

METHOD

The same apparatus and the same general procedures used in Experiment I were also used in Experiment II. In order to accelerate the testing schedule, only one-day rest periods were given between trials and between problems. Animals were tested between noon and 5 P.M.

Subjects

The F_1s were obtained by matings of the Ss previously tested in Experiment I. Twenty-one male and 26 female offspring were obtained from matings of C57BL/10 females with A/Jax males. Twelve male and 8 female offspring were obtained from matings of A/Jax females with C57BL/10 males. These groups will subsequently be designated, respectively, C57BL/10 × A/Jax and A/Jax × C57BL/10, with the strain of the female parent named first.

Because an accelerated testing schedule was used, it was necessary to obtain comparative measures on new Ss from the parental strains. Ten male and 12 female C57BL/10 and 10 male and 10 female A/Jax were obtained from Jackson Memorial Laboratory. All Ss were approximately 60 days of age at the beginning of testing.

RESULTS

Reproducibility of Results of Parental Strains

Table 4 summarizes the data for the first trials of the various tests. With the exception of the Barrier test, the present

TABLE 4

First-Trial Results by Groups on All Tests

Strain	N	Apparatus and Measure: Arena: Mdn entries	Hole in Wall: Mdn sec.	Barrier: per cent climbing	Open Field: Mdn crossings
C57BL/10	22	252	15	36	9
C57BL/10 x A/Jax	47	194	18	67	5
A/Jax x C57BL/10	20	195	10	55	6
A/Jax	20	22	180	5	1

results on C57BL/10 and A/Jax very closely parallel those of the previous study. The compressing of the testing schedule, or the change in time of day of testing, might have resulted in the greatly diminished percentage of C57BL/10 animals climbing barriers, or the difference might represent sampling fluctuations. It is clear, however, that, for the other tests, the reduced intertrial and intertest rest interval did not affect the behavior.

Comparison of F_1s with Parental Strains

All differences reported as significant are at the .01 level, unless otherwise specified.

Arena. The cumulative percentage curves for all groups are shown in Figure 1. The reciprocal F_1s do not differ, and are intermediate between the parental strains, though displaced somewhat toward the C57BL/10. By Mann-Whitney tests, the parental strains differ from each other and also differ from the F_1s. The F_1 variability appears greater than that of either parental strain.

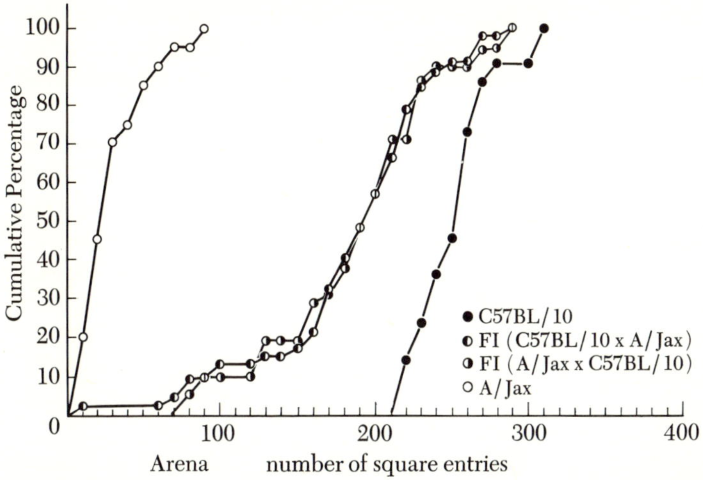

Fig. 1. Cumulative distributions for parental strains and reciprocal F1s for square entries in Arena.

Hole in Wall. Figure 2 presents the cumulative percentage of animals entering the goal chamber within the time interval

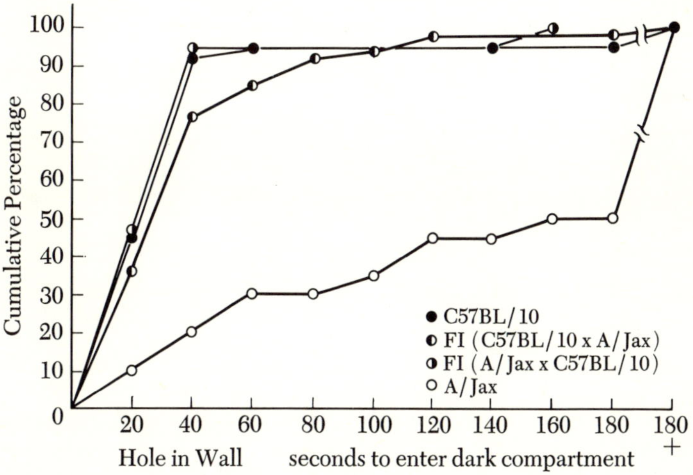

Fig. 2. Cumulative distributions for parental strains and reciprocal F1s for time to enter dark compartment in Hole in Wall.

indicated on the abscissa. In this test, the F_1 distributions are essentially indistinguishable from that of the C57BL/10, while all of these are clearly different from A/Jax. This is confirmed by Median test.

Open Field. The curves of Figure 3, for the Open Field line crossings, are similar to those for the Arena in that the F_1s do not differ from each other, are intermediate to the parental strains, and are significantly different from each of them.

Barrier. It will be recalled that the C57BL/10 of Experiment II did not climb barriers as frequently as did those of Experiment I. The results of a chi-square test, betwen C57BL/10 and A/Jax, based on a climb–no-climb dichotomy, gave a $p < .05$. By the same test, both F_1s differed from A/Jax with $p < .01$. Comparisons of C57BL/10 with A/Jax x C57BL/10 and with C57BL/10 x A/Jax gave p values of $> .30$ and $< .05$, respectively.

Because of the differences in testing schedule, combining Experiment I and Experiment II data for A/Jax and for C57BL-/10 can be, at best, suggestive. Comparisons using the pooled

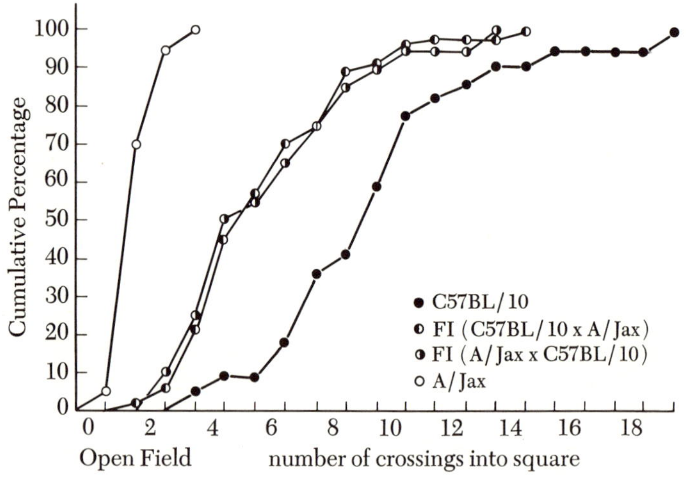

Fig. 3. Cumulative distributions for parental strains and reciprocal F_1s for number of crossings into square in Open Field.

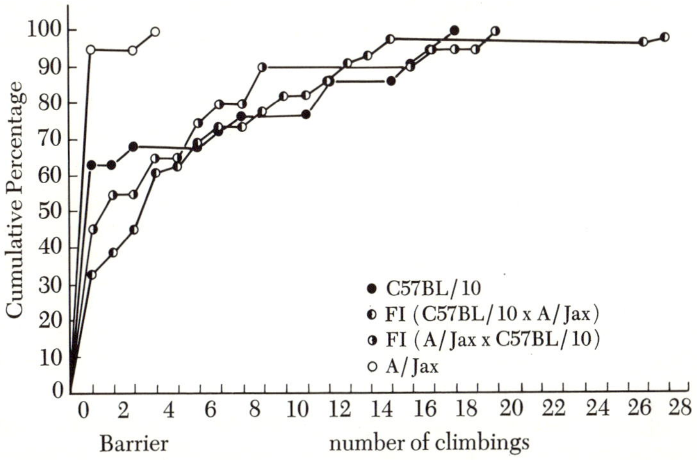

Fig. 4. Cumulative distributions for parental strains and reciprocal F_1s for number of climbings in Barrier.

data show that C57BL/10 and A/Jax differ, and that the F_1s differ from A/Jax, but not from C57BL/10.

Discussion

The principal results of Experiment I are, first, the successful general replication of Thompson's results in Arena, and, second, the demonstration that strain differences in activity in novel situations are not specific to any one technique of measurement. These conclusions are supported by further work of Thompson (1956) in which similar strain rankings were found in a **Y** maze and in an Arenalike situation. The situational generality of the behavior suggests that a fundamental mouse "personality" dimension is being measured.

In view of the essentially continuous array of strain medians in this study, and the similar earlier results, a polygenic (multiple-factor) model appears more adequate than a single-locus model, which would require an unreasonable number of supplementary assumptions. Furthermore, the high coefficient of concordance among the tests indicates that a common genetic system is involved in determining the strain differences on the various tests.

Whereas it is undoubtedly only an approximation, the initial working hypothesis assumes that highly inbred animals are homozygous at all loci. The F_1s of Experiment II, therefore, will be heterozygotic at all loci at which the C57BL/10 and A/Jax strains differ, and the F_1 genotypic value will be midway between the genotypic values of the parental strains. The genotypic intermediacy is accompanied by phenotypic intermediacy in Arena and Open Field, but not in Hole in Wall and probably not in Barrier. This outcome suggests that "exploratory behavior," as it is currently conceived, is made up of at least two subcharacters. The postulation of subcharacters is not inconsistent with the above conclusion concerning a common genetic system. Mather (1949, p. 28) has pointed out that ". . . we must expect that some genes will affect both of any pair of sub-characters while others may affect only one of the pair. The variation of the two will then be correlated but not completely so, and the degree of correlation cannot be predicted. It will depend on the sub-characters in question and on the genes which are contributing to the variation."

Finally, it must be noted that, although it has been convenient to refer to the behavior investigated as "exploratory behavior," it may be equally accurate to consider it to be a measure of timidity, fear, spontaneous activity, sensory acuity, or perhaps even intelligence (running about in a strange situation may prove to be either quite intelligent or quite stupid in terms of natural selection). Further investigations of the genetic mechanism, including study of the physiological link between the genotype and the phenotype, should contribute substantially to the understanding of the biological nature of the behavior dimension.

Summary

Data were obtained on the behavior of six inbred strains of mice, and of the F_1s between two widely differing strains, in four different situations. Two situations involved running, one around barriers, one in an open field. Two situations involved climbing, one through a hole in a wall, one over a barrier.

Wide differences exist among the strains, and there is high consistency in strain rank over the four tests. It is proposed that a polygenic system is common to all tests but that there may also be loci of specific function.

References

Committee on Standardized Nomenclature for Inbred Strains of Mice. Standardized nomenclature for inbred strains of mice. *Cancer Res.*, 1952, *12*, 602-613.

Fredericson, E. The wall-seeking tendency in three inbred mouse strains (*Mus musculus*). *J. genet. Psychol.*, 1953, *82*, 143-146.

Mather, K. *Biometrical genetics.* London: Methuen, 1949.

Siegel, S. *Nonparametric statistics for the behavioral sciences.* New York: McGraw-Hill, 1956.

Thompson, W. R. The inheritance of behaviour: Behavioural differences in fifteen mouse strains. *Canad. J. Psychol.*, 1953, *7*, 145-155.

Thompson, W. R. The inheritance of behavior: Activity differences in five inbred mouse strains. *J. Hered.*, 1956, *47*, 147-148.

Inheritance of Avoidance Conditioning in Mice: A Diallel Study

ROBERT L. COLLINS

Although most genetic research has been concerned with the study of morphological and physiological characteristics, there is considerable current interest in the inheritance of behavior

From Science, 1964, *140*, 1188-90. Reprinted by permission of the author and publisher.

characteristics. Behavioral differences have been demonstrated among highly spective F_1 crosses in areas of locomotion activity, exploration of novel environments, and other relatively simple classes of behavior (*1*). Whereas some recent evidence implicates the importance of hereditary differences in learning behavior (*2*), most previous studies considered only a few pure strains, or two pure strains and their crosses. Consequently, the findings may be applicable only to the strains chosen and may not be representative of genetic relationships in differing strains. The objective of this investigation was to study the range of genetic influence on learning in mice by using a diallel mating system in which all intercrosses among a set of inbred strains were tested for the acquisition of a learned avoidance habit.

Ten males and 10 females from each of the 25 mating combinations of the following 5 pure strains of mice were tested for acquisition of avoidance conditioning: A/JAX, BALB/c, C3H, C57BL/10, and DBA/1. A total of 500 mice were tested. At approximately 100 days of age, each subject was placed in a shuttle-box, measuring 26.5 by 5 by 4 cm, and received 200 consecutive avoidance-conditioning trials. The conditioning stimuli were a muffled buzzer and a 3000 cy/sec sine-wave tone which began 2.5 seconds after the onset of the buzzer. Electric shock was programmed to be delivered to the grid floors 5 seconds after the onset of the first cue. A jumping response across a continuously charged pit during the first 5 seconds of each trial activated a photosensitive control system which postponed shock for that trial and was considered an avoidance response. A pit-crossing after 5 seconds terminated the cues and shock and was considered an escape response. The interval between trials was fixed at 30 seconds. Four mice were trained simultaneously according to this schedule in separate automatic testing chambers. The score for each subject was the number of avoidances in 200 trials.

The data were arranged in a 5 by 5 square design, in which the rows represented the genotype of sires, and the columns, the genotype of dams. The 25 squares thus formed represented the 25 genotypic combinations from all matings of the 5 pure strains, and included 20 F_1 crosses and the 5 pure strains. Two complementary statistical models were used to assess the data. Model 1 represented a factorial analysis of variance by which the genetic effects (gametes) of dam and sire could be evaluated independently of each other. Model 2 represented an extension of Griffing's combining abilities analysis of variance, which enabled both sexes of offspring and resulting interactions to be included (3). By this model the effect of the five pure strain gametes upon learning was assessed simultaneously in sires and dams. Table 1 presents a summary of the statistical tests.

Wide differences in learning ability were evident among the 25 genotypes. Figure 1 illustrates the avoidance conditioning scores for each of the 20 hybrid crosses in comparison with the scores of the respective parental strains. The avoidance means of the highest scoring genotypes, C3H × DBA/1, DBA/1 × C3H, and C57BL/10 × C3H, were approximately eight times the mean of the lowest scoring pure strain, BALB/c. The highly significant dam and sire effects, as well as the significant general combining abilities effect in Table 1, demonstrate that the strain of the parents considerably influenced the level of performance of the F_1 crosses. The performance score of hybrids in which there was a C3H parent was approximately 12 percent higher than the mean of all crosses. Conversely, hybrids in which there was an A/JAX parent performed 7 percent below the mean of all crosses.

The highly significant dams × sires interaction and the specific combining abilities effect indicate that the scores of certain genotypes were statistically different from an expectation based on the observed additive effects of the parental genes. This interaction was primarily due to the low scores of the five pure strains and the unusually high performance of C3H × DBA/1 hybrids.

Female mice were superior to males; females averaged 26.9 percent avoidances while males averaged 19.3 percent. The absence of any significant genetic interactions with sex of offspring indicates that essentially identical conclusions could have been drawn from studying either sex of offspring alone. In particular, the absence of the reciprocal × sex interaction indicates that sex-linkage did not affect avoidance acquisition.

However, other findings at this laboratory indicate that sex-linked genes in C3H × DBA/1 hybrids do have some effect on avoidance extinction as well as on the weight of the adrenal gland (*4*).

To test the effect of possible influences of the early maternal environment upon learning uncomplicated by genetic variables, orthogonal contrasts were formed between individual pure strain dam and sire total scores (row and column totals). This test indicated that no significant differential maternal effects were associated with rearing differences among the five pure strain mothers ($p > 0.10$; $F = 1.94$, $df = 4{:}350$). However, cross-fostering techniques would be needed to evaluate maternal influences in greater detail.

An indication of the mode of inheritance underlying a quantitative characteristic may be obtained by comparing the mean for each hybrid against the average of the respective parental scores (midparent average). If the hybrid score is the same as the midparent value, or lies between it and the higher scoring parent, the mode of inheritance is termed intermediate or partially dominant, respectively. If the hybrid score is equal to, or exceeds the score of the higher scoring parent, the mode of inheritance is called dominant or over-dominant, respectively. In 15 of the 20 comparisons in Fig. 1, the learning performance of F_1 crosses was higher than the midparent average, and in 14 cases, the hybrids exceeded the score of the higher-scoring parent. A statistical evaluation of the 20 comparisons by orthogonal contrasts demonstrated that F_1 crosses learned to avoid shock to a higher degree than would be expected on the basis of the predicted additive effects of the parental genes ($p < 0.0001$; $F = 51.9$, $df = 1{:}450$). Thus, the overall mode of inheritance of avoidance conditioning in F_1 crosses was dominant, and in most cases was overdominant.

The superiority of the F_1 cross compared to the midparent expectation, usually called heterosis, or hybrid vigor, has been repeatedly demonstrated in hybrid progeny obtained from the crossing of highly inbred lines. Whether this heterosis arises from interlocular interactions, the complementary gene hypothesis, or from intralocular interactions, the hypothesis of overdominance, cannot be settled from these data (*5*). Since

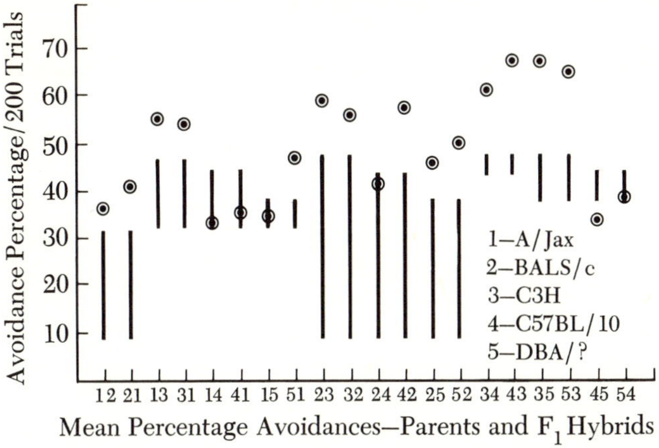

Fig. 1. Average scores for individual hybrid groups compared with the mean scores of the respective pure-strain parents. The circled dot represents the score of ten male and ten female hybrid mice; the endpoints of each line represent the learning scores of the respective parental strains. The first number of each F_1 designation indicates the strain of dam; the second number refers to the strain of sire. For example, 12 is the F_1 cross from the mating of an A/JAX female to a BALB/c male.

quantitative behavioral differences most probably arise from polygenic origin rather than from single-gene action, either hypothesis or both may account for the observed heterotic effect. Whatever the mechanism or genetic route, the important finding is that heterozygotes are clearly superior to their inbred parents in avoidance learning. Aside from the practical consideration of breeding faster-learning mice, the observed hybrid vigor suggests that avoidance learning may be of significant adaptive value to a species. Bruell has theorized that only quantitative polygenic traits, which were biologically adaptive and under past selection pressure, would result in heterotic inheritance, and that traits which possessed negligible adaptive value would be inherited in an intermediate manner (*6*). This suggests that the degree of heterosis induced by crossing inbred lines may be a useful index of the biological adaptiveness of a quantitative phenotype.

This study suggests that behavior-genetic research in con-

TABLE 1

Summary of statistical evaluations of avoidance conditioning performance among 25 genotypes of mice. Analyses according to models 1 and 2.

Source	*df*	Mean square	F-ratio
	Model 1		
Dams	4	5,839	16.82*
Sires	4	5,131	14.78*
Sex	1	25,891	74.57*
Dams × sires	16	2,923	8.42*
Dams × sex	4	93	0.27
Sires × sex	4	625	1.80
Dams × sires × sex	16	401	1.16
	Model 2		
GCA†	4	10,310	29.70*
SCA‡	10	4,366	12.58*
Reciprocals	10	574	1.65
Sex	1	25,891	74.57*
GCA × sex	4	408	1.18
SCA × sex	10	437	1.26
Reciprocals × sex	10	328	0.95
Error	450	347	
Total	499		

* $P < .0001$. † General combining ability.
‡ Specific combining ability.

junction with morphological and physiological studies may be of greater value in the investigation of evolutionary theory than morphological and physiological research alone.

REFERENCES AND NOTES

1. J. H. Bruell, in *Roots of Behavior*, E. L. Bliss, Ed. (Hoeber, New York, 1962); J. L. Fuller and W. R. Thomson, *Behavior Genetics* (Wiley, New York, 1960).

2. J. R. Joyce and M. Covington, *J. Comp. Physiol. Psychol.* 53, 197 (1960); G. Lindzey and H. Winston, *ibid.* 55, 748 1962); J. A. King and A. Mavromatis, *ibid.* 49, 465 (1956); C. Y. Nakamura and N. H. Anderson, *ibid.* 55, 740 (1962).
3. B. Griffing, *Australian J. Biol. Sci.* 9, 463 (1956); R. L. Collins, thesis, Western Reserve Univ., 1963.
4. R. J. Meckler and R. L. Collins, in preparation.
5. G. W. Gowan, *Heterosis* (Iowa State College Press, Ames, 1952).
6. J. H. Bruell, *Am. Zoologist,* in press.
7. This report is a portion of my Ph.D. dissertation at Western Reserve University and was supported in part by a U.S. Public Health Service predoctoral research fellowship. National Institutes of Health grant HE-7216-02, and National Science Foundation grant G-14410. I thank Jan H. Bruell for advice and encouragement during this investigation.

Genetic Aspects of Learning and Memory in Mice

DANIEL BOVET, FILOMENA BOVET-NITTI,
ALBERTO OLIVERIO

One of the most controversial issues in biology since the time of Locke and Leibniz has been that of the role exerted by heredity and environment—what is inborn and what is acquired. Two factions arose in attempting to explain this delicate balance. One minimized the role of genetics as a determinant of behavior, and the other (mainly represented by ethologists and geneticists) emphasized the importance of heredity (*1*).

As a matter of fact, the genetic determinants of behavior have been confused with, or identified with, fixed action patterns currently classified as instinctive. Heredity, however,

From *Science,* 1969, *163,* 139–49. Reprinted by permission of the author and publisher.

also affects many forms of behavior less organized or less systematic than nest building, sexual behavior, or some forms of highly organized aggressiveness.

Each individual or species is provided with genetic baggage responsible for the so-called spontaneous motor activity, for the search for new stimuli (which are the grounds of the exploratory behavior and curiosity), and, more generally, for all types of motivational and emotional behavior. Besides, any type of adaptive behavior, like the aptitude to respond to a conditioned stimulus, to learn, and to remember, is genetically determined. Heredity plays an important role in different forms of adaptive behavior, since individual reactions to environmental influences and the ability of man to benefit from cultural heritages depend on hereditary factors.

Due to the difficulties of controlling the genetic and environmental determinants and the role played by cultural factors in human societies, it seems necessary to start with experimentation on animals in order to confront the problem of "nature and nurture" in men. The use of other animal species involves a less-complex approach. The problem of nature and nurture constitutes the target of interest in today's psychobiology, a new branch of biology that affords a tentative bridge between neurophysiology and comparative psychology, and ethology. Psychobiology attempts to study whether, and to what extent, learning behavior is genetically determined.

Behavioral Genetics

One of the first extensive experiments in the field of psychogenetics was Tryon's selective breeding program (2). In mating rats which displayed characteristic traits, Tryon selected those among their offspring with the highest or lowest performance in a standardized maze and bred their off-spring. Bright maze learners produced bright progeny, whereas dull maze learners produced dull progeny. Tryon's experiments, based on carefully controlled breeding and environmental factors, established the inherited nature of this adaptive behavioral trait. In light of these experiments it seemed advisable to use animals bred selectively, from genetically homogeneous strains to assess the effects of heredity on behavioral aptitudes (*3*).

In agreement with Hall (*4*), the main objectives of psychogenetics are (i) to discover whether a given behavior pattern is transmitted from generation to generation, (ii) to determine the number and nature of the genetic factors (*5*) involved in the trait, and (iii) to locate the genes on the chromosomes. We discuss here the first of the above problems.

Scott, Frederickson, and Thompson (*6*) demonstrated that inbred strains of mice differ not only with respect to certain morphological traits but also with regard to behavioral traits. As far as adaptive behavior is concerned, the acquired behavior of inbred mise has been studied by the use of a reward as motivation; in a few investigations, avoidance conditioning has been used (*7*).

Individual and Strain Performances in Avoidance Behavior

Our experiments with different conditioning tests for screening the action of psychotropic agents demonstrated that the large individual variability of a heterogeneous population was an obstacle to the assessment of the factors affecting the patterns of learning and retention, characteristic of a given species. By the use of different strains of inbred mice, we developed an avoidance technique similar to one which we had used for the rat. In this paper we describe results obtained by studying the avoidance behavior of different strains of mice in a shuttle box.

Double-Compartment Grill-Box Technique

The apparatus used for the study (*8*) of escape and avoidance responses is a two-compartment cage adapted from Warner's model (*9*). Each apparatus consists of a rectangular plexiglass box divided into two equal compartments which are connected by a small opening (Fig. 1). The floor is a tilting platform of stainless steel rods. On each trial the conditioned stimulus (a

Fig. 1. Shuttle box used for the study of avoidance learning in mice.

lamp which lights the compartment where the animal is) is preceded by a constant interval (5 seconds), and the unconditioned stimulus is represented by a continuous electric shock administered through the grid floor. An unconditioned escape response is recorded when the mouse shuttles into the adjacent compartment after the onset of the shock. A conditioned avoidance response is recorded when the mouse avoids the shock by running into the other compartment within 5 seconds after the onset of the light. A series of eight automated shuttle boxes driven by the same programming device was used.

An advantage of using the shuttle box in experiments on avoidance lies in the type and the constancy of the motivation adopted. Staying alert to avoid and escape from predators is one of the basic requirements for survival for a wild species (*10*). A technique based on an escape reaction gives reliable and homogeneous results in that it appeals to a type of inborn behavior.

Other reasons support the study of escape and avoidance behavior. (i) In contrast to what happens in operative behavior, and typically in a lever-pressing apparatus, it is possible to study the initial learning phases in a shuttle box, because escape and avoidance reactions are closely related. (ii) The anticipatory nature of the acquired responses makes this type of learning the prototype of an intentional "forward looking" behavior different from other types of "contiguity" learning (*11*). The complexity of this task is also reflected in the large number of trials required: this is in contrast to the "single-trial" tasks used to measure other types of learning. This complex task allows analysis of the different phases of the learning processes.

The use of an aversive stimulus in the Warner cage has been widely criticized. The use of electric shock gives the advantage of a constant and adjustable motivation but might also be responsible for emotional components which can be variously interpreted.

Harlow (*12*) claimed that "the double compartment grill box is without doubt the most efficient torture chamber which is still legal." Actually, the intensity of the electric shock generally used for testing small laboratory rodents is only sufficient to keep the animal alert. A real state of anxiety is evident only during the first trials, and the experimenter can select conditions which minimize this phenomenon. Fear and anxiety, which may be an emotional reaction of freezing, are largely dependent on species and strain. One of the reasons for using mice is that in most of the strains this emotional component can almost be ignored.

In an amusing experiment we showed that when a "bridge" was placed between the mouse pen and the shuttle box where the animal had been trained, the mouse spontaneously returned to the shuttle box, presumably because of curiosity, in spite of the risk of receiving an electric shock.

Behavioral Variability of a Common Noninbred Strain

In establishing a learning curve, the experimenter has the same difficulties as a man who is training a dog or taming a beast. These troubles have often been matters of perplexity or jokes. For instance, Scott (*13*) says that "there is an old joke among biologists known as the Harvard law of animal behavior: when a stimulation is repeatedly applied under conditions in which environmental factors are precisely controlled the animal will react exactly as it pleases."

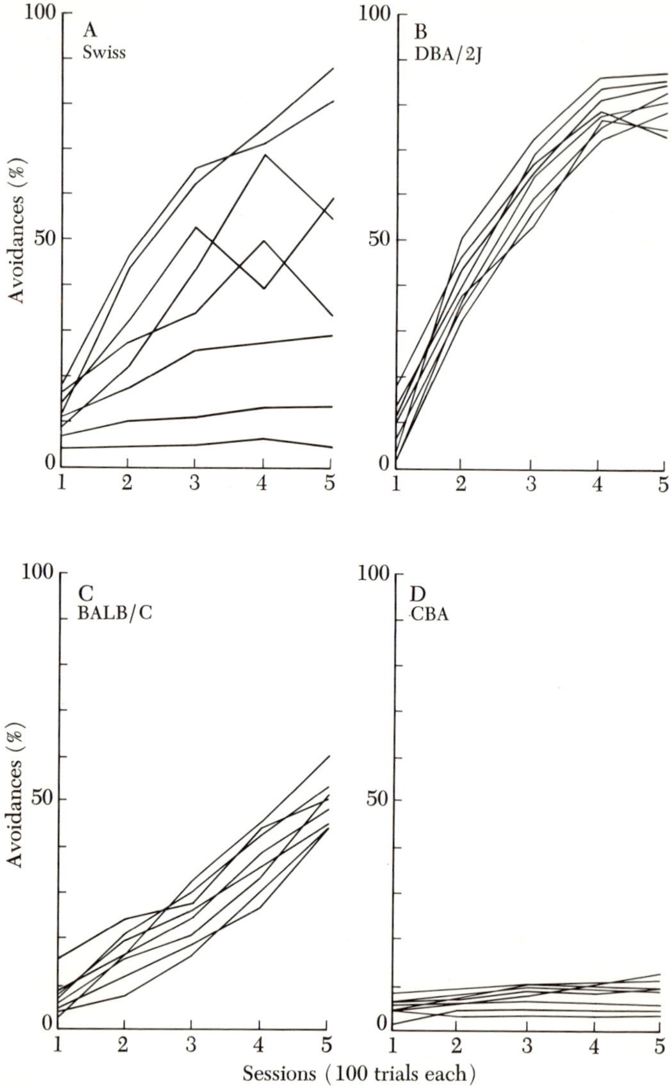

Fig. 2. Avoidance learning in (A) a heterogeneous population of Swiss mice; (B) DBA/2J mice; (C) BALB/c mice; (D) CBA mice. Each curve represents the individual performance of a mouse during five avoidance sessions of 100 trials each.

When Warner (*9*) described his technique for testing avoidance behavior he was confident that it would become an "ideal and universal yardstick for the measurement of the learning capacity." However, the results did not always support his hopes. Also, with animals of different species trained in a shuttle box, individual differences between subjects of the same species and even of the same colony represent a bias and sometimes make interpretation of the results hazardous.

Even when the techniques used are not considered, individuals may appear refractory to any training because the behavior may differ as widely between two animals of the same colony as among fish, birds, rats, or even monkeys. In this respect, our findings agree with numerous reports on conditioning, maze learning, or operative behavior.

A specific example (*14*) shows the extent of the individual differences in groups of Swiss Webster mice from one colony and trained in a shuttle box (Fig. 2). A comparison between the individual response curves shows that different levels of performance are attained by each subject. Some animals reach a high rate of response during the first session, whereas other

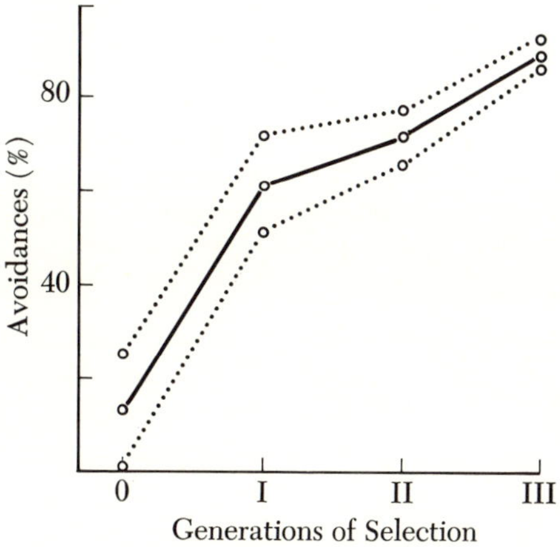

Fig. 3. Selective breeding for avoidance learning in the mouse (Swiss Webster). Each point represents the performance reached at the end of five 100-trial sessions (solid line). Dotted lines, fiducial limits of the mean.

mice seem to be unable to associate the conditioned stimulus with the escape response (*15*).

The level of performance of a given animal shows sharp fluctuations from session to session. This last finding, that is, a sharp decrement in performance between consecutive sessions, can be interpreted in different ways. From an ethological standpoint, if the behavior of a laboratory animal is compared to that of the species, we might ascribe this type of inhibitory behavior to the reaction of immobility usual in the presence of danger, a type of instinctive behavior well known among the naturalists. From a psychological viewpoint, freezing behavior has generally been interpreted as a type of emotional behavior associated with experimental neuroses. In our experiments we used strains of mice in which this type of emotional behavior is absent or at a minimum.

The extreme variability in a heterogeneous population of Swiss Webster mice led us to initiate a selective breeding experiment to obtain a line of mice with high performance levels in the shuttle box. Previous experiments had shown that within the span of a few generations it was possible to breed two lines of rats characterized by patterns of high or low avoidance (*16*).

From Fig. 3 we note that a large group of Swiss mice (32 subjects) attains a rather low performance at the end of five 100-trial sessions. If the high avoiders are mated together performance improves within a generation from the original 15 percent to 60 percent. Selective breeding, continued until the third generation, produced a population of mice with a performance level of 85 percent.

Selective breeding not only produced a line of high avoiders, but was also accompanied by progressively reduced individual variability. The fiducial limits of the mean decreased from the original level of 23 percent to 4 percent in the third generation.

Fig. 4 (opposite page, top). Avoidance conditioning of nine strains of inbred mice during five consecutive daily sessions of 100 trials. Fig. 5 (opposite page, bottom). Maze learning (Lashley III maze) in nine strains of inbred mice. Each point represents the mean errors of 16 mice given one daily trial for ten consecutive days.

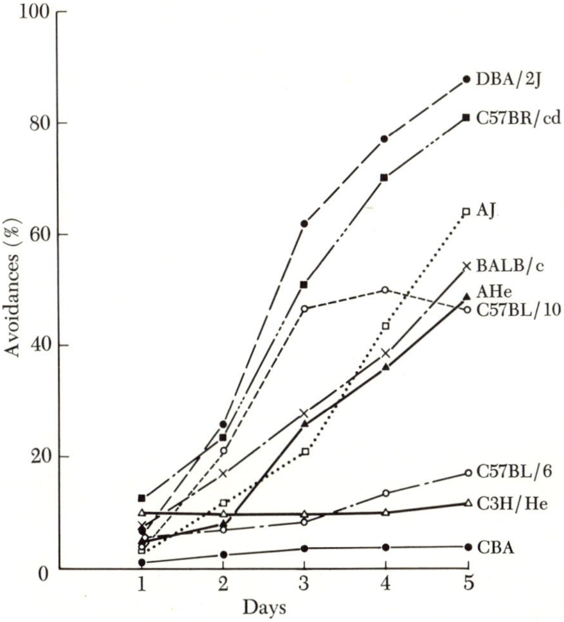

100
80
60
40
20
0
Avoidances (%)
1
2
3
4
5
Days
DBA/2J
C57BR/cd
AJ
BALB/c
AHe
C57BL/10
C57BL/6
C3H/He
CBA

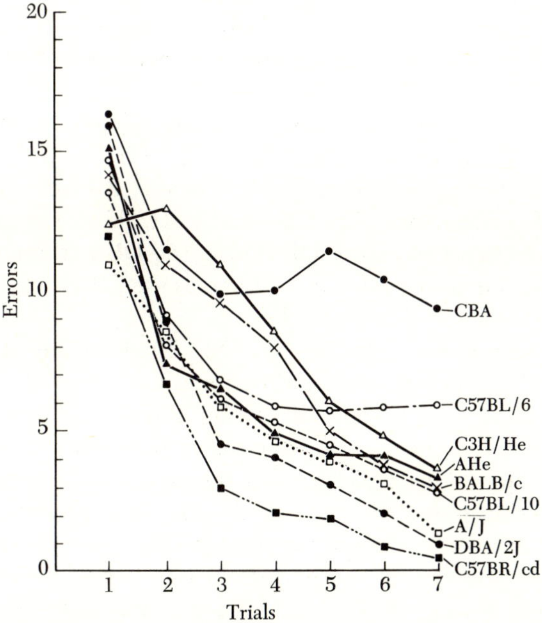

20
15
10
5
0
Errors
1
2
3
4
5
6
7
Trials
CBA
C57BL/6
C3H/He
AHe
BALB/c
C57BL/10
A/J
DBA/2J
C57BR/cd

Strain Differences in Avoidance Acquisition

Similar to what we witnessed in experiments based on the selection of a given trait, a comparison between the performances of different strains of mice shows that avoidance learning is genetically determined. The advantage of using mice comes from the many strains now available for research.

Figure 4 gives the avoidance performance of nine strains assessed by training different groups of naive animals. A comparison between the shuttle-box performance attained during a cycle of five daily sessions under rigid control of the experimental conditions shows the following.

1) Three strains (CBA, C3H/He, and C57BL/6) were quite poor in acquisition of avoidance learning. The mean of the responses during the last session was lower than 20 percent (*17*).

2) In four strains (A/He, A/J, BALB/c, and C57BL/10) the mean percent of avoidances had reached, at the end of the experiment, values ranging between 40 and 70 percent.

3) Finally, two strains (C57BR/cd and DBA/2J) attained a very high level and during the last session reached an 80-percent level of performance.

The results obtained with genetically related strains show that the three strains from the group C57, although they belong to the same "strain family," attain strikingly different levels of performance. The CBA and C3H/He strains are both derived from a crossing of DBA mice; the DBA strain reached a high percentage of avoidances whereas the CBA and C3H/He remained at a very low level (*18*).

Our results for C3H/He and DBA/2J mice are in agreement with those reported by Royce and Covington, by Royce, and by Meier and Foshee for solving ability in a water maze (*19*). There is, however, a discrepancy between the performance of the strains tested in this research and that reported by Schlesinger and Wimer, which can be explained by the fact that different techniques were used (*19*).

The importance of the genetic factors is evident from the differences among strains and from the homogeneity of per-

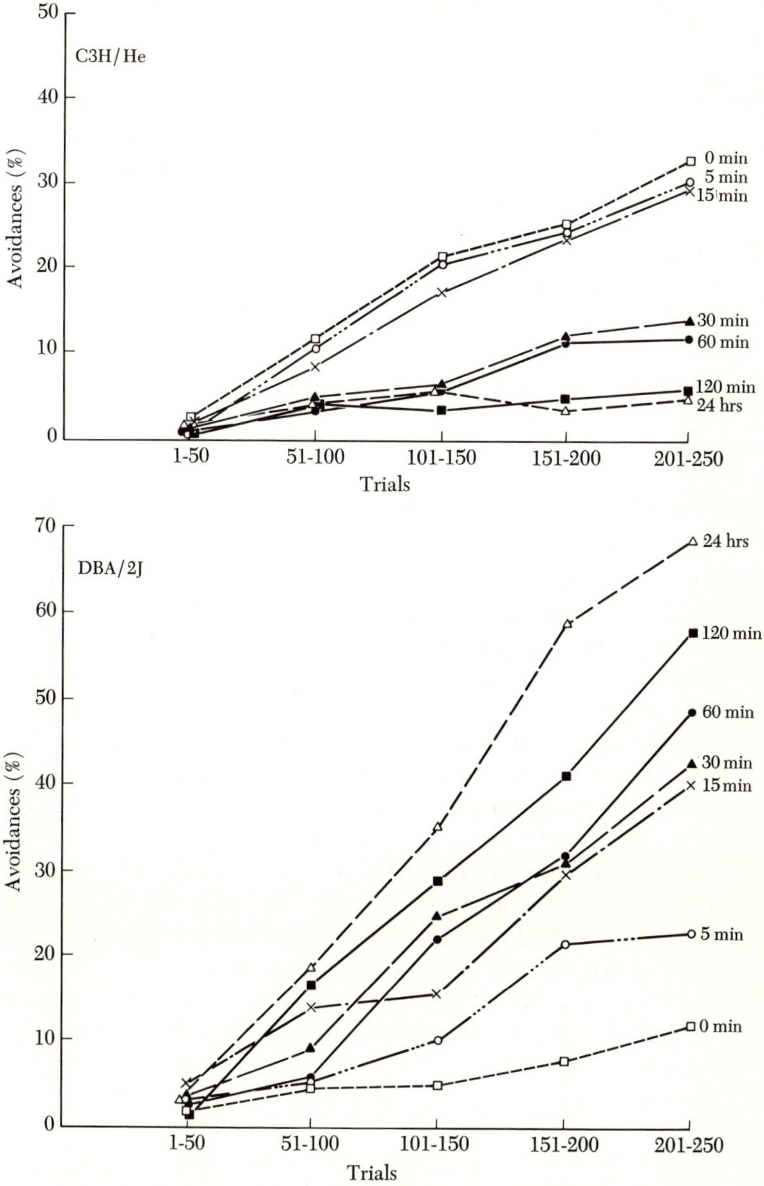

Fig. 6. Comparison of the performances of the strains C3H/He and DBA/2J when various time intervals between five sessions of 50 trials are used. The numbers at the end of the curves represent the length of the interval.

formance in animals of the same strain. The homogeneous behavior of inbred mice contrasts with the heterogeneous performance of randomly bred mice. The fiducial limits of the mean are in fact very low (< 10 percent) in the strains DBA/2J, C57BR/cd, and BALB/c, and lower than 20 percent in the remaining six strains (Fig. 2).

The wide variability of adaptive behavior of animals in a homogeneous population which, for almost a century, have been selectively bred for their activity or absence of aggressivity, is astonishing. It is to be expected that in a wild species the variability would be even greater (*20*).

Relations between Avoidance and Maze-Solving Behaviors

Different strains of inbred mice attain different levels of performance when trained in a shuttle box. On the basis of these results and of previous findings showing sharp differences among inbred strains with respect to ability as judged by performance in a maze (*6*), we trained nine strains of mice in a small Lashley III maze. Figure 5 shows the performance of the different groups of animals given one trial a day for ten consecutive days. In agreement with previous results (*19*), ability to get through a maze also appears to be genetically determined, in that some strains (DBA/2J and BALB/c) of mice quickly attain good performance, whereas other strains (C57BL/6 and C3H/He) are poorer learners. The agreement between our findings and those of the authors quoted above (*19*) is particularly interesting in that those strains of good maze learners also attained a better performance when trained in the shuttle box. However, though these findings appear suggestive, it seems premature to extend the meaning of this particular problem-solving or learning ability.

Strain Differences in Learning Patterns

Can behavioral data be used to enable the synthesis of a concept of the process of retention? Having observed the striking differences in the performance of learning tasks among strains of inbred mice, we analyzed the nature of these different types of behavior by studying the effect of variable schedules of conditioning. The rather surprising conclusion was that the differences between the various strains were not only quantitative but also qualitative, and that at least two different mechanisms of retention seem to be involved in learning processes of the various strains.

Dependence of Learning Differences on Massed or Distributed Practice

We demonstrated that different strains of mice are characterized by different levels of performance. Further analysis revealed that these differences arc not only quantitative but also qualitative as evidenced by the shape of the curve representing acquisition. An example of this fact comes from comparing the response of strain DBA/2J to that of strain C3H/He (*21*).

Let us start by asking whether our mice prefer a long uninterrupted training session or a series of short ones, and, in case they choose the second solution, how long the break should be to take advantage from the practice. Actually this is one of the most controversial and debated problems in social psychology and experimental pedagogy.

The experiments were conducted under comparable, rigid conditions. The interval between each trial consisted of 30 seconds; a total of 250 trials was massed in a single uninterrupted session (interval 0 minute) or spaced in five sessions (each session consisting of 50 trials) the interval between each session being 5, 15, 30, 60, 120 minutes, or 24 hours. The performances achieved by the two strains under these conditions are completely opposite (Fig. 6). Strain C3H/He attains good performance only when the trials are massed in a single (125-minute) session, whereas performance is very poor when the interval between each session is 2 or 24 hours. By contrast, strain DBA/2J performs better under distributed practice when the interval between each session is increased.

Another way to space or mass the trials is by increasing the

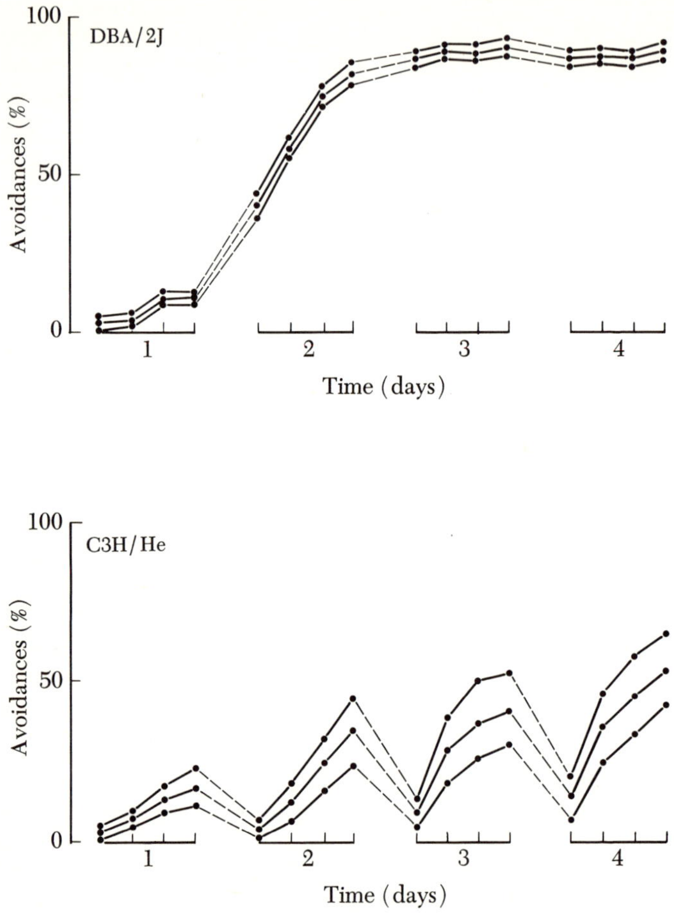

Fig. 7. Progression of the performance within each session and in the interval between consecutive sessions in C3H/He and DBA/2J mice. Solid lines, mean percentage of avoidances in groups of 32 mice during four daily sessions of 200 trials each. Dotted lines, fiducial limits of the means.

length of the interval between trials. By this means we can compare performances of two groups of mice subjected to a single 200-trial session in which the trials were spaced by 30 or 120 seconds. In this case (by giving in the same period of time a different number of trials), C3H/He mice perform better when the interval between trials is shorter (30 seconds), whereas DBA/2J mice attain a higher level of response when the trials are more widely spaced (interval between intertrials, 120 seconds).

Differences between strains can be analyzed by studying the variations of performance during each session and during the interval between consecutive sessions (*21*). Figure 7 gives the results of an experiment in which mice of two strains (C3H/He and DBA/2J) were given four sessions of 200 trials each with intersession intervals of 24 hours. The results show different patterns of behavior. If the variations of performance during the interval between successive sessions are considered (that is, between the level observed at the end of a given session and that observed at the beginning of the following session) the results show: (i) a decrement of performance in the strain C3H/He; and (ii) a marked improvement of performance between the last trials of session 1 and the first trials of session 2 in DBA/2J mice.

The type of learning pattern evident in DBA/2J mice is the one most frequently observed in the other strains studied. An "intermediate" pattern was evident in BALB/c mice. The first results of experiments conducted on strains of rats and guinea pigs show that avoidance learning in these two species often takes place in a "DBA/2J-type" (*21*).

Performance after Rest between Long Avoidance Sessions

A third experiment was conducted in which recovery of the performance at the end of a prolonged session was compared in the two strains, in the context of research on the effects of reactive inhibition (fatigue) and rest on avoidance performance of mice subjected to prolonged avoidance sessions (*22*). In a first experiment groups of trained mice were given a 2000-trial session (16 hours and 30 minutes). A decrement of the performance was evident in both C3H/He and DBA/2J mice after the first 3 to 4 hours (400 to 500 trials). This decre-

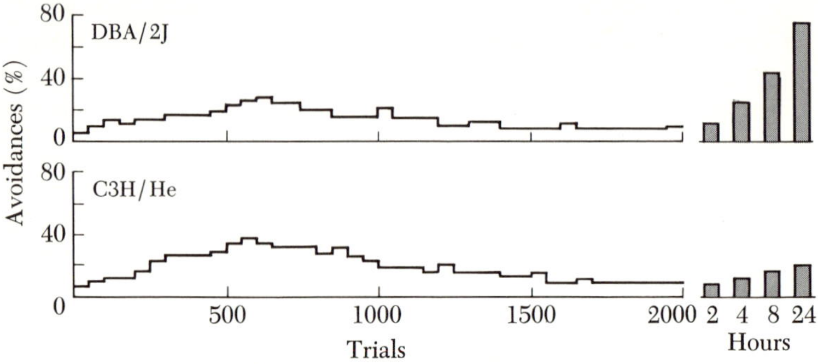

Fig. 8. Performance of two different strains of mice during an uninterrupted (16.5-hour) avoidance session consisting of 2000 trials. Columns at the right represent the performance of four different groups of mice given an additional 100-trial session after different intervals of rest from the end of the previous session.

ment progressively became more pronounced; at the end of the session the rate of avoidance responses was very low. For both strains a rest period of at least 16 hours was needed before they could again perform at their initial level. Figure 8 shows the difference between C3H/He and DBA/2J naive mice when both strains were subjected to the same type of session.

A progressive increment in the rate of responding (higher in C3H/He mice) is evident during the first part of the session. As far as trained mice are concerned a progressive decrement in performance, evident 4 to 5 hours after the beginning of the session, became more evident until the end of the session.

A second difference in the behavior of the two strains (besides that observed in conditioning rates) was in their performance after different rest intervals after the end of the 2000-trial session. After 16 hours, the performance of the C3H/He mice returned to levels similar to, or lower than, the highest (peak) level reached during the uninterrupted session. It was rather surprising to observe that after a similar rest period DBA/2J mice attained high performance (about 70 percent), a highly significant improvement as compared to the highest performance attained during the previous session.

In the most general terms the results show that after the rest period, the responding was a function of the retention mechanisms characteristic of the strain. Under these experimental conditions, as in the previous experiment, consolidation mechanisms were active in DBA/2J mice and defective in strain C3H/He.

Strain Differences in Inhibitory One-Trial Task for Avoidance Learning

The results obtained for the two strains of mice show that there are two time-dependent processes in memory storage. However, we were concerned about the possible dependency of this phnomenon on the technique used. In other words, did the differences observed in DBA/2J and C3H/He mice with respect to patterns of learning and consolidation result only because a particular active avoidance task was presented?

In order to investigate this problem, groups of ten mice belonging to the two strains received a single "training" trial on a task involving inhibitory avoidance (*23*) and a "retention" trial given at time intervals from 2 minutes to 24 hours

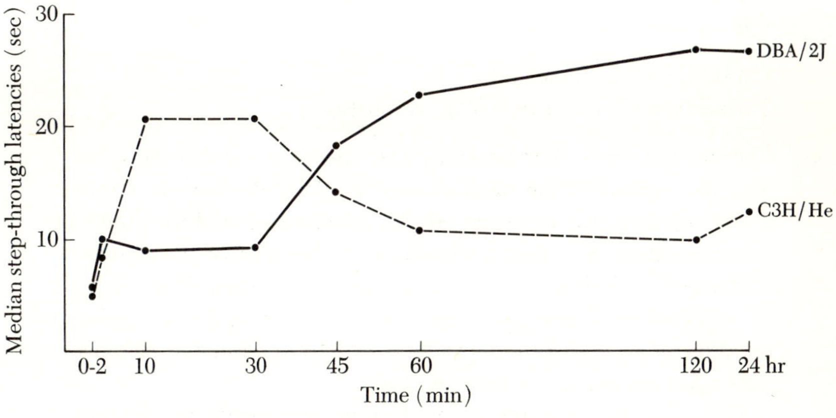

Fig. 9. One-trial passive avoidance learning in two different strains of mice.

after the first trial (*24*). On each trial a mouse was placed on a small metal platform leading to a dark chamber. As the mouse stepped through the hole into the box he received a foot shock of about 3 milliamperes. The amount of time the mouse spent on the platform before he stepped into the box was recorded. Figure 9 shows that when the "retention" trial was given 0 or 2 minutes after the training trial, the median latencies to step through the hole were not different in the two strains. However, when longer intervals of time are considered, differences are evident between the two groups of mice. An increment in median latencies is evident in C3H/He mice when the "retention" trial is given 10 or 30 minutes after the "training" trial. After this time a decrement in latencies is evident, and the performance at 24 hours is lower than that at 10 or 30 minutes. This indicates a decrement of memory in time. In DBA/2J mice a poor retention of the "training" experience is evident when short intervals of time separate the "training" from the "retention" trial (2, 10, or 30 minutes). For longer intervals the latencies increase—indicating an improvement of retention. These findings therefore seem to be consistent with the recent interpretation given by McGaugh and by Jarvik on the multiple mechanisms involved in passive avoidance learning (*25*).

Interpretation Based on Two-Stage Concept of Memory Storage

Our results show that not only the speed of conditioning but also the nature of the mechanisms involved depend on strain, and are genetically determined. An interpretation of the studies on reminiscence is often rather complex because of the number of variables which frequently are not controlled. Two types of explanations have been put forward to describe the usual superiority of distributed practice and to explain the reminiscence phenomenon in motor learning.

The first explanation probably does not apply to the type of avoidance learning used in our experiment. It postulates that during the rest pauses there is a recovery from the effects of practice that are detrimental to performance. Key words in these various hypotheses have been "fatigue," "work decrement," "disinhibition," "maturation of anxiety." Hull, in his classic theory, viewed learning as the acquisition of a "reaction potential" at the same time as "inhibitory" elements are building up. He felt that, with distributed practice, inhibitory processes would by and large disappear. Thus, he concluded, more improvement should result when practice periods were spaced (*26*).

A second and more probable interpretation postulates the role of an active consolidation process in reminiscence. The hypothesis of two-phase or multiple-stage retention is to be considered. During an early phase, the labile trace of the process of sensorial or ideational registration process corresponds to the so-called short-term memory. It is followed by the formation of a stable latent trace, potentially capable of subsequent evocation, which corresponds to the so-called long-term phase of memory. The hypothesis has been advanced that dynamic processes such as reverberation consolidation or perseveration, integration, and organization could act during the phase of storage. During this process of storage the labile trace may go through an autonomous decay or may be modified by new experiences such as a learning or interference (*27*).

A neurophysiological model of a dual mechanism of memory storage based on neurophysiological circuits (during the first phase) and on a molecular coding (during a second phase) has been suggested by Hebb, Gerard, and Young (*28*). Biochemical and pharmacological studies provide support for molecular coding (*29*). The experimental data shown above concerning the discrepancies of the learning curves of two strains of inbred mice strongly support the hypothesis of such a dual mechanism. In agreement with this concept, the strain C3H/He should be characterized by a good short-term memory and by autonomous decay, or by a rather ineffective consolidation mechanism. However, the processes of consolidation are a determinant in avoidance learning of DBA/2J mice. This interpretation at the same time explains the differences of performance that depend on massed or distributed practice (such as the patterns of improved or impaired performance evident during the intersession interval), the latent learning in long sessions, and the strain differences in the passive avoidance task.

Effects of Electroconvulsive Shock, Age, and Drugs on Consolidation

The interpretation already proposed concerning the nature of the differences in the learning patterns of C3H/He and DBA/2J mice is further supported by experiments dealing with the action of electroconvulsive shock, aging, and administration of convulsant agents after the trial. Electroconvulsive shock, a therapy used in psychiatric treatment (*30*), reportedly produces transient impairments of memory. Retrograde amnesia affects from 50 to 75 percent of the patients treated with shock, although amnesia may not be complete.

Clinical implications aside, the use of electroconvulsive shock opened a new area of research in the psychobiological sciences. The first findings are from Duncan (*31*), who reported that administration of shock after a trial impaired consolidation in rats. This effect was no longer evident if time elapsed between the end of the training session and the application of shock. Analogous experiments were undertaken by the use of other species of animals and other training procedures. Although most of these experiments dealt with "one-trial" learning procedures, we adopted longer training schedules consisting of sessions of 50 trials with intervals of 24 hours between each session (*32*). The results showed that retrograde amnesia occurred if shock treatment was applied immediately after the end of each training session (Fig. 10). The disruptive effects were only evident if shock was administered within 2 minutes after the end of the session. The time courses of the consolidation interval shows a linear regression after log-probit transformation in agreement with results of one-trial learning experiments plotted by Cherkin (*33*).

The analysis of the learning curves within each session enables a distinction to be made between the effects of shock on learning (short-term memory) and on the consolidation

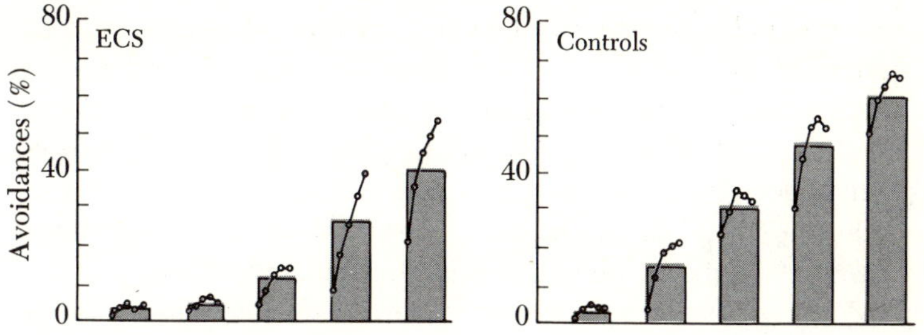

Fig. 10. Effects of posttrial electroconvulsive shock (ECS) on consolidation of shuttle box learning. Shock (15 ma, 100 msec) was administered to DBA/2J mice immediately after the end of each 50-trial session. In additional experiments shock treatment was ineffective if administered 15 minutes after the end of each session. The curves within each session represent the mean avoidances (percent) during five blocks of ten trials each.

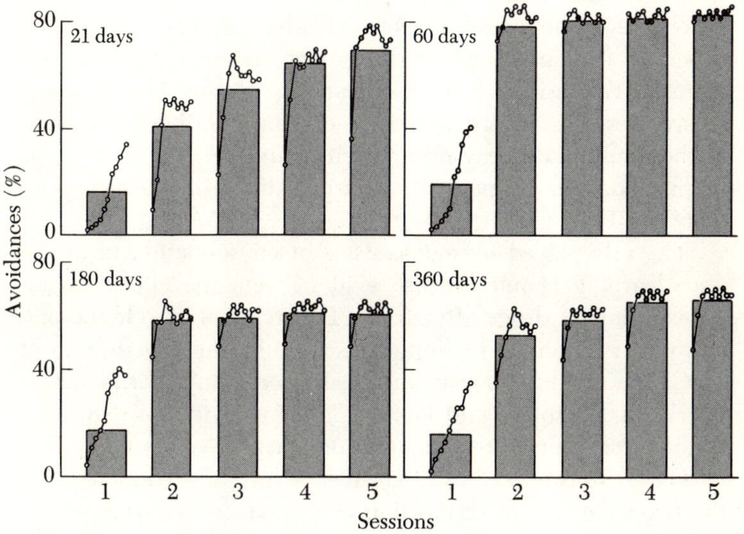

Fig. 11. Differences in the learning and retention curves in mice of different ages during sessions of 400 trials. The curves represent the patterns of performance within each session; each point represents the mean of 40 trials.

processes which occur in the interval between two consecutive sessions (long-term memory). In fact, although the performance of control mice improves between the end of one session and the beginning of the following session, a decrement is evident during the same period in animals treated with shock. With the exception of the initial level of performance, the shape of the learning curves shows no sharp differences between controls and animals treated with shock.

We now examine a second group of results relating to the performance of different groups of animals belonging to the same strain (DBA/2J) but tested at different ages (*34*).

As shown by Fig. 11, four groups of mice (21, 60, 180, and 360 days old) were tested in a task involving avoidance learning during five consecutive days. In agreement with data collected on rats by Verzar-McDougall and others and on mice by Meier (*35*), our adult mice showed the best performance when compared to younger (21 days) or older (180 to 360 days) animals. However, the most striking result was not the difference in the overall performance but the variations of responding (enhancement or decrement) evident between two consecutive sessions. Briefly, during the interval between successive sessions, active processes of consolidation took place in the adult, and decay occurred in young or old mice. These findings further support the interpretations of Kirby and of Doty and Doty (*36*).

The role played by mechanisms of consolidation in avoidance learning of mice is also evident from the effects of administration of drugs after trials. Injections of strychnine and picrotoxine facilitate learning, as shown by the experiments of McGaugh and his associates (*37*). By contrast, experiments by Pearlman, Sharpless, and Jarvik (*38*) gave evidence that anesthesia after the trial results in an impairment of learning.

Our results (*39*)show that the effects of treatment with drugs on the rate of acquisition of avoidance behavior is particularly evident when performance at the beginning of the second session is compared to performance at the end of the previous session. In the group of animals injected after a trial with strychnine or picrotoxine, the percentage of avoidances during the first trials of session 2 are much higher than the level during the last trials of session 1. The enhancement, due to the effect of convulsants on consolidation mechanisms, is higher than that in control mice. Similarly, an impairment of the storage was observed when nitrous oxide was administered immediately after a short training session.

The results of two series of experiments, under identical conditions and in which stimulants and depressants were used on the central nervous system, are consistent with a view of memory storage as a process involving several traces. Furthermore, these results are in agreement with the findings that learning patterns are affected by aging and that retrograde amnesia occurs when shock is administered after a trial.

Pharmacogenetics of Centrally Active Drugs

There is a large body of results on the different reactivity of different species or strains of animals to the action of psychotropic agents. Various strains of mice have different pharmacological responses to a wide variety of drugs such as *d*-amphetamine, iproniazid, or chlorpromazine (*40*).

On the basis of the electrophysiological and pharmacological evidence of a central cholinergic transmission and on the hypothesis that there is a cholinergic system regulating arousal (*41*), the effects of nicotine and arecoline on learning and retention were tested. During investigations on drugs which facilitate elementary forms of learning, we observed, in 1963, that treatment with nicotine improved avoidance learning of naive rats. We used inbred mice (*42*) to extend these results and to study the pharmacological implications of the genetic differences in avoidance learning.

The effect of the same dose of nicotine (0.5 milligram per kilogram) was compared in nine inbred strains subjected to the same avoidance schedule in a shuttle box. The results show that under similar conditions, the various strains attain quite different performance, and that there are important differences between the effects of nicotine in the various strains. Nicotine had a facilitating effect on six out of nine inbred strains: the incidence of avoidance increased to 35 percent above the level of the control (C3H/He) mice. The same dose had a smaller effect in DBA/2J mice (15 percent), and an impairing effect on the performance of two other strains (Fig. 12).

In contrast to nicotine (a typical cholinergic agent), arecoline did not facilitate avoidance learning. On the contrary, a decrement of performance was evident after treatment with

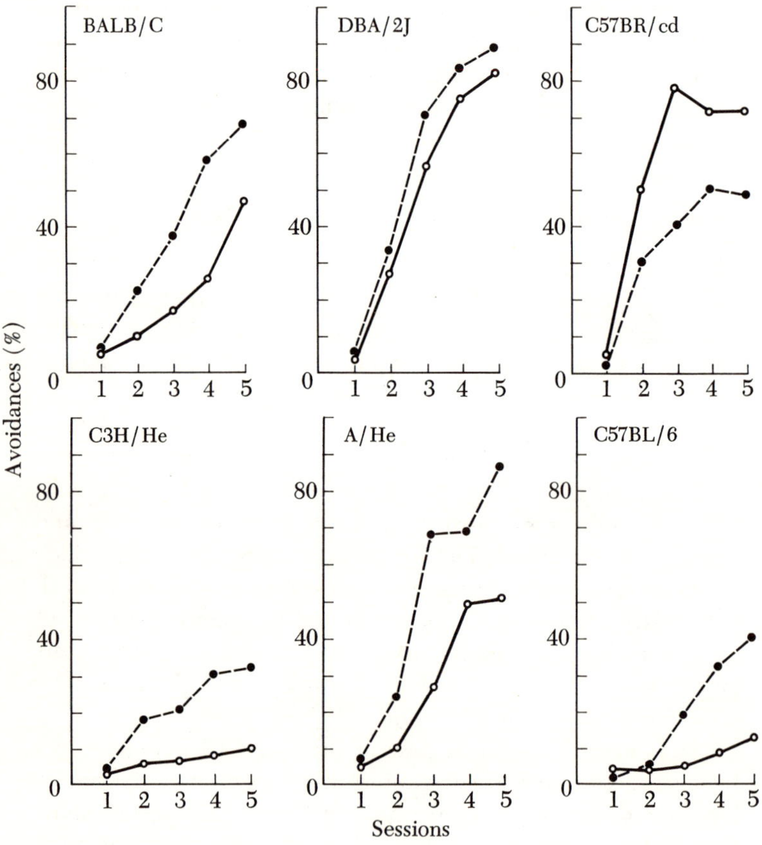

Fig. 12. Effects of nicotine (0.5 mg/kg) on avoidance conditioning of six inbred strains of mice. Dotted lines, nicotine-injected groups; solid lines, control groups.

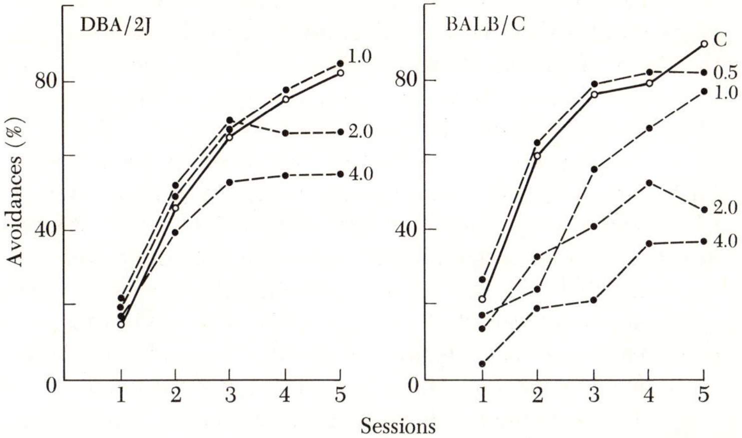

Fig. 13. Effects of arecoline on avoidance learning of BALB/c and DBA/2J mice. Each point represents the mean of eight mice during five consecutive daily sessions of 100 trials. *C* = controls. The numbers indicate the dose of the drug.

various dosages of this drug. This impairment was strain dependent, as shown by a comparison between the effects of arecoline on DBA/2J and BALB/c mice (Fig. 13).

A correlation between the pharmacological effects of these cholinergic agents and the significance of the short- and long-term memory balance in the strains of the mice tested is still hypothetical. However, the facilitating effect of nicotine is generally higher in the strains characterized by low performance levels.

Effects of Environment on Learning and Memory

How can we determine whether a given type of behavior depends on heredity or environment? The interdependence of these factors shed light on studies in which we attempted to find out to what extent either of these elements could be responsible for any behavioral variation.

The role of genetic patterns in the determination of simple behavioral traits has been determined by studies on selective breeding and the comparative behavior of inbred strains of laboratory rodents. However, relatively little is known about the limits of the effects of the environment on learning aptitudes and memory. Here are a few examples of the advances of the research in this field.

1) Concerning sexual behavior, the effects of environmental factors on gonadotropic hormones and the nervous system have been widely studied. Light, diet, and olfactory stimuli are some of the most important nonsocial patterns modifying the sexual behavior of the male and the estrus cycle and the spontaneous activity of the female (*43*). The implications of these and other social factors on the dynamics of population of confined house mice have been extensively studied (*44*).

2) Many experiments have been conducted by Rosenzweig and his associates (*45*) on the effects of an enriched or impoverished environment in the rat. The most striking results deal with modifications of the weight and morphology of the brain and with changes in cerebral chemistry. Some evidence has been furnished that an enriched environment leads to an improvement of ability to get through a maze.

3) Imprinting is, according to the ethologists, a species-specific type of learning that occurs (mostly in birds) within a limited period of time, early in the life of the organism, and is relatively unmodifiable thereafter. Although this concept does not seem immediately applicable to the behavior of small rodents, many experiments have dealt with the effects of early experience. As demonstrated by Sluckin (*46*), a distinction may be drawn between the last effects of early learning in a form of training and the effects of other kinds of early experience. The value of a very early training is that it may be, according to Hebb's theory "half-transferable" from early to later learning.

The fact that the lack of any training often precludes the acquisition of skills and habits, or may adversely affect the learning ability, was shown by studies dealing with the nature of free environmental experience and with the effects of early visual and motor experience as determinants of the rat's ability to solve problems in a maze (*47*). Various experimental factors, such as the exposure to high and low temperatures, bright and dim light, noise, electric shock, and drugs, have been ma-

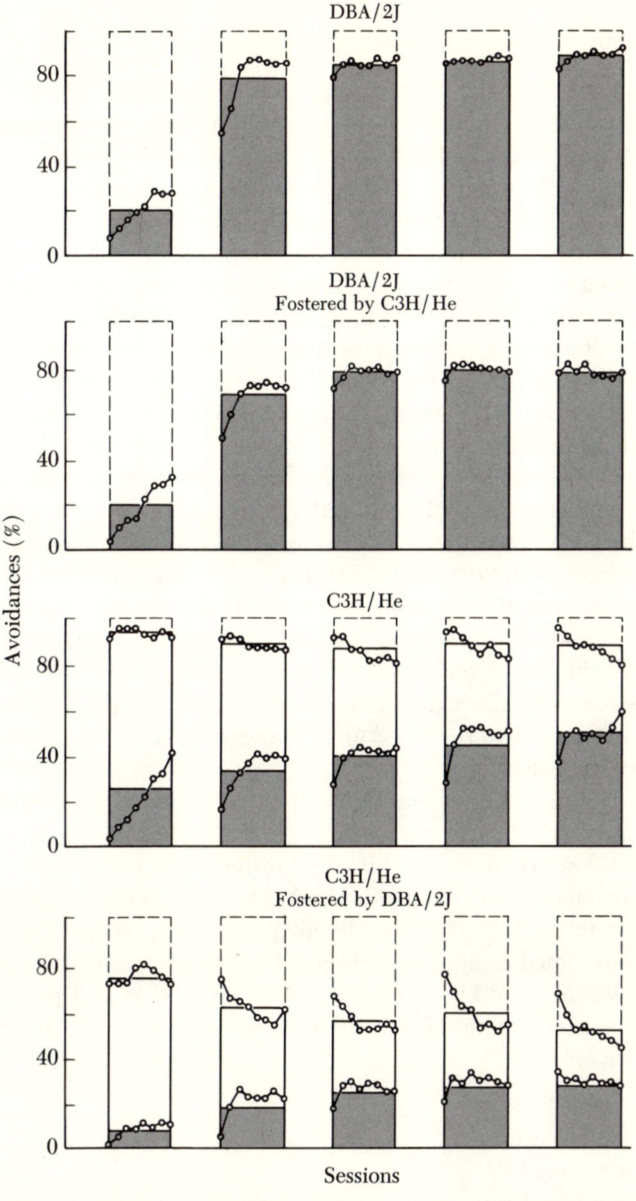

Fig. 14. Mean percentage of avoidance (dark columns) and freezing responses (white columns) of control and cross-fostered DBA/2J and C3H/He mice during five sessions of 400 trials each. The curves within each session represent the responses in eight blocks of 50 trials each.

nipulated to discover their long-term effects (*48*). Levine (*49*), for example, reported that stimulation such as momentary handling could facilitate the rat's ability with regard to conditioning. Denenberg (*50*) reported that the shocking of infant mice furthered subsequent classical conditioning.

Of particular interest are the studies on the effects of the preweaning environment on behavioral patterns; it was demonstrated that behavior of mice reared by alien foster mothers was affected by fostering, or that it resembled behavior of the alien strain (*51*). The results of these experiments support findings (*52*) that emotionality of the offspring is affected by maternal environment.

By the use of a shuttle box in experiments on avoidance, we demonstrated (*53*) that the usual patterns of avoidance behavior were not affected by cross-fostering two inbred strains of mice (DBA/2J and C3H/He), whereas some emotional components of behavior were strongly affected. In particular, we found, by measuring avoidance and "freezing" behavior with the same technique, that modifications of preweaning environment affect the emotional behavior, whereas variations of learning patterns are due to the concomitant appearance of freezing behavior.

Figure 14 shows that DBA/2J mice, which in control conditions reach a high performance and do not present freezing behavior, when reared by alien foster mothers appear to be more emotive. On the other side, the emotivity is enhanced by fostering C3H/He mice by alien mothers in which this type of emotional behavior is absent. In general terms, the results show that a modification of the environment, as that produced by cross-fostering, affects the emotional but not the adaptive behavior (*54*).

Conclusions

Among laboratory mammals the mouse is used in most genetic studies and is bred in many inbred strains which are currently available. The findings reported here suggest that there are at least two reasons for using inbred strains in psychobiology:

(i) the extreme behavioral homogeneity of the individuals belonging to the same strain, and (ii) the characteristic differences in behavioral traits of each strain. These inbred strains provide the psychobiologist with unlimited groups of individuals presenting a homogeneous adaptive behavior. This availability is particularly important in view of previous difficulties and problems surrounding the establishment of learning and retention curves in laboratory animals.

Particularly in the field of learning, the average individual is, as Tryon stated (2), "a man made fiction." The individuals in an inbred strain form a kind of family—a multitude of one-egg twins. Psychobiologists studying the comparative psychology of individuals take advantage of this anomalous and almost unnatural situation.

It is also important to point out the "artificial" nature of the inbred strains and the profound differences among inbred strains and the general concept or "race."

Of his classic experiments on the dancing mouse Yerkes (55) wrote: "All dancing (mice) are alike in certain important respects but to the trained observer of animal behavior, their individual peculiarities are quite as evident and even more interesting than their points of resemblance."

We have seen how the use of different strains of inbred mice is a useful tool in the study of the mechanisms of retention. The findings obtained by analyzing the learning curves of two strains of mice support the two-stage concept of memory storage. In agreement with this theory, a type of short-term memory is peculiar to the C3H/He strain, whereas DBA/2J mice have long-term memory storage.

It is probable that the individual differences observed in the "adaptive" behavior of a heterogeneous population are not only quantitative, and that they derive from the different mechanisms of memory that characterize each subject. A similar line of reasoning might explain the different effects exerted by some psychotropic agents in different strains.

The findings reported until now in the literature and in the present paper suggest that there can be no clear-cut determination of the role of the environmental factor modifying learning and retention aptitudes. In a general way our findings show that adaptive behavior, less than emotional behavior, is subject to the influences of environment (*56*).

As an answer to the question which Hull raised concerning the role exerted by the genetic factors in the determination of the individual variability in adaptive behavior, the findings reported here illuminate the important role exerted by heredity in the determination of learning and retention mechanisms (*53*).

References and Notes

1. A. Anastasi, *Psychol. Rev. 65,* 197 (1958); T. Dobzhansky, *Science 111,* 161 (1950); J. P. Guilford, *The Nature of Human Intelligence* (McGraw-Hill, New York, 1967), pp. 347-386; J. B. S. Haldane, *Ann. Eugen. 13,* 197 (1946); R. S. Woodworth, *Soc. Sci. Res. Coun. Bull.* No. 47, pp. 1-50 (1941).
2. R. C. Tryon, in *Comparative Psychology,* F. A. Moss, Ed. (Prentice-Hall, Englewood Cliffs, N.J., 1934).
3. J. Hirsch, in *Roots of Behavior,* E. L. Bliss, Ed. (Harper, New York, 1962), pp. 3-23; C. S. Hall, in *Handbook of Experimental Psychology,* S. S. Stevens, Ed. (Wiley, New York, 1951), pp. 304-329; G. E. McClearn and W. Meredith, *Ann. Rev. Psychol. 17,* 515 (1966); R. Robinson, *Genetics of the Norway Rat* (Pergamon, Oxford, 1965).
4. C. S. Hall, in *Handbook of Experimental Psychology,* S. S. Stevens, Ed. (Wiley, New York, 1951), pp. 304-329.
5. E. L. Green, Ed., *Biology of the Laboratory Mouse* (McGraw-Hill, New York, 1966); J. H. Bruell, in *Roots of Behavior,* E. L. Bliss, Ed. (Harper & Row, New York, 1962); R. E. Wimer and J. L. Fuller, in *Biology of the Laboratory Mouse,* E. L. Green, Ed. (McGraw-Hill, New York, 1966).
6. E. Fredricson, C. D. Fink, J. R. Parker, *J. Genet. Psychol. 86,* 131 (1955); J. P. Scott and E. Fredricson, *Physiol. Zool. 24,* 273 (1951); W. R. Thompson, *Can. J. Psychol. 7,* 144 (1953).
7. J. R. Royce and M. Covington, *J. Comp. Physiol. Psychol. 53,* 197 (1960); R. L. Collins, *Science 143,* 1188 (1964); K. Schlesinger and R. E. Wimer, *J. Comp. Physiol. Psychol. 63,* 139 (1967).

8. D. Bovet, G. L. Gatti, J. Pecori-Giraldi, M. Frank, in *Neuropsychopharmacology,* E. Rothlin, Ed. (Elsevier, Amsterdam, 1961), p. 142; D. Bovet, F. Bovet-Nitti, A. Oliverio, *Life Sci. 5,* 415 (1966); F. Robustelli, *Atti Accad. Naz. Lincei Rend. 33,* 565 (1965).
9. L. H. Warner, *J. Genet. Psychol. 41,* 57 (1932).
10. H. Hediger, *Studies of the Psychology and Behavior of Captive Animals in Zoos and Circuses* (Butterworth, London, 1955).
11. R. L. Solomon and E. S. Brush, *Neb. Symp. Motiv., 4,* 212 (1956); M. E. Bitterman, *Psychol. Bull. 59,* 81 (1962); F. S. Keller and W. Schoenfeld, *Principles of Behavior* (Appleton-Century-Croft, New York, 1950); N. E. Miller, in *Handbook of Experimental Psychology,* S. S. Stevens, Ed. (Wiley, New York, 1951).
12. H. F. Harlow, *Psychol. Rev. 60,* 23 (1953).
13. J. P. Scott, *Animal Behavior* (Univ. of Chicago Press, Chicago, 1958).
14. D. Bovet, F. Bovet-Nitti, A. Oliverio, *Brain Res.,* in press.
15. The fiducial limits of the mean observed in five groups of Swiss Webster mice coming from different breeding colonies of the United States and Italy were generally so high that a learning curve representing the mean performance was meaningless.
16. G. Bignami and D. Bovet, *C. R. Hebd. Seances Acad. Sci.* Paris *260,* 1239 (1965).
17. Genetic abnormalities in vision have been reported in some strains of mice with retinal degeneration and particularly in some strains deriving by crosses between C3H/He and other strains [R. L. Sidman and M. C. Green, *J. Hered. 56,* 23 (1965)]. This problem has been widely discussed by J. L. Fuller and R. E. Wimer [in *Biology of the Laboratory Mouse,* E. L. Green, Ed. (McGraw-Hill, New York, 1966), p. 609]. Different findings prove that animals belonging to the strains used in this experiment respond to luminous stimuli, and that similar patterns of avoidance responding were elicited by visual or auditive conditioned stimulus.
18. D. Bovet, F. Bovet-Nitti, A. Oliverio, *Psychopharmacologia 10,* 1 (1966); *Ann. N.Y. Acad. Sci. 142,* 261 (1967); *Brain Res. 10,* 168 (1968).
19. J. R. Royce and M. Covington, *J. Comp. Physiol. Psychol. 53,* 197 (1960); J. R. Royce, *Multiv. Behav. Res. 1,* 209 (1966); G. W. Meier and D. P. Foshee, *J. Genet. Psych. 102,* 267

(1963); K. Schlesinger and R. E. Wimer, *J. Comp. Physiol. Psychol. 63*, 139 (1967); R. E. Wimer, L. Symington, H. Farmer, P. Schwartzkroin, *ibid. 65*, 126 (1968).

20. S. A. Barnett, *The Rat: A Study in Behavior* (McGibbon and Kee, London, 1960); *"Instinct"* and *"Intelligence"* (McGibbon and Kee, London, 1967); J. L. Kavanau, *Science 155*, 1623 (1967).
21. D. Bovet, F. Bovet-Nitti, A. Oliverio, *Life Sci. 5*, 415 (1966); *Brain Res. 10*, 168 (1968); M. Sansone and D. Bovet, *Quart. J. Exp. Psychol.*, in press.
22. D. Bovet and A. Oliverio, *J. Psychol. 68*, 45 (1967); A. Oliverio, *Il Farmaco Ed. Sci. 22*, 159 (1967).
23. W. B. Essman and H. Halpern, *Psychol. Rep. 14*, 731 (1964).
24. This research was conducted in collaboration with J. L. McGaugh, Department of Psychobiology, University of California, Irvine.
25. J. L. McGaugh, *Science 153*, 1351 (1966); M. E. Jarvik, in *Recent Advances in Learning and Retention*, D. Bovet *et al.*, Eds. (Accad. Nazionale Lincei, Rome, 1968), Suppl. 109.
26. C. L. Hull, *Principles of Behavior* (Appleton-Century-Croft, New York, 1943).
27. D. P. Kimble, Ed., *The Anatomy of Memory* (Science and Behavior Books, Palo Alto, Calif., 1963); F. O. Schmit, Ed. *Horizons in Biochemistry* (Academic Press, New York, 1962).
28. D. O. Hebb, *The Organization of Behavior* (Wiley, New York, 1949); R. W. Gerard, *Science 122*, 225 (1955); J. Z. Young, *The Memory System of the Brain* (University of California Press, Berkeley, 1966); A. Cherkin, *Proc. Nat. Acad. Sci. U.S. 55*, 88 (1966).
29. B. W. Agranoff, R. E. Davis, J. J. Brink, *Brain Res. 1*, 303 (1966); S. H. Barondes and M. E. Jarvik, *J. Neurochem. 11*, 187 (1964); L. Cook, *Proc. Annu. Mtg. Amer. Coll. Neuropsychopharmacol. 6th, Puerto Rico (1967)* (PHS Publication No. 1836, 1968); H. Hydén and E. Egyhàzi, *Proc. Nat. Acad. Sci. U.S. 49*, 618 (1963).
30. U. Cerletti, *Riv. Sper. Freniatria 64*, 209 (1940).
31. C. P. Duncan, *J. Comp. Physiol. Psychol. 42*, 32 (1949).
32. D. Bovet and A. Oliverio, *Atti Accad. Naz. Lincei Rend. 41*, 18 (1966).
33. A. Cherkin, *Psychonom. Sci. 4*, 169 (1966).
34. A. Oliverio and D. Bovet, *Life Sci. 5*, 1317 (1966).

35. E. J. Verzar-McDougall, *Gerontologia 1,* 65 (1957); K. Bättig and E. Granjean, *ibid. 3,* 226 (1959); V. H. Denenberg and N. J. Kline, *J. Comp. Physiol. Psychol. 51,* 488 (1958); G. W. Meier, *J. Comp. Physiol. Psychol. 58,* 418 (1964).
36. B. A. Doty and L. A. Doty, *J. Comp. Physiol. Psychol. 57,* 331 (1964); R. H. Kirby, *ibid. 56,* 158 (1963).
37. J. L. McGaugh and L. Petrinovich, *Amer. J. Psychol. 72,* 99 (1959); *Int. Rev. Neurobiol. 8,* 189 (1965).
38. M. E. Jarvik, in *Animal Behavior and Drug Action,* H. Steinberg, Ed. (Churchill, London, 1964); C. Pearlman, S. K. Sharpless, M. E. Jarvik, *J. Comp. Physiol. Psychol. 54,* 109 (1961).
39. D. Bovet, J. L. McGaugh, A. Oliverio, *Life Sci. 5,* 1309 (1966); F. Bovet-Nitti, in *Recent Advances in Learning and Retention,* D. Bovet, *et al.* Eds. (Accad. Nazionale Lincei, Rome, Suppl. 109, 1968), p. 205.
40. H. Meier, in *Experimental Pharmacogenetics* (Academic Press, New York, 1963); ——— and J. L. Fuller, in *Biology of the Laboratory Mouse,* E. L. Green, Ed. (McGraw-Hill, New York, 1966).
41. B. Cardo, *J. Physiol. (Paris) 53,* 1 (1961); P. L. Carlton, *Psychol. Rev. 70,* 19 (1963); A. Oliverio, F. Bovet-Nitti, D. Bovet, *Proc. Int. Congr. Coll. Int. Neuropsychopharmacol. 5th Excerpta Med. Int. Congr. Ser. 129,* 213 (1967); A. Oliverio, *Psychopharmacologia 11,* 39 (1967); *ibid. 12,* 214 (1968); A. Soulairac and M. L. Soulairac, *Ann. Endocrinol. 6,* 731 (1956); L. Stein, *Canad. Psychiat. Ass. J. 11,* 34 (1966); M. J. Michelson, Ed., *Physiologitsheskja Rol Acetylcholina i Isykanije Novych Lekarstvennych Vestshetv* (Leningrad, 1957); A. Oliverio, *Proc. Annu. Mtg. Amer. Coll. Neuropsychopharmacol. 6th, Puerto Rico* (1967) (PHS Publ. No. 1836, 1968), p. 867.
42. D. Bovet and F. Bovet-Nitti, in *Tobacco Alkaloids and Related Compounds,* U. S. vonEuler, Ed. (Pergamon Press, Oxford, 1965), pp. 125, 137; D. Bovet, F. Bovet-Nitti, A. Oliverio, *Psychopharmacologia 10,* 1 (1966); *Ann. N.Y. Acad. Sci. 142,* 261 (1967).
43. R. A. Gorski and R. Whalen, in *Brain and Behavior,* M. Brazier, Ed. (University of California Press, Berkeley, 1966), vol. 3; C. H. Sawyer, in *The Hypothalamus,* A. Anderson *et al.,* Eds. in press; D. Mainardi, *Atti Accad. Naz. Lincei Rend. 37,* 484 (1964).

44. R. Z. Brown, *Ecol. Monogr. 23,* 217 (1953); P. Crowfkoft and F. P. Rowe, *Proc. Zool. Soc. London 140,* 517 (1963); C. H. Southwick, *Ecology 36,* 627 (1955).
45. M. R. Rosenzweig, *Amer. Psychol. 21,* 321 (1966); ———, E. L. Bennett, D. Krech, *J. Comp. Physiol. Psychol. 57,* 438 (1964).
46. W. Sluckin, *Imprinting and Early Learning* (Aldine, Chicago, 1965).
47. D. G. Forgays and J. W. Forgays, *J. Comp. Physiol. Psychol. 45,* 322 (1952).
48. F. A. Beach and J. Jaynes, *Psychol. Bull. 51,* 239 (1954); J. A. King, *ibid.* 55, 46 (1959); W. Sluckin, *Imprinting and Early Learning* (Aldine, Chicago, 1965).
49. S. Levine, *Science 126,* 405 (1957).
50. V. H. Denenberg, D. H. Ottinger, M. W. Stephens, *Child Devel. 33,* 65 (1962).
51. R. H. Ressler, *J. Comp. Physiol. Psychol. 56,* 882 (1963); A. J. Reading, *ibid. 62,* 437 (1966).
52. V. H. Denenberg and G. G. Karas, *Psychol. Rep. 7,* 313 (1960).
53. F. Bovet-Nitti, A. Oliverio, D. Bovet, *Life Sci. 7,* 791 (1968).
54. Wistar and Long Evans rats were reared from birth until weaning by foster mothers of their own or the other strain. During avoidance training in a shuttle box both performance and emotional (freezing) behavior were significantly influenced by cross-fostering. However, modifications of learning patterns seemed to be dependent on the appearance of the freezing behavior. In addition to that the emotional behavior of cross-fostered males of either strain was enhanced by foster rearing in a consistently higher extent than that of females. These findings seem to agree with results showing a higher reactivity of male pups to environmental changes in infancy [A. Oliverio, M. Satta, D. Bovet, *Life Sci. 7,* 799 (1968)].
55. R. M. Yerkes, *The Dancing Mouse* (Macmillan, New York, 1907).
56. To what extent will our results and future results solve the equation formulated by Hirsh [J. Hirsh, in *Roots of Behavior,* E. L. Bliss, Ed. (Harper, New York, 1962)] in interpreting heritability (h^2) as the proportion of trait variance due to genetic and environmental variances? The results obtained allow a comparison between the findings deriving from an inbred and a heterogeneous population and a quan-

tification of the indexes (σ, σ^2, σ^2 $100/x$) referring to the variability due to genetic factors. It is difficult to predict whether it will be possible to define heritability in mammals without a restrictive and arbitrary definition of the multiple elements that form the environment.

57. The experiments described here were carried out in the Department of Pharmacology of the University of Sassari and the University of California School of Medicine, Los Angeles. Supported by grants from the Italian Consiglio Nazionale delle Ricerche, from the American Medical Association, ERF Commitee for Research on Tobacco and Health,. and from a USPHS general research support grant to the University of California, Los Angeles. Many citations were furnished by the Brain Information Service, Biomedical Library, University of California at Los Angeles.

Since Learned Behavior Is Innate, and Vice Versa, What Now?[1]

WILLIAM S. VERPLANCK

My text for today may be found in the behavioral apocrypha; it pertains to the courtship of the three-spined stickleback. When the male stickleback, with nest completed, and in full bloom of red belly and bright blue eye, is confronted with a silvery, deep-bosomed, egg-laden female, he courts her. He zigs toward her, zags back, and toward and back again, approaching nearer and nearer with successive passes. After his closest approach, he swims directly toward the entrance to his nest. If she fails to follow, he turns away and repeats his game —and again and again, until she follows. And when she ac-

[1] A paper presented at the symposium titled "Unlearned Behavior: Concepts and Findings," sponsored jointly by Divisions 3 and 7 of the American Psychological Association, in September, 1954. It was made possible by a grant from the American Philosophical Society, which enabled the writer to gain first-hand familiarity with the work of ethologists.

From *Psychological Review,* 1955, *62,* 139-44. Reprinted by permission of the author and the American Psychological Association.

quiesces, when she does follow, his behavior changes; this time, he reaches the nest entrance, he swims into it and straight through the nest, leading the female.

One studies this behavior by presenting the male with a captive female, one that cannot follow because the experimenter has confined her in a glass test tube. After a series of presentations of the female in this way, with repeated elicitations of the zigzag dance of the male, it is possible to perform a kind of control experiment in which the male is presented with an empty test tube. As a control, this usually proves disappointing; the male, as like as not, will proceed to zigzag, to court the test-tube. But what is more interesting, I am told that if the experimenter is compliant, and moves the test tube like a female in pursuit of the male as he swims directly toward the nest, the male will proceed with his usual behavior toward the female; he will swim into the nest, and through it.

This is one of the kinds of things the ethologists[2] bring to us. With their background in biology, and their strictly behavioristic approach, they have over the past few years staggered us with a mass of data (10, 11; see also *Behaviour* and the *British Journal of Animal Behaviour*) on the behavior of inframammalians, and have provoked once again discussion of an issue that from time to time drags psychologists out of the laboratory and into the forum—nature or nurture, heredity or environment, learning or maturation.

Ethologists, of course, stress "innate" behavior; they find differences between the behaviors they study, and those learning-theory psychologists work on. These they attribute to innateness. That such innateness as a source of differences is hopelessly confounded with differences in the species studied, and with the degree of experimental restriction on the behavior investigated, goes without saying.

[2] Ethology, as it is defined by ethologists, is the science of behavior. And so it is: the methods and concepts of ethology bear a very close resemblance to their counterparts in the modern behaviorism psychologists are familiar with. This is true despite a difference in subject matter; ethologists, trained in the intellectual tradition of zoology, have concerned themselves largely with species-specific behavior in birds and fishes, and the individual ("acquired") behavior of insects (especially wasps). One ethologist explained ethology this way: "Ethologists are behaviorists who *like* their animals."

Ethologists have, however, stirred up the argument, and one of the features of our cultural heritage appears to be a very strong tendency to take one or another side in it. The discussion shifts its ground, changes its dominant features, but it seems to remain.

Fortunately, there are signs that progress is being made toward a solution to the endless chatter. Experimental animals, busily engaged in coping with their environments, seem to be coping with this bit of human activity as well. They seem to be telling us that there is no possible reasonable answer to questions such as "Is this bit of behavior innate?" The dilemma of innate or acquired seems to be one of those categorical pseudo problems that the philosopher Ryle (9) has concerned himself with. Analysis of the variables controlling behavior, such as that presented by Hebb (4), tells us the same thing: it is a logical impossibility to study the innate before studying learning.

As we list the criteria proposed for distinguishing the innate from the learned, many examples of learned behavior can be found that satisfy one or more—or most—of the criteria for the innate. Stereotypy, universality of appearance, orderliness, adaptivity, resistance to modification—all these fail, and only one criterion remains: execution of the behavior on its first opportunity to occur, *without* the possibility of previous learning. *How* does one rule this out?

We are forced into the position of acknowledging that the only criterion for distinguishing between innate and acquired behavior is one that requires us to accept the null hypothesis as proven.

We are forced into the position that no meaningful distinction can be drawn between learned and innate behavior, that is, between the stereotyped and highly predictable behavior studied in inframammalians by the ethologists, and the more variable behavior studied in the T maze and Skinner box. We can no more distinguish between behavior that is innate and behavior that is learned than physicists can distinguish between light that is made up of corpuscles and light that is made up of transverse vibrations. Innate and learned must always be assumed to have quotation marks about them.

Given this position—the assumption that much the same

behavioral laws apply throughout the vertebrate realm (if not further), together with a rigorous experimental approach to behavior—anumber of consequences follow. I shall note and illustrate some of these consequences.

First, the same classes of experimental variables should control both learned and unlearned behavior,[3] and in similar ways. Hence, the same kinds of difficulties should arise in experimental designs aimed at untangling the effects of these variables, whether they are applied to learned or innate behavior.

Secondly, the structure of the theory of unlearned behavior and that of learned behavior must prove to be similar if not identical. It should not matter which kind of behavior yields a theory—it should apply to both. That is to say, one theoretical account of causal variables should be found forced by the data, and it should prove applicable to the data of both. And perhaps the same classes of theoretical variables should give the same kind of theoretical difficulties.

Thirdly, the behavioral phenomena observed by ethologists in the field should be observable in the **T** maze and similar apparatus, and those observed in the **T** maze should be observable in the field. Those who study innate behavior are well advised to look for and control the effects of the variables studied in learning. Conversely, those who study learning may facilitate the accumulation of relevant data by looking for and controlling the effects of the variables that have been considered the preserve of the ethologist. I am not saying that ethologists and psychologists should borrow each other's concepts to apply to their own fields, but rather that there is *only one* field of investigation, and failure to examine all the concepts developed in that field may serve to delay the development of an ordered and comprehensive body of data.

If there were time, we could develop fully each of these points. We could draw the revealing parallels between the series of experiments on generalization gradients in the CR (e.g., 5, 6) and those on the sign stimuli releasing pecking in the gull chick (12). The differences seem to lie in the number

[3] From this point on, by learned behavior, we will refer to behavior studied by learning theory men; and by unlearned or innate behavior, to behavior studied by ethologists, and so termed by these men.

of dimensions of the stimulus that must be investigated. Hovland restricts himself to one: metric distance from the point of original conditioning along the body surface. Others have explored response strength along other single stimulus dimensions, such as frequency of a pure tone. Tinbergen and Perdeck, however, are forced by the animal, as it were, to vary the complex stimulus for the pecking response along a series of dimensions that includes those relating to the size, position, and optical properties of the spot on the bill of an adult sea gull, as well as to similar measures of the bill itself and of the remainder of the adult's head. Along each of these dimensions they find what is recognizable as a generalization gradient. To determine these gradients, careful counterbalancing of the order of test stimuli was essential, lest, as in the CR, the response extinguish.

We could point out that it does no abuse to the theories now current in ethology to change a term here and there and thereby convert ethological theory into something in large measure almost indistinguishable from 1943 Hullian theory (7), except in its treatment of the variables defining $_{S}H_{R}$ (some of which it ignores). We could expand the difficulties that ethological and learning theorists have with the concepts of stimulus and response. Both tend to commit a series of equivocations or, more charitably, both tend to gloss over some of the possible distinctions to be drawn between movement-defined and consequence-defined responses (3, 11). More profitably, we may look at some of the contributions that the work of the ethologists may make to behavioral psychology today.

After Beach's classic paper, "The Snark Was a Boojum" (2), it may seem a bit unnecessary to stress once again that behavior is a function of species membership, and that we do not have as yet more than the beginnings of a comparative psychology. The neglect of this field has served to let the work of psychologists drift out of the context of that general evolutionary theory that unifies all the biological sciences. We have little to say about the survival value[4] and the taxonomic distribution of behavior. This may or may not be a bad state of

[4] Most psychologists are surprised when they learn that the adaptiveness or survival value of a structure or of a behavior is now subject to experimental verification in biology.

affairs. But clearly, in forgetting the comparative approach, we are very sharply limiting the inductive basis of our theories of behavior. To restrict ourselves to the study of one or two species and a short list of environmental variables ensures that miniature theories will remain miniature.

The penalty we pay can be evaluated very readily indeed when once we shift from species to species. Armed with our knowledge of rats, we may attempt, for example, to do an experiment on the guinea pig. As soon as we do this, we discover that guinea pigs are not rats. This is not very startling, perhaps, but significant enough when we discover that the food-deprivation–feeding-behavior relationships[5] in the guinea pig are sufficiently different from those in the rat that it is all but impossible to recover from the guinea pig the types of relationships between bar pressing and deprivation that we are familiar with in the rat. Here is one area of comparative psychology that has hardly been touched, but one that has many implications, both practical and theoretical. The laws of learning operate within limits determined by the genetic characteristics of the animal. The behavior repertoire and the functional relationships existing among members of the repertoire and between them and the environment are set by the genetic characteristics of the animal, whether they are learned or not.

Secondly, we might look toward incorporating into our theories of learning a variable that has been remarkably neglected—in theories if not in practice. This variable is the *age* of the organism, and the general topic is "the critical period for learning." But with D. O. Hebb and Eckhard Hess among the participants in this symposium, it seems hardly necessary to stress the fact that the acquisition of particular modes of behavior is very much a function of the animal's age, of behavior acquired before the behavior we are studying is acquired, and of other behavior being acquired, or appearing, at the same time.

Thirdly, psychologists have largely set problems to their animals, and examined the incidence of particular pieces of behavior. Starting out with limited interests, we have exam-

[5] This and several other experimental results described have not yet been reported or published elsewhere.

ined limited sets of variables, both dependent and independent. Ethologists, on the contrary, have often examined somewhat broader causal chains of events, and have not restricted themselves to observing the effects of *our* sets of independent variables on the responses *we* like to deal with. To put it another way, ethologists watch everything that animals do, and everything that is happening when they do it. Given the opportunity, animals will emit behavior that is highly adaptive, and integrate it into longer causal chains, still under the control of specific motivating conditions and contingencies of reinforcement. Quite interesting behavioral effects will occur if we but let them. Let me take two examples.

In the first case, worked on by O. R. Lindsley, 20 rats were put on a deprivation schedule that allowed them access to solid Purina chow pellets for just one hour a day. Lindsley placed a number of pellets in the individual cages at the beginning of the hour, and at the end of the hour he took the remaining ones out. The first day was uneventful. There was the rat, there were the pellets, and in went the hand to remove the pellets. On the second day, when the experimenter's hand moved into the cage of *one* animal, the rat seized a pellet and retreated to the rear of the cage with it. There he held it. He hung on. After taking out the other pellets, the experimenter had to take the pellet from the rat forcibly. On the next day, two or three rats showed this behavior, and by the end of the week, not only were all 20 showing *at least* this behavior, but some had gone on to seize and carry two or more pellets to the rear, and in many the behavior was set off by simply unlatching the cage door. Reinforced? Interesting? Irrelevant? Some data we now have suggest that this behavior is associated with an increased tendency to run in a running wheel.

In the second case, 15 rats were trained to bar press in simple wire mesh Skinner boxes. After they were placed on a reinforcement schedule, discrimination training was begun, with five-minute periods of presentation of S_D, the positive discriminandum, and five minutes of S_Δ, the negative discriminandum. S_Δ was a two-inch equilateral black cardboard triangle that could be fixed by paper clips to the wire mesh behind the bar of the Skinner box. Under S_D conditions, this black triangle was absent. During the first discrimination period with the first failures of reinforcement, most of the animals seized, shook, and bit the bar, and the black triangle too. In most cases this aggressive behavior continued for several minutes; it continued in fact until the triangle was removed from behind the bar, usually by the rat. As soon as the triangle was gone, the animals resumed bar pressing at their normal rates. At the end of a series of such cycles of S_D and S_Δ, the majority of the rats, as soon as S_Δ was fixed in place, would approach it, and neatly but vigorously remove it—to get back to work. The effect appeared whether or not reinforcements were forthcoming when S_Δ was removed, although it is not surprising that the behavior was not so neat and efficient when the recurrence of S_D did not restore the previous reinforcement schedule. Altogether there was a startling shift from "violent," "aggressive," "emotional" assault to neat, smooth, precise manipulation.

Both these cases are the kinds of behavior that ethologists look for, and with which they deal. Both these cases are the kind of behavior that psychologists learn to ignore, or learn to prevent. Perhaps I may be forgiven if I suggest that they should be of some concern and interest to the learning theorist.

The next area that might be examined is that of displacement activities—responses usually associated with one drive state that appear when behavior associated with another drive is prevented from occurring, and in conflict situations (1). One mode of behavior that might be classified as displacement activity by the ethologist is the bar biting and bar shaking of our last example. Others are the face washing and general body cleaning—preening—that rats so often vigorously show during extinction, and the cooing and wing beating of the pigeon that occur under the same circumstances. These modes of behavior are familiar enough, but they are very largely associated for us with the extinction of behavior that has led to food and water. It is clear that our animals have behaviors that are not associated with reinforcement by food and water getting or by shock avoidance. It has now been shown that mice will learn to press bars when the reinforcement is access to a submissive male (which is then attacked), that rats will learn to press bars that release brakes on their running wheels (8) or that permit a female in oestrus to join them in the Skinner box. What behaviors, if any, appear when extinction is carried out with respect to such reinforcements? The ethological work would suggest strongly that we look and see. Perhaps some things that have puzzled us will be clarified.

Again, much work has been done with the hoarding of food and water pellets, and of other objects, under experimental conditions involving deprivation of these substances. But what of rats that hoard yellow pencils, pens, paper clips, and envelopes, when they have not had a history of deprivation of food or water, and when such objects have not been associated with food or water? And what of satiated rats running freely about familiar spaces that transport large pieces of favored foods to preferred places, and there eat them, although they eat small pieces on the spot when they get them? And that do the same with their standard food if they are hungry? And that do not hoard it? The psychologist's fixation on a few easily manipulated and already identified experimental variables may seriously limit the power of his generalizations.

Let me complete the circle, and close by returning to the behavior described in the beginning, that is, to the stickleback that was conditioned to zigzag to a test tube. When this conditioning was effective, he showed other behavior toward the test tube—behavior that had *not* been conditioned to it. He led it, you will recall, to his nest, and there he did what he usually does only when a female is behind him: he swam through the nest. I cannot think of a clear-cut case where such a transfer effect occurs in the animals run by psychologists. What is disturbing is that I cannot think of a clear-cut case where such a transfer effect *could* have been observed in psychologists' animals.

Some rather similar behaviors occur to me, and these are ones that have puzzled me as an S-R psychologist. Let us take rats that have run across a grid to food many times, so that their running is asymptotic. If, for the first time, they are now shocked as they cross the grid, they may show no change in their running when the shock occurs. But on the *next* trial, they may stop short of the grid, and there sniff and poke and peer.

The ethologists have much to teach us. . . .

They suggest that we must *observe* more species, that we must find new organizations of behavior (that is, new drives), that we must permit more manipulation of the environment by the animal, giving opportunities for phenomena not already in the textbooks to occur—in sum, that we must broaden the inductive basis, the experimental underpinnings, of our theories of behavior.

References

1. Bastock, M., Morris, D., & Moynihan, M. Some comments on conflict and thwarting in animals. *Behaviour,* 1953, 6, 66-84.
2. Beach, F. A. The snark was a boojum. *Amer. Psychologist,* 1950, *5,* 115-124.
3. Estes, W. K., *et al. Modern learning theory.* New York: Appleton-Century-Crofts, 1954.
4. Hebb, D. O. Heredity and environment. *Brit. J. Anim. Behav.,* 1953, *1,* 43-47.
5. Hovland, C. I. The generalization of conditioned responses: I. The sensory generalization of conditioned responses with varying frequencies of tone. *J. gen. Psychol.,* 1937, *17,* 125-148.
6. Hovland, C. I. The generalization of conditioned responses: II. The sensory generalization of conditioned responses with varying intensities of tone. *J. genet. Psychol.,* 1937, *51,* 279-291.
7. Hull, C. L. *Principles of behavior.* New York: D. Appleton-Century, 1943.
8. Kagan, J., & Berkun, M. The reward value of running activity. *J. comp. physiol. Psychol.,* 1954, *47,* 108.
9. Ryle, G. *Dilemmas.* Cambridge: Cambridge Univer. Press, 1954.
10. Society for Experimental Biology. *Physiological mechanisms in animal behaviour.* New York: Academic Press, 1950.
11. Tinbergen, N. *The study of instinct.* Oxford: Oxford Univer. Press, 1951.
12. Tinbergen, N., & Perdeck, A. C. On the stimulus situation releasing the begging response in the newly hatched Herring Gull chick (*Larus argentatus argentatus Pont.*). *Behaviour,* 1950, *3,* 1-39.

SECTION **II**

Heredity and Personality

In his presidential address to the American Psychological Association, Paul Meehl advanced this view of the experienced psychotherapist:

> Suppose that you were required to write down a procedure for selecting an individual from the population who would be diagnosed as schizophrenic by a psychiatric staff; you have to wager $1000 on being right; you may not include in your selection procedure any behavioral fact, such as a symptom or trait, manifested by the individual. What would you write down? So far as I have been able to ascertain, there is only one thing you could write down that would give you a better than even chance of winning such a bet—namely, "Find an individual X who has a schizophrenic twin."[1]

Many studies have confirmed the suspicion that emotional characteristics "run in families," but, of course, these same families provide the environment in which learning proceeds. Assessing the heritability of personality would require complete control of environmental variables—a condition not pos-

[1] P. Meehl. "Schizotaxia, schizotypy and schizophrenia." *Amer. Psychol.*, 1962, *12*, p. 827.

sible in studies of man. In addition, it would be necessary to isolate those phenotypic properties to be included under "personality," e.g., most personality theorists would include "optimism" in their list of personality traits, but would not include "intelligence." What, in fact, is *personality*?

As with all psychological concepts, personality is defined according to the theoretical biases, the current conventions, and the peculiar settings which prevail in the psychiatric and psychological circles concerned with the phenomenon. To the behavioral scientist, personality is a rather stereotyped and predictable pattern of *learned* actions occurring in a wide range of environmental circumstances. To the psychoanalytically oriented, personality is the expression of inborn tendencies which have been modified during the various (fixed) "stages" of "psychosexual" development. Conventionally, the term implies rather stable reaction patterns usually described as falling somewhere on a continuum, e.g., aggressive–submissive, hostile–pacific, selfish–altruistic, realistic–deluded, optimistic–pessimistic, egocentric–self-deprecating. Various questionnaires, "profiles," tests, and inventories have been invented to sample such characteristics of the individual. They boast a fair reliability in that the same individuals do seem to get similar "scores" on repeated administrations of the tests. Whether this means that personality is stable or that test-taking behavior is stable (or both) is, of course, moot.

The determination of personality is not limited to tests, however. Interviews, judgments by teachers, supervisors and relatives, comments of acquaintances—all of these may be used in arriving at some general picture of an individual's personality. At the root of these assessments is an attempt to ascertain the individual's attitudes about self and about the world; his personal perceptions and conceptions of the people and institutions in the world around him.

This section contains three recent papers on the subject of heredity and personality. The first paper, which ostensibly is not a genetically oriented article, is an address given by the great personality psychologist, Gordon Allport, shortly before he died. It is included because it reinforces the concept of individuality, because it flies in the face of extreme environmentalism, and because it sets the stage for the two subsequent articles.

The second paper is by D. Rosenthal, and in it he furnishes support for a "genetic predisposition" to schizophrenia. He also summarizes research which indicates a higher concordance among female twins than among male twins in the incidence of schizophrenia. Rosenthal is sensitive to the problems of interpretation and his own hypotheses are carefully worded.

The third paper, by I. I. Gottesman, is a model of methodological control. Gottesman discusses the difficulty of establishing the true zygosity of human twins and he exercises commendable rigor in selecting twins for his own research. His *method* sections should be read by anyone planning research in this difficult area or planning to review critically other work in this field. He acknowledges that the reliabilities of the tests he employed "are on much firmer ground than their validities" and, therefore, tacitly communicates the equivocal nature of any instrument which purports to "measure" personality. Finally, his paper is highly instructive in the method of computing heritability, the limitations of twin research, and the arduous task of interpreting results.

Both Rosenthal and Gottesman conclude that personality characteristics are highly heritable, if one accepts current measuring devices as valid techniques for assessing personality.

Suggested Readings

Kallman, F. J. *Heredity in Health and Mental Disorders.* W. W. Norton and Co., New York (1953).

McClelland, D. C. *Personality.* Dryden Press, New York (1951).

Pastore, N. *The Nature-Nurture Controversy.* Kings Crown Press, New York (1949).

Sorsby, A. (ed.). *Clinical Genetics.* Butterworth and Co., London (1951).

Traits Revisited

GORDON W. ALLPORT

Years ago I ventured to present a paper before the Ninth International Congress at New Haven (G. W. Allport, 1931). It was entitled "What Is a Trait of Personality?" For me to return to the same topic on this honorific occasion is partly a sentimental indulgence, but partly too it is a self-imposed task to discover whether during the past 36 years I have learned anything new about this central problem in personality theory.

In my earlier paper I made eight bold assertions. A trait, I said,

1. Has more than nominal existence.
2. Is more generalized than a habit.
3. Is dynamic, or at least determinative, in behavior.
4. May be established empirically.
5. Is only relatively independent of other traits.
6. Is not synonymous with moral or social judgment.
7. May be viewed either in the light of the personality which contains it, or in the light of its distribution in the population at large.

To these criteria I added one more:

8. Acts, and even habits, that are inconsistent with a trait are not proof of the nonexistence of the trait.

While these propositions still seem to me defensible they were originally framed in an age of psychological innocence. They now need reexamination in the light of subsequent criticism and research.

Criticism of the Concept of Trait

Some critics have challenged the whole concept of trait. Carr and Kingsbury (1938) point out the danger of reification. Our initial observation of behavior is only in terms of adverbs of action: John behaves aggressively. Then an adjective creeps

From *American Psychologist,* 1966, *21,* 1-10. Reprinted by permission of the American Psychological Association.

in: John has an aggressive disposition. Soon a heavy substantive arrives, like William James' cow on the doormat: John has a trait of aggression. The result is the fallacy of misplaced concreteness.

The general positivist cleanup starting in the 1930s went even further. It swept out (or tried to sweep out) all entities, regarding them as question-begging redundancies. Thus Skinner (1953) writes:

> When we say that a man eats *because* he is hungry, smokes a great deal *because* he has the tobacco habit, fights *because* of the instinct of pugnacity, behaves brilliantly *because* of his intelligence, or plays the piano well *because* of his musical ability, we seem to be referring to causes. But on analysis these phrases prove to be merely redundant descriptions [p. 31].

It is clear that this line of attack is an assault not only upon the concept of trait, but upon all intervening variables whether they be conceived in terms of expectancies, attitudes, motives, capacities, sentiments, or traits. The resulting postulate of the "empty organism" is by now familiar to us all, and is the scientific credo of some. Carried to its logical extreme this reasoning would scrap the concept of personality itself—an eventuality that seems merely absurd to me.

More serious, to my mind, is the argument against what Block and Bennett (1955) called "traitology" arising from many studies of the variability of a person's behavior as it changes from situation to situation. Every parent knows that an offspring may be a hellion at home and an angel when he goes visiting. A businessman may be hardheaded in the office and a mere marshmallow in the hands of his pretty daughter.

Years ago the famous experiment by La Piere (1934) demonstrated than an innkeeper's prejudice seems to come and go according to the situation confronting him.

In recent months Hunt (1965) has listed various theories of personality that to his mind require revision in the light of recent evidence. Among them he questions the belief that personality traits are the major sources of behavior variance. He, like Miller (1963), advocates that we shift attention from traits to interactions among people, and look for consistency in behavior chiefly in situationally defined roles. Helson (1964) regards trait as the residual effect of previous stimulation, and thus subordinates it to the organism's present adaptation level.

Scepticism is likewise reflected in many investigations of "person perception." To try to discover the traits residing within a personality is regarded as either naive or impossible. Studies, therefore, concentrate only on the *process* of perceiving or judging, and reject the problem of validating the perception and judgment (Cf. Tagiuri & Petrullo, 1958.)

Studies too numerous to list have ascribed chief variance in behavior to situational factors, leaving only a mild residue to be accounted for in terms of idiosyncratic attitudes and traits. A prime example is Stouffer's study of *The American Soldier* (Stouffer et al., 1949). Differing opinions and preferences are ascribed so far as possible to the GI's age, marital status, educational level, location of residence, length of service, and the like. What remains is ascribed to "attitude." By this procedure personality becomes an appendage to demography (see G. W. Allport, 1950). It is not the integrated structure within the skin that determines behavior, but membership in a group, the person's assigned roles—in short, the prevailing situation. It is especially the sociologists and anthropologists who have this preference for explanations in terms of the "outside structure" rather than the "inside structure" (cf. F. H. Allport, 1955, Ch. 21).

I have mentioned only a few of the many varieties of situationism that flourish today. While not denying any of the evidence adduced I would point to their common error of interpretation. If a child is a hellion at home, an angel outside, he obviously has two contradictory tendencies in his nature, or perhaps a deeper genotype that would explain the opposing phenotypes. If in studies of person perception the process turns out to be complex and subtle, still there would be no perception at all unless there were something out there to perceive and to judge. If, as in Stouffer's studies, soldiers' opinions vary with their marital status or length of service, these opinions are still their own. The fact that my age, sex, social status help form my outlook on life does not change the fact that the outlook is a functioning part of me. Demography deals with distal forces—personality study with proximal forces. The fact that the innkeeper's behavior varies according to whether he is, or is not, physically confronted with Chinese applicants for hospitality tells nothing about his attitude structure, except that it is complex, and that several attitudes may converge into a given act of behavior.

Nor does it solve the problem to explain the variance in terms of statistical interaction effects. Whatever tendencies exist reside in a person, for a person is the sole possessor of the energy that leads to action. Admittedly different situations elicit differing tendencies from my repertoire. I do not perspire except in the heat, nor shiver except in the cold; but the outside temperature is not the mechanism of perspiring or shivering. My capacities and my tendencies lie within.

To the situationist I concede that our theory of traits cannot be so simpleminded as it once was. We are now challenged to untangle the complex web of tendencies that constitute a person, however contradictory they may seem to be when activated differentially in various situations.

On the Other Hand

In spite of gunfire from positivism and situationism, traits are still very much alive. Gibson (1941) has pointed out that the "concept of set or attitude is nearly universal in psychological thinking." And in an important but neglected paper—perhaps the last he ever wrote—McDougall (1937) argued that *tendencies* are the "indispensable postulates of all psychology." The concept of *trait* falls into this genre. As Walker (1964) says trait, however else defined, always connotes an enduring tendency of some sort. It is the structural counterpart of such functional concepts as "expectancy," and "goal-directedness."

After facing all the difficulties of situational and mood variations, also many of the methodological hazards such as response set, halo, and social desirability, Vernon (1964) concludes, "We could go a long way towards predicting behavior if we could assess these stable features in which people differ from one another [p. 181]." The powerful contributions of Thurstone, Guilford, Cattell, and Eysenck, based on factor analysis, agree that the search for traits should provide eventually a satisfactory taxonomy of personality and of its hierarchical structure. The witness of these and other thoughtful writers helps us withstand the pessimistic attacks of positivism and situationism.

It is clear that I am using "trait" as a generic term, to cover all the "permanent possibilities for action" of a generalized order. Traits are cortical, subcortical, or postural dispositions having the capacity to gate or guide specific phasic reactions. It is only the phasic aspect that is visible; the tonic is carried somehow in the still mysterious realm of neurodynamic structure. Traits, as I am here using the term, include long-range sets and attitudes, as well as such variables as "perceptual response dispositions," "personal constructs," and "cognitive styles."

Unlike McClelland (1951) I myself would regard traits (i.e., some traits) as motivational (others being merely stylistic). I would also insist that traits may be studied at two levels: (*a*) dimensionally, that is as an aspect of the psychology of individual differences, and (*b*) individually, in terms of *personal dispositions*. (Cf. G. W. Allport, 1961, Ch. 15.) It is the latter approach that brings us closest to the person we are studying.

As for factors, I regard them as a mixed blessing. In the investigations I shall soon report, factorial analysis, I find, has proved both helpful and unhelpful. My principal question is whether the factorial unit is idiomatic enough to reflect the structure of personality as the clinician, the counselor, or the man in the street apprehends it. Or are factorial dimensions screened so extensively and so widely attenuated—through item selection, correlation, axis manipulation, homogenization, and alphabetical labeling—that they impose an artifact of method upon the personal neural network as it exists in nature?

A Heuristic Realism

This question leads me to propose an epistemological position for research in personality. Most of us, I suspect, hold this position although we seldom formulate it even to ourselves. It can be called a *heuristic realism.*

Heuristic realism, as applied to our problem, holds that the person who confronts us possesses inside his skin generalized action tendencies (or traits) and that it is our job scientifically to discover what they are. Any form of realism assumes the existence of an external structure ("out there") regardless of our shortcomings in comprehending it. Since traits, like all intervening variables, are never directly observed but only inferred, we must expect difficulties and errors in the process of discovering their nature.

The incredible complexity of the structure we seek to understand is enough to discourage the realist, and to tempt him to play some form of positivistic gamesmanship. He is tempted to settle for such elusive formulations as: "If we knew enough about the situation we wouldn't need the concept of personality"; or "One's personality is merely the way other people see one"; or "There is no structure in personality but only varying degrees of consistency in the environment."

Yet the truly persistent realist prefers not to abandon his commitment to find out what the other fellow is really like. He knows that his attempt will not wholly succeed, owing partly to the complexity of the object studied, and partly to the inadequacy of present methods. But unlike Kant who held that the *Ding an Sich* is doomed to remain unknowable, he prefers to believe that it is at least partly or approximately knowable.

I have chosen to speak of *heuristic* realism, because to me special emphasis should be placed on empirical methods of discovery. In this respect heuristic realism goes beyond naive realism.

Taking this epistemological point of view, the psychologist first focuses his attention on some limited slice of personality that he wishes to study. He then selects or creates methods appropriate to the empirical testing of his hypothesis that the cleavage he has in mind is a trait (either a dimensional trait or a personal disposition). He knows that his present purposes and the methods chosen will set limitations upon his discovery. If, however, the investigation achieves acceptable standards of validation he will have progressed far toward his identification of traits. Please note, as with any heuristic procedure the process of discovery may lead to important corrections of the hypothesis as originally stated.

Empirical testing is thus an important aspect of heuristic realism, but it is an empiricism restrained throughout by rational considerations. Galloping empiricism, which is our present occupational disease, dashes forth like a headless horseman. It has no rational objective; uses no rational method other than mathematical; reaches no rational conclusion. It lets the discordant data sing for themselves. By contrast heuristic realism says, "While we are willing to rest our case for traits on empirical evidence, the area we carve out for study should be rationally conceived, tested by rational methods; and the findings should be rationally interpreted."

Three Illustrative Studies

It is now time for me to illustrate my argument with sample studies. I have chosen three in which I myself have been involved. They differ in the areas of personality carved out for study, in the methods employed, and in the type of traits established. They are alike, however, in proceeding from the standpoint of heuristic realism. The presentation of each study must of necessity be woefully brief. The first illustrates what might be called *meaningful dimensionalism;* the second *meaningful covariation;* the third *meaningful morphogenesis.*

Dimensions of Values

The first illustration is drawn from a familiar instrument, dating almost from the stone age, *The Study of Values* (Allport & Vernon, 1931). While some of you have approved it over the years, and some disapproved, I use it to illustrate two important points of my argument.

First, the instrument rests on an a priori analysis of one large region of human personality, namely, the region of generic evaluative tendencies. It seemed to me 40 years ago, and seems to me now, that Eduard Spranger (1922) made a persuasive case for the existence of six fundamental types of subjective evaluation of *Lebensformen.* Adopting this rational starting point we ourselves took the second step, to put the hypothesis to empirical test. We asked: Are the six dimensions proposed—the *theoretic,* the *economic,* the *esthetic, social, political,* and *religious*—measurable on a multidimensional scale? Are they reliable and valid? Spranger defined the six ways of looking at life in terms of separate and distinct ideal types, although he did not imply that a given person belongs exclusively to one and only one type.

It did not take long to discover that when confronted with a forced-choice technique people do in fact subscribe to all six values, but in widely varying degrees. Within any pair of values, or any quartet of values, their forced choices indicate a reliable pattern. Viewed then as empirical continua, rather than as types, the six value directions prove to be measurable, reproducible, and consistent. But are they valid? Can we obtain external validation for this particular a priori conception of traits? The test's *Manual* (Allport & Vernon, 1931) contains much such evidence. Here I would add a bit more, drawn from occupational studies with women subjects. (The evidence for men is equally good.) The data in Table 1 are de-

TABLE 1

Mean Scores for Occupational Groups of Women: Study of Values

	Female collegiate norms $N = 2{,}475$	Graduate nurses training for teaching $N = 328$	Graduate students of business administration $N = 77$	Peace Corps teachers $N = 131$
Theoretical	36.5	40.2	37.3	40.6
Economic	36.8	32.9	40.4	29.9
Esthetic	43.7	43.1	46.8	49.3
Social	41.6	40.9	35.0	41.2
Political	38.0	37.2	41.8	39.7
Religious	43.1	45.7	38.7	39.2

rived partly from the *Manual,* partly from Guthrie and McKendry (1963) and partly from an unpublished study by Elizabeth Moses.

For present purposes it is sufficient to glance at the last three columns. For the *theoretic* value we note that the two groups of teachers or teachers in preparation select this value significantly more often than do graduate students of business administration. Conversely the young ladies of business are relatively more *economic* in their choices. The results for the *esthetic* value probably reflect the higher level of liberal arts background for the last two groups. The *social* (philanthropic) value is relatively low for the business group, whereas the *political* (power) value is relatively high. Just why nurses should more often endorse the *religious* value is not immediately clear.

TABLE 2

Significant Deviations of Scores on the Study of Values for Occupational Groups of Wellesley Alumni from Wellesley Mean Scores

Occupational groups	N	Theoretical	Economic	Esthetic	Social	Political	Religious
Business workers	64	Lower	Higher				
Medical workers	42	Higher	Lower			Lower	
Literary workers	40	Higher	Lower	Higher			
Artistic workers	37			Higher	Lower		
Scientific workers	28	Higher		Lower	Lower		
Government workers	24	Higher			Higher		Lower
Social workers	26						
Religious workers	11					Lower	Higher

Another study of external validation, showing the long-range predictive power of the test is an unpublished investigation by Betty Mawardi. It is based on a follow-up of Wellesley graduates 15 years after taking the Study of Values.

Table 2 reports the significant deviations (at the 5% level or better) of various occupational groups from the mean scores of Wellesley students. In virtually every case we find the deviation meaningful (even necessary) for the occupation in question. Thus women in business are significantly high in *economic* interests; medical, government, and scientific workers in *theoretical;* literary and artistic workers in *esthetic;* social workers in *social;* and religious workers in *religious* values.

One must remember that to achieve a relatively high score on one value, one must deliberately slight others. For this reason it is interesting to note in the table the values that are systematically slighted in order to achieve a higher score on the occupationally relevant value. (In the case of social workers it appears that they "take away" more or less uniformly from other values in order to achieve a high social value.)

Thus, even at the college age it is possible to forecast in a general way modal vocational activity 15 years hence. As Newcomb, Turner, and Converse (1965) say this test clearly deals with "inclusive values" or with "basic value postures" whose generality is strikingly broad. An evaluative posture toward life saturates, or guides, or gates (choose your own metaphor) specific daily choices over a long expanse of years.

One reason I have used this illustration of trait research is to raise an important methodological issue. The six values are not wholly independent. There is a slight tendency for theoretic and esthetic values to covary; likewise for economic and political values; and so too with social and religious. Immediately the thought arises, "Let's factor the whole matrix and see what orthogonal dimensions emerge." This step has been taken several times (see *Manual*); but always with confusing results. Some investigators discover that fewer than six factors are needed—some that we need more. And in all cases the clusters that emerge seem strange and unnamable. Here is a case, I believe, where our empiricism should submit to rational restraint. The traits as defined are meaningful, reliably measured, and validated. Why sacrifice them to galloping gamesmanship?

Covariation: Religion and Prejudice

Speaking of covariation I do not mean to imply that in restraining our empirical excesses we should fail to explore the patterns that underlie covariation when it seems reasonable to do so.

Take, for example, the following problem. Many investigations show conclusively that on the broad average church attenders harbor more ethnic prejudice than nonattenders. (Some of the relevant studies are listed by Argyle, 1959, and by Wilson, 1960.) At the same time many ardent workers for civil rights are religiously motivated. From Christ to Gandhi and to Martin Luther King we note that equimindedness has been associated with religious dovoutness. Here then is a paradox: Religion makes prejudice; it also unmakes prejudice.

First we tackle the problem rationally and form a hypothesis to account for what seems to be a curvilinear relation. A hint for the needed hypothesis comes from *The Authoritarian Personality* (Adorno, Frenkel-Brunswik, Levinson, & Sanford, 1950) which suggests that acceptance of institutional religion is not as important as the *way* in which it is accepted. Argyle (1959) sharpens the hypothesis. He says, "It is not the genuinely devout who are prejudiced but the conventionally religious [p. 84]."

In our own studies we have tentatively assumed that two contrasting but measurable forms of religious orientation exist. The first form we call the *extrinsic* orientation, meaning that for the churchgoer religious devotion is not a value in its own right, but is an instrumental value serving the motives of personal comfort, security, or social status. (One man said he went to church because it was the best place to sell insurance.) Elsewhere I have defined this utilitarian orientation toward religion more fully (G. W. Allport, 1960, 1963). Here I shall simply mention two items from our scale, agreement with which we assume indicates the extrinsic attitude:

> What religion offers me most is comfort when sorrows and misfortune strike.

> One reason for my being a church member is that such membership helps to establish a person in the community.

By contrast the *intrinsic* orientation regards faith as a supreme value in its own right. Such faith strives to transcend self-centered needs, takes seriously the commandment of brotherhood that is found in all religions, and seeks a unification of being. Agreement with the following items indicates an intrinsic orientation:

> My religious beliefs are what really lie behind my whole approach to life.

> If not prevented by unavoidable circumstances, I attend church, on the average (more than once a week) (once a week) (two or three times a month) (less than once a month).

This second item is of considerable interest, for many studies have found that it is the irregular attenders who are by far the most prejudiced (e.g., Holtzmann, 1956; Williams, 1964). They take their religion in convenient doses and do not let it regulate their lives.

Now for a few illustrative results in Table 3. If we correlate the extrinsicness of orientation with various prejudice scales we find the hypothesis confirmed. Likewise, as predicted, intrinsicness of orientation is negatively correlated with prejudice.

In view of the difficulty of tapping the two complex traits in question, it is clear from these studies that our rationally derived hypothesis gains strong support. We note that the trend is the same when different denominations are studied in relation to differing targets for prejudice.

Previously I have said that empirical testing has the ability to correct or extend our rational analysis of patterns. In this particular research the following unexpected fact emerges. While those who approach the intrinsic pole of our continuum are on the average less prejudiced than those who approach the extrinsic pole, a number of subjects show themselves to be disconcertingly illogical. They accept both intrinsically worded

TABLE 3

Correlations between Measures of Religious Orientation among Churchgoers and Various Prejudice Scales

Denominational sample	*N*	*r*
Unitarian	50	
Extrinsic–anti-Catholicism		.56
Intrinsic–anti-Catholicism		—.36
Extrinsic–anti-Mexican		.54
Intrinsic–anti-Mexican		—.42
Catholic	66	
Extrinsic–anti-Negro		.36
Intrinsic–anti-Negro		—.49
Nazarene	39	
Extrinsic–anti-Negro		.41
Intrinsic–anti-Negro		—.44
Mixed[a]	207	
Extrinsic–anti-Semitic		.65

[a] From Wilson (1960).

items and extrinsically worded items, even when these are contradictory, such as:

> My religious beliefs are what really lie behind my whole approach to life.
>
> Though I believe in my religion, I feel there are many more important things in my life.

It is necessary, therefore, to inspect this sizable group of muddleheads who refuse to conform to our neat religious logic. We call them "inconsistently proreligious." They simply like religion; for them it has "social desirability" (cf. Edwards, 1957).

The importance of recognizing this third mode of religious orientation is seen by comparing the prejudice scores for the groups presented in Table 4. In the instruments employed the lowest possible prejudice score is 12, the highest possible, 48. We note that the mean prejudice score rises steadily and sig-

TABLE 4

Types of Religious Orientation and Mean Prejudice Scores

	Mean prejudice scores			
	Consistently intrinsic	Consistently extrinsic	Moderately inconsistent (proreligion)	Extremely inconsistent (proreligion)
Anti-Negro	28.7	33.0	35.4	37.9
Anti-Semitic	22.6	24.6	28.0	30.1

Note.—$N = 309$, mixed denominations. All differences significant at .01 level.

nificantly from the intrinsically consistent to the inconsistently proreligious. Thus subjects with an undiscriminated proreligious response set are on the average most prejudiced of all.

Having discovered the covariation of prejudice with both the extrinsic orientation and the "pro" response set, we are faced with the task of rational explanation. One may, I think, properly argue that these particular religious attitudes are instrumental in nature; they provide safety, security, and status —all within a self-serving frame. Prejudice, we know, performs much the same function within some personalities. The needs for status, security, comfort, and a feeling of self-rightness are served by both ethnic hostility and by tailoring one's religious orientation to one's convenience. The economy of other lives is precisely the reverse: It is their religion that centers their existence, and the only ethnic attitude compatible with this intrinsic orientation is one of brotherhood, not of bigotry.

This work, along with the related investigations of Lenski (1963), Williams (1964), and others, signifies that we gain important insights when we refine our conception of the nature of the religious sentiment and its functions. Its patterning properties in the economy of a life are diverse. It can fuse with bigotry or with brotherhood according to its nature.

As unfinished business I must leave the problem of nonattenders. From data available it seems that the unchurched are less prejudiced on the average than either the extrinsic or the inconsistent churchgoers, although apparently more prejudiced on the average than those whose religious orientation is intrinsic. Why this should be so must form the topic of future research.

Personal Dispositions: An Idiomorphic Approach

The final illustration of heuristic realism has to do with the search for the natural cleavages that mark an individual life. In this procedure there is no reference to common dimensions, no comparison with other people, except as is implied by the use of the English language. If, as Allport and Odbert (1936) have found, there are over 17,000 available trait names, and if these may be used in combinations, there is no real point in arguing that the use of the available lexicon of a language necessarily makes all trait studies purely nomothetic (dimensional).

A series of 172 published *Letters from Jenny* (G. W. Allport, 1965) contains enough material for a rather close clinical characterization of Jenny's personality, as well as for careful quantitative and computational analysis. While there is no possibility in this case of obtaining external validation for the diagnosis reached by either method, still by employing both procedures an internal agreement is found which constitutes a type of empirical validation for the traits that emerge.

The *clinical* method in this case is close to common sense. Thirty-nine judges listed the essential characteristics of Jenny as they saw them. The result was a series of descriptive adjectives, 198 in number. Many of the selected trait names were obviously synonymous; and nearly all fell readily into eight clusters.

The *quantitative* analysis consisted of coding the letters in terms of 99 tag words provided by the lexicon of the General Inquirer (Stone, Bales, Namenwirth, & Ogilvie, 1962). The frequency with which these basic tag words are associated with one another in each letter forms the basis for a factor analysis (see G. W. Allport, 1965, p. 200).

TABLE 5

Central Traits in Jenny's Personality as Determined by Two Methods

Common-sense traits	Factorial traits
Quarrelsome-suspicious Aggressive	Aggression
Self-centered (possessive)	Possessiveness
Sentimental	Need for affiliation Need for family acceptance
Independent-autonomous	Need for autonomy
Esthetic-artistic	Sentience
Self-centered (self-pitying)	Martyrdom
(No parallel)	Sexuality
Cynical-morbid	(No parallel)
Dramatic-intense	("Overstate")

Table 5 lists in parallel fashion the clusters obtained by clinical judgment based on a careful reading of the series, along with the factors obtained by Jeffrey Paige in his unpublished factorial study.

In spite of the differences in terminology the general paralleling of the two lists establishes some degree of empirical check on both of them. We can say that the direct commonsense perception of Jenny's nature is validated by quantification, coding, and factoring. (Please note that in this case factor analysis does not stand alone, but is tied to a parallel rational analysis.)

While this meaningful validation is clearly present, we gain (as almost always) additional insights from our attempts at empirical validation of the traits we initially hypothesize. I shall point to one instance of such serendipity. The tag words (i.e., the particular coding system employed) are chiefly substantives. For this reason, I suspect, *sexuality* can be identified by coding as a minor factor; but it is not perceived as an independent quality by the clinical judges. On the other hand, the judges, it seems, gain much from the running style of the letters. Since the style is constant it would not appear in a factorial analysis which deals only with variance within the whole. Thus the common-sense traits *cynical-morbid* and *dramatic-intense* are judgments of a pervading expressive style in Jenny's personality and seem to be missed by factoring procedure.

Here, however, the computor partially redeems itself. Its program assigns the tag "overstate" to strong words such as *always, never, impossible,* etc., while words tagged by "understate" indicate reserve, caution, qualification. Jenny's letters score exceedingly high on overstate and exceedingly low on understate, and so in a skeletonized way the method does in part detect the trait of dramatic intensity.

One final observation concerning this essentially idiomorphic trait study. Elsewhere I have reported a small investigation (G. W. Allport, 1958) showing that when asked to list the "essential characteristics" of some friend, 90% of the judges employ between 3 and 10 trait names, the average number being 7.2. An "essential characteristic" is defined as "any trait, quality, tendency, interest, that you regard as of major importance to a description of the person you select." There is, I submit, food for thought in the fact that in these two separate studies of Jenny, the common-sense and the factorial, only 8 or 9 central traits appear. May it not be that the essential traits of a person are few in number if only we can identify them?

The case of Jenny has another important bearing on theory. In general our besetting sin in personality study is irrelevance, by which I mean that we frequently impose dimensions upon persons when the dimensions fail to apply. (I am reminded of the student who was told to interview women patients concerning their mothers. One patient said that her mother had no part in her problem and no influence on her life; but that her aunt was very important. The student answered, "I'm sorry, but our method requires that you tell about your mother." The *method* required it, but the *life* did not.)

In ascribing a list of traits to Jenny we may seem to have used a dimensional method, but such is not the case. Jenny's traits emerge from her own personal structure. They are not imposed by predetermined but largely irrelevant schedules.

Conclusion

What then have I learned about traits in the last 4 decades? Well, I have learned that the problem cannot be avoided—neither by escape through positivism or situationism, nor through statistical interaction effects. Tendencies, as McDougall (1937) insisted, remain the "indispensable postulates of all psychology."

Furthermore, I have learned that much of our research on traits is overweighted with methodological preoccupation; and that we have too few restraints holding us to the structure of a life as it is lived. We find ourselves confused by our intemperate empiricism which often yields unnamable factors, arbitrary codes, unintelligible interaction effects, and sheer flatulence from our computors.

As a safeguard I propose the restraints of "heuristic realism" which accepts the common-sense assumption that persons are real beings, that each has a real neuropsychic organization, and that our job is to comprehend this organization as well as we can. At the same time our profession uniquely demands that we go beyond common-sense data and either establish their validity or else—more frequently—correct their errors. To do so requires that we be guided by theory in selecting our trait slices for study, that we employ rationally relevant methods, and be strictly bound by empirical verification. In the end we return to fit our findings to an improved view of the person. Along the way we regard him as an objectively real being whose tendencies we can succeed in knowing—at least in part—beyond the level of unaided common sense. In some respects this recommended procedure resembles what Cronbach and Meehl (1955) call "construct validation," with perhaps a dash more stress on external validation.

I have also learned that while the major foci of organization in a life may be few in number, the network of organization, which includes both minor and contradictory tendencies, is still elusively complex.

One reason for the complexity, of course, is the need for the "inside" system to mesh with the "outside" system—in other words, with the situation. While I do not believe that traits can be defined in terms of interaction effects (since all tendencies draw their energy from within the person), still the vast variability of behavior cannot be overlooked. In this respect I have learned that my earlier views seemed to neglect the variability induced by ecological, social, and situational factors. This oversight needs to be repaired through an adequate theory that will relate the inside and outside systems more accurately.

The fact that my three illustrative studies are so divers in type leads me to a second concession: that trait studies depend in part upon the investigator's own purposes. He himself constitutes a situation for his respondents, and what he obtains from them will be limited by his purpose and his method. But this fact need not destroy our belief that, so far as our method and purpose allow, we can elicit real tendencies.

Finally, there are several problems connected with traits that I have not here attempted to revisit. There are, for example, refinements of difference between trait, attitude, habit, sentiment, need, etc. Since these are all inside tendencies of some sort, they are for the present occasion all "traits" to me. Nor am I here exploring the question to what extent traits are motivational, cognitive, affective, or expressive. Last of all, and with special restraint, I avoid hammering on the distinction between common (dimensional, nomothetic) traits such as we find in any standard profile, and individual traits (personal dispositions) such as we find in single lives, e.g., Jenny's. (Cf. G. W. Allport, 1961, Ch. 15, also 1962). Nevitt Sanford (1963) has written that by and large psychologists are "unimpressed" by my insisting on this distinction. Well, if this is so in spite of 4 decades of labor on my part, and in spite of my efforts in the present paper—I suppose I should in all decency cry "uncle" and retire to my corner.

References

Adorno, T. W., Frenkel-Brunswik, Else, Levinson, D. J., & Sanford, R. N. *The authoritarian personality*. New York: Harpers, 1950.

Allport, F. H. *Theories of perception and the concept of structure.* New York: Wiley, 1955.

Allport, G. W. What is a trait of personality? *Journal of Abnormal and Social Psychology,* 1931, *25,* 368-372.

Allport, G. W. Reviw of S. A. Stouffer et al., *The American soldier. Journal of Abnormal and Social Psychology,* 1950, *45,* 168-172.

Allport, G. W. What units shall we employ? In G. Lindzey (Ed.), *Assessment of human motives.* New York: Rinehart, 1958.

Allport, G. W. Religion and prejudice. In, *Personality and social encounter.* Boston: Beacon Press, 1960. Ch. 16.

Allport, G. W. *Pattern and growth in personality.* New York: Holt, Rinehart & Winston, 1961.

Allport, G. W. The general and the unique in psychological science. *Journal of Personality,* 1962, *30,* 405-422.

Allport, G. W. Behavioral science, religion and mental health. *Journal of Religion and Health,* 1963, *2,* 187-197.

Allport, G. W. (Ed.) *Letters from Jenny.* New York: Harcourt, Brace & World, 1965.

Allport, G. W., & Odbert, H. S. Trait-names: A psycholexical study. *Psychological Monographs,* 1936, *47*(1, Whole No. 211).

Allport, G. W., & Vernon, P. E. *A study of values.* Boston: Houghton-Mifflin, 1931. (Reprinted: With G. Lindzey, 3rd ed., 1960.)

Argyle, M. *Religious behaviour.* Glencoe, Ill.: Free Press, 1959.

Block, J., & Bennett, Lillian. The assessment of communication. *Human Relations,* 1955, *8,* 317-325.

Carr, H. A., & Kingsbury, F. A. The concept of trait. *Psychological Review,* 1938, *45,* 497-524.

Cronbach, L. J. & Meehl, P. E. Construct validity in psychological tests. *Psychological Bulletin,* 1955, *52,* 281-302.

Edwards, A. L. *The social desirability variable in personality assessment and research.* New York: Dryden Press, 1957.

Gibson, J. J. A critical review of the concept of set in contemporary experimental psychology. *Psychological Bulletin,* 1941, *38,* 781-817.

Guthrie, G. M., & McKendry, Margaret S. Interest patterns of Peace Corps volunteers in a teaching project. *Journal of Educational Psychology,* 1963, *54,* 261-267.

Helson, H. *Adaptation-level theory.* New York: Harper & Row, 1964.

Holtzman, W. H. Attitudes of college men toward nonsegregation in Texas schools. *Public Opinion Quarterly*, 1956, *20*, 559-569.

Hunt, J. McV. Traditional personality theory in the light of recent evidence. *American Scientist*, 1965, *53*, 80-96.

La Piere, R. Attitudes *vs.* actions. *Social Forces*, 1934, 230-237.

Lenski, G. *The religious factor.* Garden City, N.Y.: Doubleday, 1961.

McClelland, D. C. *Personality.* New York: Dryden Press, 1951.

McDougall, W. Tendencies as indispensable postulates of all psychology. In, *Proceedings of the XI International Congress on Psychology: 1937.* Paris: Alcan, 1938. Pp. 157-170.

Miller, D. R. The study of social relationships: Situation, identity, and social interaction. In S. Koch (Ed.), *Psychology: A study of a science.* Vol. 5. *The process areas, the person, and some applied fields: Their place in psychology and the social sciences.* New York: McGraw-Hill, 1963. Pp. 639-737.

Newcomb, T. M., Turner, H. H., & Converse, P. E. *Social psychology: The study of human interaction.* New York: Holt, Rinehart & Winston, 1965.

Sanford, N. Personality: Its place in psychology. In S. Koch (Ed.), *Psychology: A study of a science.* Vol. 5. *The process areas, the person, and some applied fields: Their place in psychology and in science.* New York: McGraw-Hill, 1963. Pp. 488-592.

Skinner, B. F. *Science and human behavior.* New York: Macmillan, 1953.

Spranger, E. *Lebensformen.* (3rd ed.) Halle: Niemeyer, 1922. (Translated: P. Pigors. *Types of men.* Halle: Niemeyer, 1928.)

Stone, P. J., Bales, R. F., Namenwirth, J. Z., & Ogilvie, D. M. The general inquirer: A computer system for content analysis and retrieval based on the sentence as a unit of information. *Behavioral Science*, 1962, 7(4), 484-498.

Stouffer, S. A., et al. *The American soldier.* Princeton: Princeton Univer. Press, 1949. 2 vols.

Tagiuri, R., & Petrullo, L. *Person perception and interpersonal behavior.* Stanford: Stanford Univer. Press, 1958.

Vernon, P. E. *Personality assessment: A critical survey.* London: Methuen, 1964.

Walker, E. L. Psychological complexity as a basis for a theory of motivation and choice. In D. Levine (Ed.), *Nebraska sym-*

posium on motivation: 1964. Lincoln: Univer. Nebraska Press, 1964.
Williams, R. M., Jr. *Strangers next door.* Englewood Cliffs, N.J.: Prentice-Hall, 1964.
Wilson, W. C. Extrinsic religious values and prejudice. *Journal of Abnormal and Social Psychology,* 1960, *60,* 286-288.

Familial Concordance by Sex with Respect to Schizophrenia

DAVID ROSENTHAL

A half century ago, Mott (1910) published a statistical compilation of pairs of relatives admitted to London County Council mental hospitals and noted a preponderance of female pairs as against male pairs, and of same-sexed as against opposite-sexed pairs of relatives. Myerson (1925) made a similar study in Massachusetts' Taunton State Hospital and replicated Mott's findings. Penrose (1942) postulated two autosomal genes to account for these different concordance rates regarding mental illness in the two sexes. His theoretical orientation stemmed from Freud's observation that sexual inversion predisposed to mental illness, as illustrated in the case of Schreber (Freud, 1911).

Penrose's postulated Genes A and B were assumed to augment and accentuate femaleness and maleness, respectively. If a father had Gene A and passed it on to his son, both would have the inherited tendency to sexual inversion which would predispose each to mental illness. If Gene A, however, was passed on to a daughter, her femininity would be enhanced or exaggerated, and she would escape the tendency to sexual inversion. Similarly, if mothers passed on Gene B to their daughters both would tend toward sexual inversion whereas their sons inheriting Gene B would be exceedingly masculine and would be less prone to develop mental illness.

From *Psychological Bulletin,* 1962, 59, 401-21. Reprinted by permission of the author and the American Psychological Association.

Slater (1944) believed that higher familial concordance in same-sexed as against opposite-sexed pairs had to be accepted as a well-established fact, but he objected to Penrose's theory to account for it. He pointed out that expressivity and penetrance were influenced by genetic, biochemical, and environmental factors, and that it was only necessary to suppose that some of these factors were more likely to be the same for persons of the same sex than for persons of the opposite sex. However, he made no attempt to indicate what these other genetic, biochemical, or environmental factors might be, so that his theory lacked the relative specificity of Penrose's theory.

Slater (1953b) subsequently considered the possibility of a sex-linked recessive gene in mental illness. He analyzed the frequencies of same-sexed and opposite-sexed pairs among the avuncular and sibling relationships in Mott's series of cases, and concluded that a genetic explanation was not satisfying because of inconsistencies in the frequencies when both kinds of familial relationship were considered. Also, if mental illness were due to a sex-linked recessive gene, an excess of maternal uncle-nephew pairs should have been found as compared to all other avuncular pairs, but no such excess occurred. However, he still considered the possibility that a sex-linked recessive gene might be playing a causative role with respect to a tendency for mental deficiency and paranoid schizophrenia to occur in the same families among Mott's cases.

Rosenthal (1959), in his analysis of Slater's (1953a) schizophrenic twin series, pointed out several differences between the sexes with respect to clinical features of the illness associated with concordance. Premorbid history, age of onset, deteriorative outcome, and subtype diagnosis distinguished concordant from discordant monozygotic (MZ) twins if they were male, but not if they were female. Attention was also called to the greater number of index cases and the higher concordance rate among female as compared to male MZ twins in Slater's series.

In a subsequent paper, Rosenthal (1961) noted that the preponderance of female twins obtained in the three largest of the five major twin studies of schizophrenia resulted from sampling among resident hospital populations. Females were more likely to become inhabitants of the chronic wards than males. It was also pointed out that a sample obtained in large part from a chronic population showed higher concordance with respect to severity of the illness than a sample obtained primarily through consecutive admissions, so that relationships found between sex and concordance rates might be influenced by the kind of sampling procedure used.

Thus, Mott's findings of half a century ago remain without any satisfying explanation, and the issues surrounding them seem to be more complex than originally believed. In this paper, it is intended to bring together various findings in the literature which are relevant to the theoretical considerations stated above, and to indicate the conclusions toward which these findings, in the aggregate, seem to be pointing. I shall attempt to determine whether higher familial concordance for females than males, and for same-sexed than opposite-sexed members, obtains through all close blood relationships. I shall consider various possible explanations of these findings, trying to indicate the relative merits of genetic and environmentalist hypotheses intended to account for the findings. I shall focus on these matters with respect to schizophrenia primarily, and shall use them to eliminate some obscure aspects of the etiology of this mental disorder.

Findings

The closest familial relationship, probably in the interpersonal as well as the genetic sense, is to be found among MZ twins. There are, of course, no oposite-sexed MZ twins, but concordance rates can be compared for MZ male and female pairs. Of the five major twin studies which aimed at statistically representative samples, four (Essen-Möller, 1941; Luxenburger, 1928; Rosanoff, Handy, Plesset, & Brush, 1934; Slater, 1953a) gave breakdowns of their MZ twin series by sex, and one (Kallmann, 1946) did not. All concordant and discordant MZ pairs reported in the four studies have herein been summed according to sex, the combined figures shown in Table 1.

It can be seen from Table 1 that concordance rates are

TABLE 1

Number of Male and Female Pairs of Monozygotic Twins Concordant and Discordant with Respect to Schizophrenia
(Four studies)

	Sex					
	Male	Percent	Female	Percent	Total	Percent
Concordant	47	(78.3)	23	(54.8)	70	(68.6)
Discordant	13	(21.7)	19	(45.2)	32	(31.4)
Total	60		42		102	

Note.—$\chi^2 = 5.328$; at $p = .02$, $\chi^2 = 5.412$.

indeed higher for female than for male MZ twins. All four studies are in agreement on this point, although none reaches statistical significance by itself because of the small numbers of cases. Indeed, the *dis*cordance rate is more than twice as high for the male as for the female MZ twins, representing almost half of all male pairs, but less than one in four female pairs.

This marked sex difference does not occur in Kallmann's series, although the very slight difference in concordance rates which he reports is also in the direction of higher concordance among female MZ pairs.

When we attempt to analyze the concordance rates of male and female dizygotic (DZ) twins, we find that only the series of Rosanoff et al. and Slater can be combined statistically. Kallmann does not give the actual number of same-sexed male and female DZ pairs. Luxenburger in a later report (1930) still had not found any concordant DZ pairs in his series of 37 DZ index cases. Essen-Möller did not concern himself with or report on his opposite-sexed pairs. The relevant data from the Rosanoff and Slater studies are summarized in Table 2.

It can be seen from Table 2 that there are only 28 concordant DZ pairs in both studies combined. When these are divided into three sex categories, the *N*s for analysis are rather small. Nevertheless, it is clear in Slater's series that the con-

TABLE 2

Concordance with Respect to Schizophrenia in Male and Female Dizygotic Twins

Sex of pairs	Slater[a] Concordant	Slater[a] Discordant	Rosanoff et al.[b] Concordant	Rosanoff et al.[b] Discordant	Total[c] Concordant	Total[c] Discordant
Male-male	2 (9.5%)	19	3 (27.3%)	8	5 (15.6%)	27
Female-female	9 (22.5%)	31	7 (16.7%)	35	16 (19.3%)	67
Male-female	2 (3.7%)	52	5 (10.4%)	43	7 (6.9%)	95

Note.–From Slater (1951) and Rosanoff, Handy, Plesset, and Brush (1934-35).
[a] $\chi^2 = 8.18, p < .02$.
[b] $\chi^2 = 2.21, p > .30$.
[c] $\chi^2 = 6.50, p < .05$.

cordance rate for female DZ pairs is higher than the rate for male DZ pairs, and that same-sexed pairs are more concordant than opposite-sexed pairs. In the Rosanoff et al. study, the concordance rate is higher for male DZ pairs than for female DZ pairs.[1] There are only 11 male pairs in their series as compared to 42 female and 48 opposite-sexed pairs, suggesting an unusual degree of sampling bias with respect to the male pairs. It will be noticed that the female: male ratio in the Rosanoff et al. study is approximately 4:1, a most deviant distribution, while in the Slater study the ratio is approximately 2:1, which also deviates markedly from theoretical expectancy. Some reasons for and implications of such sampling biases have been discussed at length by Rosenthal (1961).

Nevertheless, the concordance rate is again higher for same-sexed than opposite-sexed pairs in the Rosanoff et al. study, although the difference does not reach statistical significance. Combining both studies (which at least provides larger *N*s for each category, and which perhaps cancels out

[1] This is the only instance in the literature where, for any given blood relationship, I have found a higher concordance rate for male than female pairs with respect to schizophrenia.

some of the sampling errors of each study) yields statistically significant differences between the different types of sex pairs. The concordance rate is somewhat higher for females than males, and the concordance rate for same-sexed pairs is 18.3% as against 6.9% for opposite-sexed pairs, an appreciable difference.

Since twins comprise a special and unusual kind of familial group, we might pause in our investigation of familial concordance among the sexes to ask if there is any information available relevant to twins which might account for the sex differences in concordance reviewed here. One finding worth noting was reported by Rosanoff et al. They examined the histories of all twin pairs, both MZ and DZ, in which only one twin was diagnosed schizophrenic, and observed that the proportion of affected cases having a "'probably traumatic or infectious etiology" seemed to be higher in the males than in the females. The relevant data are brought together and summarized in Table 3.

TABLE 3

Incidence of Case Histories Suggesting "Traumatic or Infectious Etiology" in All Twin Pairs Discordant as to Schizophrenia

Type of history	Sex of the schizophrenic twin		Total
	Male	Female	
History of trauma or infection	16 (41.0%)	6 (10.0%)	22
History negative for trauma or infection	23	54	77
Total	39	60	99

Note.—Data from Rosanoff et al., 1934. $\chi^2 = 11.43$, $p < .01$.

It can be seen from Table 3 that clinical histories suggesting traumatic or infectious etiology occurred about four times as often among the male as among the female schizophrenic twins. The sex difference in this respect is highly significant, statistically Hereditarily speaking, such a finding is consistent with the hypothesis of genetic predisposition in the affected twin, and one could infer that manifestation of the illness might not have occurred without the trauma or infection, the latter serving as environmental factors which fostered and abetted manifestation.

The following considerations are raised by such an hypothesis:

1. Has it been shown in any study heretofore that physical trauma of infection are etiologic with respect to schizophrenia generally? Reviews of the literature relevant to this point indicate that, aside from a number of isolated case reports, there is no evidence to support this view (Bellak 1947; Kety, 1959; Overholser & Werkman, 1958). We must therefore question why such factors would be relevant to twins but not to single born persons. Should such a finding be confirmed with nontwins, the hypothesis would be bolstered considerably.

2. Such a finding really explains higher *dis*cordance among male pairs rather than higher *con*cordance among females. In other words, the number of concordant pairs should be about the same for both sexes, but the number of discordant pairs should be greater among male twins, if the hypothesis is correct. The data in Tables 1 and 2 suggest that it is rather the higher frequency of concordant pairs among females which is distinguishing the two sexes. There may however be some sampling bias involved here, as noted above, so that the hypothesis cannot be discredited with complete assurance on this point alone.

3. The authors do not mention whether the frequency of trauma and infection was similarly or dissimilarly distributed among the *non*-affected twin partners. Presumably, they did not have this information since it would be so hard to come by. It would be reasonable to assume that physical trauma, if not infection, occurs more frequently among males than females generally. If this is true, one would suppose that trauma and infection occurred more frequently among the nonaffected male twins than among the nonaffected female twins. Moreover, the probably higher incidence of trauma among male twins would also imply a higher incidence of schizophrenia among males than among females generally. Although there is

some evidence to show that such a difference does occur (Landis & Page, 1938; Malzberg, 1935), the difference is not large and probably not at all proportional to the difference between the sexes with respect to the frequency of trauma.

Thus, although the Rosanoff et al. hypothesis is a suggestive one, there is little to support it and some evidence against it. Another kind of information relevant to the main problem under discussion can, however, also be found in their study. The authors searched the same clinical histories of the affected twins among discordant pairs for evidence of "etiological psychic factors." The relevant data are summarized in Table 4.

TABLE 4

Incidence of "Etiological Psychic Factors" in the Clinical Histories of Male and Female Discordant Schizophrenic Twins

	Male	Female
Psychic factors in evidence	6	21
Psychic factors not in evidence	33	39

Note.—Data from Rosanoff et al., 1934. $\chi^2 = 3.65$; at $p = .05$, $\chi^2 = 3.84$.

Table 4 shows that etiological psychic factors were in evidence among 35% of the affected female twins but among only 15.4% of the affected males. The difference falls barely short of the 5% level of significance. The authors state that it was especially those psychic factors having to do with the sphere of sex or love life which were more common in the female sex. If such "etiological" factors are more common among female twins, then they would be more likely to occur among both members of a pair of female as compared to male twins, and would thus make for a higher concordance rate for the female twins.

Against this hypothesis, it may be said that we do not know about the distribution of these factors in the *non*affected twin partners. One could well imagine that a similar distribution might have obtained among them, since problems around sex and the love life probably were at that time (and perhaps

TABLE 5

Concordance as to Schizophrenia in Same-Sexed and Opposite-Sexed Dizygotic Twins

	Concordant pairs	Discordant pairs	Total
Same sex	34	262	296
Opposite sex	13	208	221
Total	47	470	517

Note.—Data from Kallmann, 1946. $\chi^2 = 4.15$, $p < .05$.

still are) more prevalent among women than men (Offergeld, 1957). Moreover, one would normally have some reservations about inferences regarding psychological problems when these inferences are made from hospital records. It may be simply that the women were less loath to talk about psychological problems, especially sexual ones, than were the men. Gross (1959) found that women admitted to having psychopathological traits more freely than men, while men showed more conscious and unconscious denial and constriction of behavior. Moreover, if such problems are crucial etiologically with respect to schizophrenia, and if they occur more commonly among women than men, then the incidence of this disorder should be higher among females. As noted above, frequency of hospitalization for schizophrenia in this country is, if anything, higher among males.

In favor of the hypothesis is the clinical observation, so often reported that it is needless to cite references here, of morbid sexual preoccupation among schizophrenics of *both* sexes. Moreover, it is well established that good premorbid sexual adjustment is a favorable prognostic sign (Phillips, 1953; Wittman, 1941). But even if psychological disturbances in the erotic sphere could be proved to be of etiological significance in schizophrenia, it would still be necessary to integrate theoretically the findings in Table 4 with data which suggest a probably higher incidence among males generally but a higher concordance among female twins.

Kallmann (1946) has published data regarding the number of same-sexed and opposite-sexed DZ twin pairs concordant with respect to schizophrenia. These data are summarized in Table 5, in which it is again shown that the concordance rate is higher for the same-sexed pairs.

Kallmann also stated that the concordance rate was 17.7% for female DZ pairs and 17.4% for male DZ pairs, as shown in Table 6. It is difficult to evaluate these figures since there were only 34 concordant same-sexed DZ pairs in his series, and since the percentages cited represent morbidity risk estimates rather than actual concordant pairs, as in Table 2. Kallmann also published concordance rates (morbidity risk estimates) for the *siblings* of twin index cases, as shown in Table 6. When the sibling was of the same sex as the twin, the concordance rate was 16.1%. When he was of the opposite sex, the rate was 12.3%. Among those siblings who were of the same sex as the twin index cases, the concordance rate was 16.3% for females and 15.9% for males. Although again we cannot infer the number of actual pairs, and the numbers are obviously small, the slight differences reported are in the expected direction. In Slater's series, which had about two females for every male index twin, the age-corrected incidence of schizophrenia among sibs of the index cases was 3.4% for male sibs and 7.3% for female sibs, a finding which is again in the direction of higher concordance among females.

TABLE 6

Morbidity Risks as to Schizophrenia in Siblings and Dizygotic Cotwins of Twin Index Cases According to Sex

Sex of the twin index case	Siblings of twin index cases			Dizygotic cotwins		
	Male	Female	Total	Male	Female	Total
Same sex	15.9	16.3	16.1	17.4	17.7	17.6
Opposite sex	12.5	12.0	12.3	10.5	10.2	10.3
All cases	14.0	14.5	14.3	14.3	14.9	14.7

Note.—Data from Kallmann, 1946.

Concordance in sibling pairs where neither is a twin has been reported in two extensive investigations by Schulz (1932) and Zehnder (1941). The relevant findings are summarized in Table 7.

On the basis of sex distribution alone, we expect that the

TABLE 7

Sex of Siblings Concordant with Respect to Schizophrenia

Sex	Schulz (1932)[a]	Zehnder (1940)[a]	Total
Male-male	37	13	50
Female-female	38	30	68
Male-female	56	29	85
All pairs	131[b]	72	203

Note.—χ^2, same- versus opposite-sexed pairs, = 5.04, $p < .05$, (using 1 : 1 expectancy).

[a] Schulz's sample is based on siblings of a series of index cases collected by Rüdin at the University of Munich Psychiatric Clinic, whereas Zehnder simply ascertained pairs of affected sibs, without specifying either as an index case. From a genetic viewpoint, the former procedure is desirable. However, the studies agree well with respect to the proportions of same-sexed to opposite-sexed pairs.

[b] If sibs had been included of whose diagnosis Schulz was less certain, the figures would have been 43, 47, and 71 for the brother, sister, and brother-sister pairs, respectively.

number of same-sexed pairs should be approximately equal to the number of opposite-sexed pairs. If we combine the figures from both studies (which are in general agreement on this point) we find 138 same-sexed pairs as against 85 opposite-sexed pairs. The discrepancy differs significantly from the expected equal distribution. With respect to male versus female pairs, we again find a higher number of female pairs in Zehnder's sample. If we assume an expected equality of pairs by sex, then the discrepancy from equality is statistically significant ($p < .05$). In the Schulz sample, however, equality appears to be almost perfect. But in this study, there were actually 367 male index cases as against only 293 female index cases in the original sample, or 55.6% male versus 44.4% female. Moreover, male index cases had 1079.5 sibs as compared to 880 sibs of the female index cases, which would have increased the probability of having more brother pairs relative to sister pairs. But among pairs of siblings, the females still comprised over 50%. Thus, a difference of unknown magnitude in favor of female pairs occurred in the Schulz study as well.

It is relevant to point out at this time how sampling procedures may by themselves lead to such differences. Zehnder, for example, collected all pairs of siblings admitted to a single Swiss mental hospital over a 20-year period. This sounds like a total sample which should therefore be free of biasing factors. However, we need only to consider that males are generally more migrant than females, and that the differential was probably at a peak during the early years of this century when most of these cases were admitted. If two brothers were both schizophrenic, but one had emigrated before his illness, this pair would not have been included in Zehnder's sample. Sisters, however, who would probably both have been relatively more housebound, would have found their way to the same hospital if both had become ill.

In this regard it is interesting to note that there were 50 male pairs as compared to 85 opposite-sexed pairs in the Schulz and Zehnder studies combined. These values do not differ significantly from the expected values of 45 and 90 pairs for the respective groups. However, the 68 female pairs are significantly greater than one-half the number of opposite-sexed pairs. Thus, in comparing same-sexed with opposite-sexed pairs, it is the group of female pairs which makes for the higher frequency or concordant siblings among the same-sexed pairs. The possibility also exists that same-sexed siblings, male or female, are more likely to be close to one another and live in the same area than are opposite-sexed siblings. Such factors alone, if proven correct, could account for the data reported in Table 7.

With respect to research in this area, two points are worth making. The first follows directly from the immediately preceding discussion. In studies of concordance among siblings, each and every sibling in all families comprising the starting cases of the sample must be investigated. The same principle holds true for other blood relationships as well. It is not enough to report only known cases of readily available relatives. Of course, such a sampling ideal is in practice extremely difficult to achieve, for apparent reasons.

This leads us to the second point, viz., that research on twins has an advantage over familial studies in that the sampling ideal just indicated can be more readily approached, once one has defined his sample of starting cases. This is true because: (*a*) Twins are the same age. One does not have the problems of cases covering two or more generations where the antecedent generation has been excessively decimated by deaths, or the subsequent generation has not yet reached or lived through an appreciable part of the morbidity risk period. (*b*) Twins are more readily located. This is true because twins are less likely than other siblings or relatives to be separated (which can of course be a disadvantage from another point of view), and are more likely to keep in touch or know of each other's whereabouts if they are separated. (*c*) The number and type of relatives to be ascertained are more clearly specified than in the case of some other familial groupings. An individual may have many sibs or none at all, but a twin always has just one cotwin, and only that particular person needs to be found and studied. Moreover, from a genetic standpoint, the same paternity in the case of twins is always assured, but no similar guarantee obtains in the case of other siblings. (*d*) The representativeness of the sample is more easily verified since the distribution of twins in the general population according to sex and zygosity is known and can serve as a standard against which the sample can be compared.

Concordance by Sex for Primary and Collateral Familial Relationships

In pursuing our inquiry further we may ask two questions of theoretical relevance: (*a*) Does the pattern of familial concordance by sex obtain only with the primary family group, i.e., parent, child, and sibling, or does it also extend to aunts, uncles, and cousins? If genetic factors are at work, the extension to avuncular relationships should be found, but to a lesser extent. If not, hypotheses emphasizing intrafamilial influences of a psychological nature would be enhanced. (*b*) Does the pattern of familial concordance by sex obtain for mental illness generally, or is it specific to those disorders called schizophrenic?

To deal with these questions, we have recourse to large scale studies by Mott (1910), Myerson (1925), and Penrose (1945). Mott began his studies shortly after the turn of the century when he established an "Heredity Index," a card-filing system in which names, diagnoses, dates of admission, and discharge, etc. were recorded for all persons admitted to one of the London County Council Mental Hospitals who were known to have or to have had any other relative as a patient in one of these hospitals. The files were continued long after Mott's original article in 1910, and Myerson (1925) and Slater (1953b) have reported more recent compilations of the filed data, these larger figures being cited here instead of Mott's original figures. Myerson (1925) used the records of the Taunton State Hospital in Massachusetts to obtain a list of all pairs of relatives who had been admitted to that institution. For both the Mott and Myerson studies, the figures reported here include all such pairs, no attempt being made to differentiate individuals diagnostically. They are simply called mentally ill to a degree warranting certification.

Penrose's unpublished study has the largest accumulation of pairs of relatives. His survey was based on records of patients admitted to all Ontario Hospitals. Because his *N*s were sufficiently large, and becuase he separated out two groups of disorders which he called schizophrenic and affective, the figures for the two groups are listed separately.[2] Under the *schizophrenic disorders* were included all patients who had been diagnosed: schizophrenic—the simple, paranoid, catatonic, or hebephrenic subtypes; involutional paranoid or senile paranoid; dementia praecox; schizophrenic defective; arterio-

[2] Omitted here are all those cases diagnosed primarily as: arteriosclerotic, senile, paretic, other organic, epileptic, defective, and Huntington's chorea.

sclerotic paranoid; and alcoholic paranoid. Under the *affective disorders* were included all patients who had been diagnosed: manic, manic depressive; reactive depressive or psychoneurotic depressive; involutional depressive or melancholic; involutional psychosis; manic alcoholic; manic psychosis with mental defect; senile manic or depressive; arteriosclerotic manic or depressive. Clearly, these groupings will not satisfy everyone, perhaps not even many. The inclusion of known "organic" cases is questionable and distinctions such as between "involutional paranoid" and "involutional psychosis" may not be easy to make. Nevertheless, the attempt to relate cases in which a cognitive disorder predominates and cases in which an affective or mood disorder predominates is reasonable enough and is preferable to a labeling anarchy which generates a large number of discrete, idiosyncratic categories. The main relevant data of the three studies are brought together in Table 8.

With respect to the parent-child relationship, it can be seen in Table 8 that there were consistently more mothers (601) than fathers (455) and more mother-daughter (339) than father-son (209) pairs. This was true for the broad classification "mentally ill," but was most striking in Penrose's group of schizophrenic disorders. In this group there were 106 mothers versus 44 fathers, and twice as many mother-daughter pairs as father-son pairs. These data may appear to have interesting psychological implications, but can readily be explained simply on the basis that schizophrenic females are more likely to marry than schizophrenic males, and that married schizophrenic females tend to have more children than their male counterparts (Essen-Möller, 1959). There was also a preponderance of mentally ill daughters (585) as compared to sons (471). This is less readily explained, although the possibility that more sons than daughters may have emigrated from the areas covered by the respective studies may have been an important factor. The possibility also exists that daughters of mentally ill parents are more likely to be affected than sons. It is interesting to note that both the mentally ill mothers and fathers had more affected daughters than sons, the proportions of daughters to sons being 56.6:43.4 in the case of the mothers and 54.0:46.0 in the case of the fathers. The slight difference

TABLE 8

The Sexual Distribution of Pairs of Relatives Certified Mentally Ill (Three studies)

Relationship	Study	Criteria	Both male	Both female	Male-female[a]	Female-male[b]
Parent and child	Mott (1910)[c]	Mentally ill	78	137	103	96
	Myerson (1925)	Mentally ill	55	80	59	56
	Penrose (1945)	Schizophrenic disorder	28	55	16	51
	Penrose (1945)	Affective disorder	48	67	68	59
		Total	209	339	246	262
Sibling	Mott (1910)[c]	Mentally ill	140	211	212	
	Myerson (1925)	Mentally ill	57	80	110	
	Penrose (1945)	Schizophrenic disorder	106	124	192	
	Penrose (1945)	Affective disorder	69	118	140	
		Total	372	533	654	
Uncle, aunt and nephew, niece	Mott (1910)[d]	Mentally ill	67	73	53	30
	Myerson (1925)	Mentally ill	41	42	37	43
	Penrose (1945)	Schizophrenic disorder	39	39	35	42
	Penrose (1945)	Affective disorder	26	45	25	43
		Total	173	199	150	158

[a] Male patient and female relative designates father-daughter and uncle-niece combinations.

[b] Female patient and male relative indicates mother-son and aunt-nephew combinations.

[c] Figures cited by Myerson (1925).

[d] Figures cited by Slater (1953).

is in favor of mother-daughter as compared to father-daughter pairs, which is in keeping with the previously noted findings of higher concordance among female than other familial pairs.

With respect to Penrose's schizophrenic group, the differences in concordance by sex are somewhat clearer. The certified mothers had slightly more affected daughters (55) than sons (51), whereas the fathers had more affected sons (28) than daughters (16). As a measure of the sex-concordance association, we may use the tetrachoric correlation which in this case is .25 or slightly less than two times its standard error. Although it does not reach statistical significance, this finding is again consistent with the sex-concordance hypothesis.

The findings with respect to siblings are in agreement with the Schulz and Zehnder studies cited above. There were more sister than brother pairs (533:372) and more same-sexed (905) than opposite-sexed pairs (654). The differences are marked and highly significant, statistically. In Penrose's schizophrenic group, the difference was less pronounced, but in the same direction. The question of sampling procedures which could be influenced by factors like differential emigration rates in the sexes must be raised again.

Such a factor would not account, however, for the fact that the number of brother pairs (372) exceeded more than one half the number of all brother-sister pairs (654). Chi square for this discrepancy from expectancy equals 5.78 which is significant at the .02 level. Thus again, we have a fairly firm finding of higher concordance in same-sexed as against opposite-sexed pairs of relatives. It should be noted, however, that most of the discrepancy occurred in the Mott series and not at all in Penrose's group of affective disorders. Failure to replicate the main finding in the latter group may simply reflect the fact that women generally are more susceptible to affective disorders than are men (Dayton, 1940; Landis & Page, 1938; Larsson & Sjögren, 1954; Malzberg, 1935).

When we examine the avuncular groups, we find that whereas among primary familial groups (parent-child and sibling) there were 872 female pairs as against 581 male pairs, there were only 199 aunt-niece pairs as against 173 uncle-nephew pairs. This latter difference is not statistically significant from a 1:1 ratio. Moreover, almost all the difference occurred in Penrose's group of affective disorders, a finding which is again probably reflecting the higher incidence of these disorders in females than males.

[Comparing same-sexed with opposite-sexed avuncular pairs, we find a significantly higher frequency of same-sexed pairs in the Mott study but not in any of the other three groups.[3] I have no idea why Mott's data deviate from the others in this respect. With respect to Penrose's group of schizophrenic disorders, we find exactly the same number of aunt-niece as uncle-nephew pairs (39:39), and virtually the same number of same-sexed (78) as opposite-sexed (77) pairs. Thus, the higher concordance of female than male pairs and of same-sexed than opposite-sexed pairs seems to obtain in the primary family groups but most of the evidence suggests that it no longer occurs, or occurs only to a very slight degree, when the familial relationships are one step removed.

One could only wish that the sampling in these studies had been of a higher order so that one could have greater confidence in drawing conclusions on a point of such high theoretical interest. Fortunately, one such study exists and we may use it as a check on the major findings and inferences just discussed.

Penrose (1942) collected what was virtually a consecutive series of 500 male and 500 female patients who had been certified at the Ontario Hospital, London. The known relatives of each case were investigated and classified according to whether or not they had suffered from mental illness. Thus, this was more than an attempt to obtain pairs from hospital records only. Unfortunately, the classification approached completeness only in the case of parents, children, and sibs of the 1,000 starting cases, complete ascertainment of grandparents, uncles, aunts, nephews, nieces, and cousins being more difficult to achieve. Thus, although a substantial improvement, sampling still fell considerably short of ideal. Relatives were divided into three groups: (*a*) Those who had been certified (excluding cases of mental defect or epilepsy without psychosis), (*b*) those who showed signs of psychosis but who had not been

[3] It is worth noting that Slater (1953b) used the avuncular series in Mott's data to test the hypothesis of sex-linked recessives in mental illness and failed to find anything more than an infirm suggestion of a linkage between mental deficiency and paranoid schizophrenia. The Myerson (1925) and Penrose (1945) data are not encouraging with respect to finding any sex-linked recessives.

certified, (*c*) a conglomerate group designated "psychopathic" which we shall not include in our discussion.[4]

Rather than present the figures separately for each kind of familial relationship, I have arranged the pairs by sex according to whether the relationships were primary or secondary. Certified and uncertified psychotic relatives were combined. In this way, we avoid too small *N*s and yet address ourselves to our main hypotheses. The relevant data have been compiled as in Table 9.

TABLE 9

Concordance by Sex with Respect to Psychosis in Primary or Secondary Familial Relationships

Relative	Primary relationships[a]		Secondary relationships[b]	
	Patient		Patient	
	Male	Female	Male	Female
Female	59	93	57	66
Male	84	52	55	62
	$r_{tet} = .35$		$r_{tet} = .01$	

Note.—Data from Penrose, 1942.

[a] Includes parents, children, and siblings.

[b] Includes grandparents, uncles, aunts, nephews, nieces, and cousins.

It can be seen in Table 9 that there were slightly more female than male pairs in both the primary and secondary groups. The difference between the groups in this respect found in Table 8 no longer obtained. The findings of Table 8 with respect to same- and opposite-sexed pairs however receive strong support in Table 9. Among primary familial groupings, there were 177 same-sexed as against 111 opposite-

[4] In this group were relatives described as nervous, excitable, hysterical, highly strung, neurasthenic, neurotic, hypochondriacal, depressive, temperament, quarrelsome, bad tempered, unbalanced, peculiar, queer, eccentric, manneristic, bigamous, deserted family, in jail, alcoholic, or drug addict. Such a "group" clearly is not relevant to our inquiry, even if we disregard questions about the reliability of such descriptions.

sexed pairs. The tetrachoric correlation equals .35, which is 3.77 times its standard error, $p < .01$. Among secondary familial groupings there were 121 same-sexed as against 119 opposite-sexed pairs. The tetrachoric correlation equals .01, which is only about one-tenth its standard error, and of course not significantly different from zero.

From a genetic point of view, members of pairs in the primary group share one-half of a common heredity, whereas those in the secondary groupings share one-fourth (uncles, aunts, nephews, nieces, and grandparents) or one-eighth (cousins) of a common heredity. These values indicate the theoretical expectancies regarding the relative proportions of explained variance in the association between sex and concordance among different familial groupings, if genetic factors are solely accountable. Since there appears to be virtually no association at all between sex and concordance in the secondary group, one is left to conclude that factors other than genetic ones are involved in the association found in the primary familial groupings. It is assumed that a more complete ascertainment of relatives in the secondary group would not change the proportions shown in Table 9. This assumption seems tenable, since factors such as differential sex emigration should apply in all degrees of familial relationship and should be manifest in even a partial sample that is representative, but it surely warrants further investigation.

If other than genetic factors are implicated in sex differences in concordance rates, one might expect these differences to be accentuated in mental disorders defined primarily as the sharing by two people of characteristics of the illness in which heredity is probably playing less of a role. I refer to what has been known historically as "disorders of association." Diagnostically, patients with such disturbances would often be subsumed under paranoid schizophrenia, but commonly too under hypochondriasis, these disorders not always being easy to differentiate. Since 1873, the term "folie à deux" has been applied most often to such cases (Lasègue & Fabret, 1873). A scholarly review of this subject has been written in an unpublished doctoral dissertation by Greenberg (1961) submitted to the University of Sydney.

A predominance of female over male pairs with folie à deux has been found by a number of investigators who compiled series of cases, mostly from reports in the literature (Gralnick, 1942, Kröner, 1891; Marandon, 1894; Wollenberg, 1889). Since we do not know the factors making for some cases being reported and others not, we are unable to estimate the role of possible sampling bias in these reports. Recently, however, Greenberg's (1961) study has become available and it is the first systematically obtained sample of such cases, as far as I know. He states:

> From the Board of Control's central registry of admissions to mental hospitals in England and Wales, instances of the more or less simultaneous admission of two or more individuals, related either by blood or marriage, were noted for a period of five years. These, with the addition of several cases from the out-patient clinics of Guy's and St. Bartholomew's Hospitals, made up a total of 114 cases, involving 234 individuals, each of which was then investigated individually. A number of cases irrelevant to this investigation were then excluded: those involving aged sibs living together, whose increasing dementia had precipitated their admission to hospital; related individuals developing psychoses apparently quite independently of one another; and several where the available clinical records were such as to preclude adequate appraisal. . . . There remained 60 instances of folie à deux, à trois and à quatre, involving 124 individuals, and these were submitted to detailed investigation. . . . The positive criteria of selection were that the subjects should show very similar or identical clinical pictures; that they wholly or partially shared the same delusions; that there was presumptive evidence from the history that the development of these states was in some way interrelated.

The criteria for inclusion of cases and the sampling procedures are relatively specific and straightforward. The sample turned up 13 mother-daughter pairs as against 2 father-son pairs, and 18 pairs of sisters as against 3 pairs of brothers. There were 6 mother-son pairs, no father-daughter pairs, and 3 brother-sister pairs. Thus, although there was a preponderance of same-sexed as against opposite-sexed pairs, the difference was accounted for primarily by the relatively high frequency of female pairs. Although one could raise the question of differential fertility rates as a possible explanation of the higher number of mothers than fathers, nevertheless the ratio of mother-daughter to father-son pairs was almost identical with the ratio of sister to brother pairs, suggesting that the sex difference here may have been relatively independent of fertility factors. This suggestion is supported by the fact that among *all* parent-child combinations, same-sexed pairs occurred 2.5 times as often as opposite-sexed pairs. The high frequency of sister pairs, plus the concurrent finding of equal numbers of brother-sister and brother pairs (the number of course are small), pose a constellation of frequencies which strain any attempt to provide a simple genetic explanation.

Discussion

Although relevant to a specific theoretical issue, the material covered in this analysis is quite heterogeneous, especially with respect to methods of sampling and diagnosis. In the main, the data point up concordance rates with respect to schizophrenia which are higher for female than male pairs and for same-sexed than opposite-sexed pairs of relatives in primary family groups but not in familial relationships further removed.

There are three possible ways of trying to account for such findings, and each warrants separate discussion: (*a*) the findings may be artifacts produced by vagaries of sampling, (*b*) the findings may be valid and explainable on a genetic basis, (c) the findings may be valid and explainable on a psychological basis.

1. In presenting the material, I have called attention to ways in which sampling procedures alone could have led to the findings in some studies. Factors such as differential migration of the sexes, incompleteness of ascertainment, severity of illness or age of the subjects in the samples might have been contributory in various ways. Can such factors possibly account for all the findings?

Differential migration rates must be considered as a serious and relevant source of sampling bias when pairs of relatives are ascertained only from records of hospitals in a fairly circumscribed area. The findings in the studies of Zehnder (1941), Schulz (1932), Mott (1910), Myerson (1925), Penrose (1945), and possibly Greenberg (1961) could conceivably have arisen from such a bias. However, in the studies of twins such a bias could not have occurred since the sampling involved *single* index cases rather than *pairs* of relatives, and since each ascertainable cotwin of every index case was included in the evaluation, even if the cotwin had emigrated or died after reaching the age of morbidity risk. Since the concordance rates were found to be higher for female than male MZ pairs in the five major studies, and for female than male DZ pairs in Slater's study, and for same-sexed than opposite-sexed pairs of DZ twins in the studies of Rosanoff et al., Slater, and Kallmann, we have evidence that factors other than differential migration rates must have been involved. Also, higher migration rates for males than females would not by themselves have accounted for a higher incidence of brother pairs relative to brother-sister pairs in the Mott, Myerson, and Penrose (1945) studies combined. One would have to add the qualification that brothers of females might have been more likely to migrate than brothers of males. Of course the brother pairs must have had sisters in their families and opposite-sexed pairs must have had other brothers, so that the point would have to be qualified further with respect to the incidence of the sex of other sibs, straining the hypothesis even more.

Moreover, we would expect that as the geographic area encompassed in the sampling increased, migration effects would have been proportionately reduced. Thus, sex-concordance ratios should have been less in a study like Greenberg's which covered all of England and Wales than in a study like Zehnder's or Myerson's which sampled from one hospital, or than in a study like Mott's or Penrose's, which sampled from one county or province. Actually, the sex-concordance ratios were highest in Greenberg's study, and there was great similarity between the ratios in Myerson's study and those in the Mott and Penrose studies.

Lastly, in a more fully ascertained sample, Penrose (1942) found higher concordance for same-sexed as against opposite-sexed pairs of sibs. Thus, although they may well have been contributory, differential migration rates alone could not have accounted for the array of sex-concordance findings presented here.

Incompleteness of sampling does not per se constitute a serious objection to the kinds of data presented unless one can point toward specific biases which are likely to occur on a selective basis when some proportion of prospective cases is missed. One such bias, probably the most important, would be differential migration rates in the sexes, but we have just seen that this bias alone could not account for many of the findings presented. A second possible bias, selective *reporting* of cases by sex, is somewhat obviated in studies which ascertained cases from hospital records. We might, however, consider the possibility in these studies that relatives with the same family name would more likely have been found than those with different family names. If so, more brother pairs should have been found than sister pairs since many sisters had probably married, attenuating the seriousness of such a possibility. However, if it is true that concordance rates are really higher for female than male pairs among collateral relatives as well, then a disproportionate loss of aunt-niece pairs through changes of family name might have occurred in studies which ascertained cases from hospital records (Mott, 1910; Myerson, 1925; Penrose, 1945). Such loss could conceivably account for the fact that no sex difference was found among avuncular pairs in these studies. However, in the Penrose, 1942 study which ascertained cases through personal inquiry among families, such loss is much less likely and we still find that the higher concordance rate among same-sexed pairs in the primary family groups no longer obtains in the collateral groups. The latter finding would have to assume a disproportionate loss among same-sexed pairs as against opposite-sexed avuncular pairs if the difference between primary and secondary family groups is to be attributed to sampling bias, and this assumption seems improbable.

It has been shown with respect to MZ twins that when the

severity of the illness in the index case is great, concordance is likely to be considerably higher than if the index case is only mildly ill (Rosenthal, 1961). Thus, one may wonder if a similar relationship obtains among other relatives and if such a factor could explain the sex-concordances reported. However, there seems to be no reason to think that, for example in the twin studies, the males would have been less severely ill than the females. It has been pointed out in a few studies that males are actually more likely to get out of the hospital than females[5] (Lengyel, 1941; Rosenthal, 1961; Sternberg, 1948), so that even in the twin samples which were overloaded with chronic, severe cases, most of whom were women, the males ascertained as index cases would probably have been at least as severely ill as the female index cases. Thus, on this ground alone there would be no reason to expect higher concordance in the female than the male twin pairs, and it seems reasonable that this argument would extend to other pairs of relatives as well.

One might conjecture that female patients are more likely than males to be communicative about themselves and their relatives, making for a more complete ascertainment of female than male cases. Such a conjecture is plausible enough, but it would not explain those findings where there was a higher frequency of brother pairs than brother-sister pairs. All told, it seems unlikely that sampling biases alone could have accounted for the main findings in this paper.

2. Both Penrose (1942) and Slater (1944, 1953b) concerned themselves primarily with a possible genetic explanation of higher concordance for psychiatric disorder among same-sexed than opposite-sexed pairs of relatives. According to the data presented here, especially those relating to schizophrenics, such an explanation needs to be expanded to account also for the fact that this relationship no longer obtains among collateral pairs of relatives, that concordance rates are higher for female than male pairs, and that the frequency of admis-

[5] This statement is based on data obtained during the first half of this century. Since the advent of tranquilizing drugs, differences between the sexes in this regard may no longer occur, but all the studies reviewed here were done before these drugs were in widespread use.

sions for schizophrenia is as high for males as for females, and maybe higher.

Penrose's theory of two autosomal sex-augmenting genes could be modified to account for higher concordance in female than male pairs. It would only be necessary to postulate that the effect of the male-augmenting Gene B is greater (has higher expressivity) than that of the female-augmenting Gene A. Thus, sisters affected with Gene B would both be more likely to manifest aspects of sexual inversion than both brothers affected with Gene A, and the sisters would therefore be more highly predisposed to schizophrenia. However, even with this modification, one should still find these factors operating in collateral as well as primary family groups and one should also expect to find a greater number of females admitted for schizophrenia than males. Since the best evidence available goes contrary to these expectations, the expanded theory still would not be adequate.

The same point would apply to Slater's theory (1944) which was drawn along lines similar to Penrose's, but which objected to the specific "genetic" characteristics assigned a key etiological role by Penrose. Slater preferred to talk more broadly in terms of genetic, biochemical, and environmental factors which influenced expressivity and penetrance of the inherited trait (schizophrenia), supposing only that some of these were more similar for same than opposite-sexed pairs. Some of these factors could also have been said to have higher expressivity in females than males, thus accounting for the higher concordance found in female pairs. However, further hypothetical qualifications would be necessary to try to account for similar or lower admission rates for females than males and the virtual disappearance of the sex-concordance ratios among collateral relatives. These qualifications, even if conceptually possible, would strain the hypothesis considerably, making it unusually complex and unwieldy. Moreover, the theory would eventually have to specify what these factors might be and how they might facilitate or inhibit manifestation. Until it did, the theory could not be evaluated further.

3. Psychological theories to account for the above findings would not at all be hard to come by, and the task really be-

comes one of excluding from consideration loosely conceptualized formulations which seem to be able to account for findings in one direction as well as another. Even if we limit ourselves to sound experimental and statistical studies in the psychological literature, we find such an abundance of heterogeneous studies of sex differences in personality and behavior that a thorough analysis of them could not be attempted here. The findings are not always consistent, and methods, measures, source and age of subjects, and research goals differ widely among studies. Therefore, I shall selectively present a few studies which appear relevant to our inquiry, briefly examining their explanatory power with respect to our major findings. The studies chosen bear no two lines of thought, not necessarily incompatible: one is the "anxiety-generalization hypothesis" of schizophrenia and the other has been called "sex-role identification." I do not present these as "theories" nor do I wish to convey the impression that I espouse either one. I present them to show that findings already exist in the psychological literature which are at least consistent with the sex-concordance rates presented above, to illustrate in two ways how findings obtained in studying the genetics of schizophrenia may be related to some studies of personality traits, to point up some problems which such lines of thought are confronted with when they are considered as possible factors accounting for the sex-concordance rates reported, and to imply that such lines of thought are deserving of a fuller critical and theoretical exposition than can be attempted here.

If we assume that high levels of anxiety predispose to schizophrenia (Mednick, 1958), and that at pre-psychotic stages high anxiety levels are often manifested as "neurotic" traits, then, based on the above data, we should expect the correlation of such traits to be higher for female than male pairs and higher for same-sexed than opposite-sexed pairs of relatives. Olson (1929) studied the "nervous habits" of 201 pairs of siblings in elementary school. The correlations were .32, .16, and .09 for sister, brother, and sister-brother pairs, respectively. Crook (1937) administered the Bernreuter Personality Inventory to 503 pairs of college students and found correlations for "neuroticism" of .35, .22, and .13 for sister, brother, and sister-brother pairs, respectively. Carter (1933) administered the same inventory to 117 pairs of twins. Correlations for neuroticism were .61, .32, and .18 for MZ, same-sexed DZ, and opposite-sexed DZ pairs, respectively. The higher correlation for same-sexed than opposite-sexed DZ twins is consistent with the findings presented earlier, but the higher correlation for MZ than same-sexed DZ twins suggests that inherited factors may be importantly involved in neuroticism. Eysenck and Prell (1951) maintained this point of view, but a comparison of their study with Carter's indicates that each was measuring quite different traits labeled neuroticism by the respective authors.

These correlations are consistent with the main findings of this article. Two studies of parent-child pairs regarding neuroticism on the Bernreuter are not quite as consistent (Crook, 1937; Hoffeditz, 1934). In both studies, mother-daughter pairs had higher correlations (.27 and .57) than father-son pairs (.06 and .05), but opposite-sexed parent-child pairs tended to have higher correlations (.23, .01, .23, .30) than father-son pairs. There is a suggestion that the ordering of most of these correlations could be accounted for in part by the assumption of a higher incidence of neurotic traits in females than males generally. There are a number of studies which support such an assumption (Castaneda, McCandless, & Palermo, 1956; Hattwick, 1937; Jersild & Holmes, 1935; Mathews, 1923; Rosenblum & Callahan, 1958). However, if such factors have relevance for the found concordance rates by sex with respect to schizophrenia, they should also predict a higher incidence of schizophrenia in females than males, which is not the case, at least for hospital admissions. It may be possible that more females develop milder forms of the illness and are therefore less likely to be hospitalized, accounting for the discrepancy from the prediction. Supporting this possibility is the well-established fact that females are generally admitted to hospitals later in life than males (Landis & Page, 1938; Larrson & Sjögren, 1954, Malzberg, 1935) and the findings of a much higher frequency of males than females with the most severely disorganizing sub-type of the disorder, hebephrenia (Schulz, 1932).

If, however, anxiety predisposes to schizophrenia, and females are more anxious, and more males are hospitalized for the disorder, at least earlier in life, then we seem to be confronted with an apparent contradiction. It may be that (speaking in Mednick's terms) *generalization* of anxiety occurs more readily in males than females, fostering illness-inducing anxiety-generalization spirals more frequently in males. Some evidence for such a possibility may be noted in a study by Sontag (1947) who found that girls were physiologically more reactive to stress than boys, but recovered more quickly.

At least one additional line of thought needs to be presented in an accounting of the main sex-concordance findings in this paper. If the aforementioned data are valid in that the sex-concordance ratios found in primary family relationships do not obtain among collateral relatives, we are led to infer that some factors peculiar to the structure of nuclear family life are contributing to those ratios. What might these factors be? In a previous paper, I examined Jackson's (1960) "confusion of identity" hypothesis as one such possible factor and found it wanting (Rosenthal, 1960).

Psychologists have been paying increasing attention to a related concept borrowed from psychoanalysis, called "identification," but they do not always agree on its definition, either in conceptual or operational terms. One aspect of the concept which has been studied in some detail has been called sex-role identification. To have explanatory power with respect to the sex-concordance ratios above, sex-role identification should be demonstrably greater among females than males, and among same-sexed than opposite-sexed pairs of family members.

Because children of both sexes have most contact during early rearing with their mother, they first identify with her, but the male child soon shifts to a masculine identification (Lynn, 1959). By age 3, most children are able to make sex-role distinctions and this knowledge increases with age (Brown, 1956, 1958; Fauls & Smith, 1956; Rabban, 1950; Sears, Maccoby, & Levin, 1957). The girl retains her feminine identification, which is apparently not weakened even though she is likely to go through a stage of developing a preference for more masculine activities. Girls are in more direct contact with their mothers than boys are with their fathers, so that whereas the girl is more likely to identify with a specific feminine model, viz., mother, the boy tends to identify with the cultural stereotype of masculinity rather than directly with his father (Lynn, 1959; Stoke, 1950). Women tend to be more like their mothers than their fathers in areas of major interest, but the reverse is not true for men (Beier & Ratzeburg, 1953; Gray & Klaus, 1956). However, both men and women tend to perceive themselves as more like their same-sexed than their opposite-sexed parent (Beier & Ratzeburg, 1953; Crook, 1937; Gray & Klaus, 1956; Sopchak, 1952). Boys with older sisters tend to be substantially more feminine than boys with younger sisters, but sisters with older brothers show only a slight increase in masculine traits as compared to girls with younger brothers (Brim, 1958; Brown, 1956; Koch, 1955). In the main, these findings lend support to the hypothesis that identification with same-sexed family members is stronger in females than males.

Using an operational definition of identification which was then related to patterns of psychopathology, Sopchak (1952) found that male subjects who had tendencies toward abnormality on the MMPI showed a lack of identification in varying degree with their fathers, their mothers, and "most people," in that order, the lack of identification in each case being positively correlated with scores on the Schizophrenia scale. Female subjects who had tendencies toward abnormality showed positive but not significant correlations between identification with mother and all types of abnormal trends. These findings could be interpreted in ways which would make consistent with the sex-concordance ratios reported above, suggesting further lines of research.

Even though the bodies of data presented have suggestive value in an accounting of the sex-concordance ratios found in studies of schizophrenia, the role of genetic factors cannot be excluded. However, if the found sex-concordance ratios are valid, it seems reasonable to conclude that some psychological factors are influencing these ratios in good part. Such a conclusion would be in accord with a previously reported finding of a group of schizophrenic cases where the genetic contribution to etiology was either minimal or absent (Rosenthal, 1959), and with the finding that hereditary factors were not accounting for as much of the variance with respect to schizophrenia as some leading investigators had supposed (Rosenthal, 1960).

Summary

The literature regarding concordance rates with respect to schizophrenia among relatives of both sexes is reviewed. These rates are generally found to be higher for female than male pairs and higher for same-sexed than opposite-sexed pairs of relatives in primary family groups, but not among collateral relatives. The possible role of sampling errors, genetic contributions, and psychological factors in generating such sex-concordance ratios is examined. It seems reasonable to infer that psychological factors are influencing the sex-concordance ratios. Two lines of thought in the psychological literature, "anxiety-generalization" and "sex-role identification," are briefly discussed as such possible factors.

References

Beier, E. G., & Ratzeburg, F. The parental identifications of male and female college students. *J. abnorm. soc. Psychol.*, 1953, *48*, 569-572.

Bellak, L. *Dementia praecox*. New York: Grune & Stratton, 1947.

Brim, O. G., Jr. Family structure and sex role learning by children: A further analysis of Helen Koch's data. *Sociometry*, 1958, *21*, 1-16.

Brown, D. G. Sex-role preference in young children. *Psychol. Monogr.*, 1956, *70*(14, Whole No. 421).

Brown, D. G. Sex-role development in a changing culture. *Psychol. Bull.*, 1958, *55*, 232-242.

Carter, H. D. Twin similarities in personality traits. *J. genet. Psychol.*, 1933, *43*, 312-321.

Castaneda, A., McCandless, B. R., & Palermo, D. S. The children's form of the manifest anxiety scale. *Child Develpm.*, 1956, *27*, 317-326.

Crook, M. N. Intra-family relationships in personality test performance. *Psychol. Rec.*, 1937, *1*, 479-502.

Dayton, N. A. *New facts on mental disorders.* Springfield, Ill.: Charles C Thomas, 1940.

Essen-Möller, E. *Psychiatrische Untersuchungen an einer Serie von Zwillingen.* Copenhagen: Munksgaard, 1941.

Essen-Möller, E. Mating and fertility patterns in families with schizophrenia. *Eugen Quart.*, 1959, *6*, 142-147.

Eysenck, H. J., & Prell, D. B. The inheritance of neuroticism: An experimental study. *J. ment. Sci.*, 1951, *97*, 441-465.

Fauls, Lydia B., & Smith, W. D. Sex-role learning of five-year-olds. *J. genet. Psychol.*, 1956, *89*, 105-117.

Freud, S. Psycho-analytic notes upon an autobiographical account of a case of paranoia (dementia paranoides). *Jb. psychoanal. psychopath. Forsch.*, 1911, No. 3.

Gralnick, A. Folie à deux: The psychosis of association. *Psychiat. Quart.*, 1942, *16*, 230-263.

Gray, Susan W., & Klaus, R. The assessment of parental identification. *Genet. psychol. Monogr.*, 1956, *54*, 87-114.

Greenberg, H. P. Folie à deux: An historical and clinical study. Unpublished doctoral dissertation, University of Sydney, Australia, 1961.

Gross, L. R. MMPI *L-F-K* relationships with criteria on behavioral disturbance and social adjustment in a schizophrenic population. *J. consult. Psychol.*, 1959, *23*, 319-323.

Hattwick, L. A. Sex differences in behavior of nursery school children. *Child Develpm.*, 1937, *8*, 343-355.

Hoffeditz, E. Louise. Family resemblances in personality traits. *J. soc. Psychol.*, 1934, *5*, 214-227.

Jackson, D. D. A critique of the literature on the genetics of schizophrenia. In D. D. Jackson (Ed.), *The etiology of schizophrenia.* New York: Basic Books, 1960. Pp. 37-87.

Jersild, A. T., & Holmes, F. B. Children's fears. *Child develpm. Monogr.*, 1935, No. 20.

Kallmann, F. J. The genetic theory of schizophrenia: An analysis of 691 schizophrenic twin index families. *Amer. J. Psychiat.*, 1946, *103*, 309-322.

Kety, S. Biochemical theories of schizophrenia. *Science,* 1959, *129*, 1528-1532, 1590-1596.

Koch, Helen L. The relation of certain family constellation characteristics and the attitudes of children toward adults. *Child Develpm.*, 1955, *26*, 13-40.

Kröner, E. Die folie à deux. *Allg. Z. Psychiat.*, 1891, *46*, 634-662.

Landis, C., & Page, J. D. *Modern society and mental disease.* New York: Farrar & Rinehart, 1938.

Larsson, T., & Sjögren, T. A methodological, psychiatric and statistical study of a large Swedish rural population. *Acta psychiat. neurol. Scand. Kbh.*, 1954, Suppl. No. 89.

Lasègue, C., & Fabret, J. La folie à deux. *Ann. Med.-psychol.*, 1873, *10*, 483-495.

Lengyel, B. A. Remission and mortality rates in schizophrenia. *Amer. J. Hyg.*, 1941, *33*, 16-32.

Luxenburger, H. Vorläufiger Bericht über psychiatrische Serienuntersuchungen an Zwillingen. *Z. ges. Neurol. Psychiat.*, 1928, *116*, 297-326.

Luxenburger, H. Psychiatrische-neurologische Zwillingspathologie. *Zbl. ges. Neurol. Psychiat.*, 1930, *56*, 145-180.

Lynn, D. B. A note on sex differences in the development of masculine and feminine identification. *Psychol. Rev.*, 1959, *66*, 126-135.

Malzberg, B. A statistical study of age in relation to mental disease. *Ment. Hyg.*, 1935, *19*, 449-476.

Marandon de Montyel, E. Des conditions de la contagion mentale morbide. *Ann. Med.-psychol.*, 1894, *19*, 266-293.

Mathews, E. A study of emotional stability in children. *J. Delinqu.*, 1923, *8*, 1-40.

Mednick, S. A learning theory approach to research in schizophrenia. *Psychol. Bull.*, 1958, *55*, 316-327.

Mott, F. W. The Huxley Lecture on hereditary aspects of nervous and mental disease. *Brit. med. J.*, 1910, *2*, 1013-1020.

Myerson, A. *The inheritance of mental disease.* Baltimore: Williams & Wilkins, 1925.

Offergeld, H. Narkose und Psyche. *Z. Psychother. med. Psychol.*, 1957, *7*, 261-265.

Olson, W. C. *The measurement of nervous habits in normal children.* Minneapolis: Univer. Minnesota Press, 1929.

Overholser, W., & Werkman, S. L. Etiology, pathogenesis and pathology. In L. Bellak (Ed.), *Schizophrenia: A review of the syndrome.* New York: Logos Press, 1958. Pp. 82-106.

Penrose, L. S. Auxiliary genes for determining sex as contributory causes of mental illness. *J. ment. Sci.*, 1942, *88*, 308-316.

Penrose, L. S. Survey of cases of familial mental illness. *Dig. Neurol. Psychiat.*, 1945, *13*, 644.

Phillips, L. Case history data and prognosis in schizophrenia. *J. nerv. ment. Dis.*, 1953, *117*, 515-525.

Rabban, M. Sex-role identification in young children in two diverse social groups. *Genet. psychol. Monogr.*, 1950, *42*, 81-158.

Rosanoff, A. J., Handy, L. M., Plesset, I. R., & Brush, S. The etiology of so-called schizophrenic psychoses. *Amer. J. Psychiat.*, 1934-35, *91*, 247-286.

Rosenblum, S., & Callahan, R. J. The performance of high-grade retarded, emotionally disturbed children on the children's manifest anxiety scale and children's anxiety pictures. *J. clin. Psychol.*, 1958, *14*, 272-275.

Rosenthal, D. Some factors associated with concordance and discordance with respect to schizophrenia in monozygotic twins. *J. nerv. ment. Dis.*, 1959, *129*, 1-10.

Rosenthal, D. Confusion of identity and the frequency of schizophrenia in twins. *Arch. gen. Psychiat.*, 1960, *3*, 297-304.

Rosenthal, D. Sex distribution and the severity of illness among samples of schizophrenic twins. *J. psychiat. Res.*, 1961, *1*, 26-36.

Schulz, B. Zur Erbpathologie der Schizophrenie. *Z. ges. Neurol. Psychiat.*, 1932, *143*, 175-293.

Sears, R. R., Maccoby, Eleanor E., & Levin, H. *Patterns of child rearing*. Evanston, Ill.: Row, Peterson, 1957.

Slater, E. Genetics in psychiatry. *J. ment. Sci.*, 1944, *90*, 17-35.

Slater, E. *Psychotic and neurotic illnesses in twins.* London: Her Majesty's Stationery Office, 1953. (a)

Slater, E. Sex-linked recessives in mental illness. *Acta. genet.*, 1953, *4*, 273-280. (b)

Sontag, L. W. Physiological factors and personality in children. *Child Develpm.*, 1947, *18*, 185-189.

Sopchak, A. L. Parental "identification" and "tendency toward disorders" as measured by the Minnesota Multiphasic Personality Inventory. *J. abnorm. soc. Psychol.*, 1952, *47*, 159-165.

Sternberg, S. The prognosis in schizophrenia. *Acta. genet. statist. med.*, 1948, *1*, 5-76.

Stoke, S. M. An inquiry into the concept of identification. *J. genet. Psychol.*, 1950, *76*, 163-189.

Wittman, Phyllis. A scale for measuring prognosis in schizophrenic patients. *Elgin State Hosp. Pap.*, 1941, *4*, 20-33.

Wollenberg, R. Über psychische Infektion. *Arch. Psychiat. Nervenkr.*, 1889, *20*, 62-88.

Zehnder, M. Über Krankheitsbild und Krankheitsverlauf bei schizophrenen Geschwistern. *Mschr. Psychiat., Neurol.*, 1941, *103*, 231-277.

Heritability of Personality

A Demonstration[1]

IRVING I. GOTTESMAN

Ss were 68 pairs of adolescent twins from the public schools of an urban area. They comprised 60% of the same-sexed twin population. Zygosity diagnosis was determined by blood grouping for 9 groups; half of the sample was identical (MZ) and half was fraternal (DZ). Personality was assessed by the MMPI

[1] This paper is based upon a doctoral dissertation submitted in partial fulfillment of the requirements for the Ph.D. degree at the University of Minnesota, 1960.

I wish to express my appreciation to my adviser and mentor, R. D. Wirt. I was fortunate to have also had the encouragement and friendship of S. C. Reed, Director of the Dight Institute for Human Genetics, University of Minnesota, Minneapolis.

The cooperation of the Minneapolis, Saint Paul, and Robbinsdale, Minnesota, public school systems through the efforts of H. Cooper, N. C. Kearney, and F. C. Gamelin, Assistant Superintendents, is gratefully acknowledged. A number of other persons contributed their skills and support to this study. Among them are Marianne Briggs, Joan Drues, A. C. Wahl, and Jane Swanson. Conversations with D. Freedman, R. Rosenthal, and D. S. Jones helped me to clarify a number of my thoughts. I benefited from the comments of S. G. Vandenberg and W. R. Thompson on this revised manuscript. Other individuals are given credit in later pages for their specific contributions.

Funds in support of this research were provided by the Tozer Foundation of Stillwater, Minnesota, and the Dight Institute for Human Genetics. The large expense of blood typing was borne by the Minneapolis War Memorial Blood Bank through the interest of G. A. Matson. Subsequent analyses of the data for sex differences were financed by a grant from the Laboratory of Social Relations, Harvard University. Preparation of this monograph was aided in part by Grant M-5384 from the National Institutes of Mental Health.

From *Psychological Monographs,* 1963, 77 (No. 9), 21 pp. Reprinted by permission of the author and the American Psychological Association.

and Cattell's HSPQ. 24 standard scales were analyzed by intra-class correlations (R) and heritability indexes (H). Holistic analyses of MMPI profile similarity were done clinically and statistically. 6 personality scales had significant genetic components as revealed by higher MZ Rs. 11 scales had appreciable hereditary variance. Pooled accuracy of profile similarity judgments matching zygosity was 68% ($p = .005$). The general idea that psychopathology in man has a substantial genetic component, especially the psychoses, was supported. A dimension of introversion was the most heavily influenced by genetic factors.

Mankind in general and life scientists in particular have long been curious about the basic nature of man. In recent years curiosity and speculation have given way to experimentation and controlled observation. One of the eventual outcomes of such research will be a knowledge of the sources of the individual differences in human behavior so that the variation may be explained, predicted, or controlled. The genetic source of variation in human personality is the focal concern of the present research. *Genetic* is used in the strict sense to refer to that science launched by Mendel's work with peas and not, as is common in psychology, as an abbreviation for *ontogenetic*. Although it is axiomatic that the phenotypic expression of a trait is dependent upon the resultant of interaction between genotype and environment, much heat and little light has been generated by attempts to answer the question, "How much of Trait X is due to heredity and how much to environment?" Since neither agent alone can produce the observed behavior, the question as stated is meaningless. The question has precise meaning only when framed in terms of the variation between individuals. Two answerable questions should be posed in the nature-nurture issue: (*a*) How much of the *variability* observed within a group of individuals in a specified environment on a specific measure of a specific trait is attributable to genetic factors? (*b*) How modifiable by systematic environmental manipulation is the phenotypic expression of each genotype? The answers to these questions will vary according to age, sex, culture, trait, and method of assessment. The importance of genetic factors will range from almost none to overwhelming for those individuals who have been found to have either more or less than 46 chromosomes (Lejeune & Turpin, 1961).

The genes exert their influence on behavior through their effects at the molecular level of organization. Enzymes, hormones, and neurons may be considered as the sequence of complex path markers between the genes and the aspects of behavior termed personality. The inability of behavioral genetics to demonstrate this type of reduction at the present time need be no more embarrassing than the lack of information concerning the biochemical changes associated with habit formation is to the psychology of learning. It may be that measures of behavior qua behavior are the only reliable indicators of certain kinds of genetic differences (Fuller, 1957; Fuller & Thompson, 1960). For our purposes the best way to conceptualize the contribution of heredity to a personality trait is in terms of heredity's determining a norm of reaction (Dobzhansky, 1955) or of fixing a reaction range (Gottesman, 1963). Within this framework a genotype determines an indefinite but circumscribed assortment of phenotypes, each of which corresponds to one of the possible environments to which the genotype may be exposed. Allen (1961) pointed out that the most probable phenotype of some genotypes may be such a deviant one that even the most favorable of currently known environments would not suffice to bring it within the normal range.

Within the broad context of evolution the demonstration of heritable components for personality traits involves more than an academic exercise. The one nonrandom genetic process which accounts for the adaptive orientation of evolution is differential success in reproduction. If there are heritable aspects to some personality traits and if there is assortative mating for these traits, the frequencies of the associated genes will increase in the gene pool of the population. Tryon (1957) has proposed a behavior genetics model of society which suggests that relative reproductive isolation between social strata plus social mobility could account for some of the class differences observed in achievement and personality.

In the brief history of contemporary psychology the search

for genetic aspects of personality has been dominated by an understandable emphasis on mental illness and mental deficiency (Allen, 1958; Kallman, 1959). Almost all of such research has occurred within a context of classical Mendelian major gene mechanisms. The inappropriateness of this model for the observed quantitative variation in normal personality traits has been one of the inhibitors to investigations by psychologists. Although the classical study by Newman, Freeman, and Holzinger (1937) focused on intelligence and achievement, a number of the personality tests then available were included in the battery administered to their large sample of twins. From their results the authors concluded,

> The only group of traits in which identical twins are not much more alike consists of those commonly classed under the head of personality [p. 352].

With very few exceptions (Cattell, Blewett, & Beloff, 1955; Vandenberg, 1962) this conclusion appears to have been accepted as a valid statement of the relationship of genetics to normal personality. Improvements in personality measurement (Cronbach, 1960) and a new era of sophistication about the construction and application of psychometric devices (Cronbach & Meehl, 1955; Loevinger, 1957) make it possible for us to re-examine the relationship in question. This study is an effort in this direction.

The present research attempted to improve upon the methodology of previous twin studies by incorporating a number of refinements. The sample was selected after the entire population of same-sex twins had been enumerated in a large area; the representativeness of the sample to the population was then ascertained. Accuracy and objectivity of zygosity diagnosis were ensured by means of serological tests for 46 different phenotypes in nine different blood group systems and by the use of fingerprints, height, and photographs. Two types of objectively scored personality inventories were used, each purporting to be comprehensive but parsimonious in its characterization of the personality domain. Unlike the tests which historically have been used in an effort to elucidate the "nature and nurture" of personality, both types of tests in the present research have a recognized claim on construct validity (Cronbach, 1960, p. 122). One test depends upon the process of factor analysis for the derivation of "pure" trait measures while the other stems from item analyses which separate criterion groups on various empirical dimensions.

A primary goal of the present research is to answer the first question posed above about *how much* of the variability on various traits for a specified sample reared in a particular environment is attributable to genetic factors and how much to environmental. Earlier criticisms of this strategy (Anastasi, 1958; Loevinger, 1943) are well taken but do not render it obsolete. Fuller (1960) has noted that, while the question has no significance for an individual, the contribution of heredity to total variance in a population

> is still a useful object of inquiry, though with increased sophistication we have come to see that the answer to "How much?" is not a universal constant [p. 43].

PROCEDURE

This section gives the details of twin selection as well as a description of the first application in psychological research of the Smith and Penrose (1955) method for the determination of zygosity by extended blood grouping. The efficiency of this method is compared with the methods of the past. Only questionnaire measures of personality were used so as to facilitate the handling of scale scores within the context of quantitative genetics; the tests and their validity and reliability are described. A brief rationale of the twin method is presented followed by a description of the scheme of data analysis.

Selection of the Twin Sample

In sound twin methodology, it is essential that the sample be a miniature of the population of twins. This ensures proportional representation of the two kinds of twins (identical or MZ and fraternal or DZ) and allows accurate genetic analysis with the computed heritability indexes or concordance rates. It is also essential that the two groups of twins in the sample be matched on as many variables as possible so that differences in variance cannot be attributed to differences in age, sex, intelligence, socioeconomic status, or other factors which may influence personality other than the independent variable, genotype.

All class cards for the over 31,000 children in public school Grades 9 through 12 in the cities of Minneapolis, Saint Paul, and Robbinsdale, Minnesota, were examined. All pairs of children with the same last name, same sex, same address, and same birthdate were recognized as the twin population available. Opposite-sexed fraternal twins were not included in the study in order to eliminate the questionable procedure of comparing a boy with a girl on the same personality traits.

The best data available to date about the incidence of twin births in the United States "white" population are those of Strandskov and Edelen (1946) who reported that 1.29% of all births are twin births. Of these, one third are opposite-sexed fraternals, one third are same-sexed fraternals, and one third are identicals. The best known twin studies (Kallmann, 1946; Newman et al., 1937) assumed that only one fourth of all twins are identical.

A total of 163 pairs of same-sex twins were located in the schools' files of 31,307 children. Based on the incidence of twin *births*, 11.29%, the expected incidence of same-sex twins would have been 237 pairs. After calculating neonatal mortality, however, Allen (1955) found that the incidence of twins at 1 month of age had already been lowered to .87 pairs per 100 children (.87%). Based upon this incidence, the expected number of same-sex twin pairs in the entire population would have been 182. After 1 month of age, the mortality of twins is the same as that of single born survivors. By subtracting the known mortality rate in the general population for children reaching the age of 15 (5%), the final expected number of same-sex twin pairs was 173.[2] At the time the present study

[2] Evaluation of the sampling adequacy was suggested by and facilitated by E. Anderson.

was conducted, there were no adolescent twins in either the correctional or mental institutions (not counting the two housing mental defective and brain damaged cases) of the state. It would appear that virtually the entire population of same-sex adolescent twins in the public high schools of the three communities was enumerated.

The parents of all pairs in Minneapolis and Robbinsdale and of those in the largest high school in Saint Paul were sent a letter describing the project and a return postcard on which was printed a medical release authorizing blood typing. After 10 days, telephone calls were made to those who had not returned the card indicating the voluntary participation of their children. After another 10 days, a second and last, hopefully persuasive, telephone call was made. These efforts secured the initial cooperation of 26 pairs of boys and 48 pairs of girls. By the end of the study 6 pairs of twins had defaulted for various reasons: 1 pair was lost as a result of fear of the intravenous removal of the blood specimen, one member of another pair had cerebral palsy and could not take the personality tests in the standard manner, and 4 pairs were unavailable at the times provided for the tests which were Saturday afternoons and mornings.

The final study sample, then, consisted of 23 pairs of boys and 45 pairs of girls. These 68 pairs, disregarding sex, represented 60.2% of the total possible 113 pairs in the schools sampled. The sample contained, respectively, 43.4% and 75.0% of the male and female twin pairs. The sample of the present study compares favorably in size with the majority of twin studies reported in the psychological literature. In representativeness, it is superior to the majority. The simplest explanation for the preponderance of girls over boys is the reluctance of adolescent boys to volunteer their spare time for taking paper-and-pencil tests, especially on Saturdays. The children came from 13 different high schools (all that were sampled), some of which included a ninth grade, and 5 different junior high schools. Participation ranged from 8 out of 8 pairs to 3 out of 8 in the high schools. There was a tendency toward better participation as the economic level of the neighborhood increased.

TABLE 1

Descriptive Characteristics of the Sample

Character	MZ	DZ	Combined
Pairs of boys	12	11	23
Pairs of girls	22	23	45
Age			
14	0	2	2
15	15	4	19
16	9	13	22
17	7	10	17
18	3	5	8
Grade			
9	7	5	12
10	15	11	26
11	7	11	18
12	5	7	12
	Level of paternal occupation[a]		
I and II	15	9	24
III	7	13	20
V and VI	12	12	24
M Otis IQ	105	108	107
IQ *SD*	12	12	12

[a] The Minnesota Scale for Paternal Occupations, Institute of Child Welfare, University of Minnesota.

After the parents of a twin pair had returned the signed authorization for participation and blood typing, an appointment was made to drive the pair to the Minneapolis War Memorial Blood Bank. An appointment was then made for the personality tests. The children were tested in small groups ranging up to 12 pairs. At the time of testing, the children filled out a personal history data sheet, the Minnesota Multiphasic Personality Inventory (MMPI), and the High School Personality Questionnaire (HSPQ; Cattell, Beloff, & Coan,

1958); they were weighed, measured for height, fingerprinted, and photographed. The entire procedure usually took between 3 and 4 hours for each group.

John D. Douthit, Identification Officer with the Minnesota Bureau of Criminal Apprehension, fingerprinted about half of the twins and, after being tutored in the technique, the author fingerprinted the rest. The Faurot inkless method was used with acceptable results and a considerable saving of time. It makes use of a colorless fluid and chemically sensitive paper. An ordinary bathroom scale was used for weighing. Height measurement is estimated to be correct within 5 millimeters. Photography was done by the author with a 35 mm. camera; both a front view and a profile were shot of the head and shoulders.

Some descriptive characteristics of the sample are presented in Table 1. By a serological procedure described in the next section, the 68 pairs of twins were classified into 34 pairs of MZ and 34 pairs of DZ. That this split corresponds to genetic theory is a stroke of luck and an illustration of the representativeness of the sample. About 90% of the twins reported they were of Scandinavian or Western European extraction. In addition to obtaining Otis IQs on the sample, the same data were obtained for 30 more pairs of twins who had not volunteered in Minneapolis and Robbinsdale so that any selection for intelligence might be revealed. It should be noted that the total of 98 pairs accounts for information about the IQs of 86.7% of the *population* of same-sex twins in the schools used. A sampling bias was revealed by the fact that the mean IQ of the non-sample twins was 97, while those of the MZ and DZ samples were 105 and 108, respectively. A t test for the significance of these differences showed that both study samples were significantly different from the nonvolunteers ($t = 2.66$, $p < .01$; $t = 3.66$, $p < .001$).

Criteria for the Diagnosis of Zygosity and an Evaluation of their Relative Efficiencies

One of the most serious criticisms of much twin research is the inaccuracy of zygosity diagnosis. In reaching a judgment in the past, reliance has been placed on an evaluation of the type of birth membrane or the degree of physical resemblance between the twins. Diagnoses based upon the birth membranes are unreliable because while MZ twins are always monochorionic (i.e., a single membrane surrounding both fetuses), the presence of two chorions is known to occur with both MZ and DZ twins (Stern, 1960, p. 536). In addition, when studying adult or adolescent twins, it is difficult to obtain accurate information about the birth membrane. Diagnosis by means of the placenta is even more unreliable. In evaluating the extent of physical resemblance, geneticists have used such traits as sex, height, weight, eye color, hair color and form, familial appearance, and various types of fingerprint or palmprint analyses. Although there is an unavoidable subjective element in evaluating many of these characteristics, one expert has estimated the error to be no greater than 1 in 10 (Newman, 1940). That this estimate may be in error is demonstrated below.

If twins differ in sex or any other known inherited characteristic, they cannot be MZ twins. However, if the characteristics are alike, the possibility still remains that the twins are DZ. Given a number of simply inherited and widely distributed traits, it is possible to state the probability of monozygosity or dizygosity for a given pair of twins. It is to be noted, however, that all such diagnoses of monozygosity, no matter how many characteristics are identical, will always be statements of probability; that is, the probability of sharing the given number of traits in common.

Numerous criteria were examined in this study with the hope that the various suggestions in the literature for the diagnosis of zygosity might be objectively evaluated against the recognized best method of extensive blood typing recently quantified by Smith and Penrose (1955). Thus blood type alone was compared with blood type combined first with height, second with a difference in total fingerprint ridge count, and then with both height and ridge count. The accuracy of fingerprints alone and height alone was ascertained. In addition, three groups of judges, geneticists, psychologists, and artists, looked at photographs of the twins and made another form of judgment.

All blood specimens were drawn and typed by the Minneapolis War Memorial Blood Bank, Incorporated. The following blood group systems (Race & Sanger, 1958) were used: ABO, MNSs, Rh, P, Lutheran (Lu), Kell (K), Duffy (Fy), Kidd (Jk), and Lewis (Le).

Smith and Penrose (1955) tabulated the probabilities used for an objective determination of the likelihood of dizygosity based on the incidence of phenotypic sib-sib concordance for the above blood groups. A specific example will illustrate the

TABLE 2

Example of the Smith and Penrose Method for Zygosity Determination

Character	Independent relative chance
Initial odds	1.9246
Likeness in sex	.5000
Likeness in ABO	.6891
Likeness in MNSs	.4556
Likeness in Rh	.5021
Likeness in Le	.8681
Likeness in K	.9485
Likeness in Fy	.8036
Likeness in Jk	.8531
Likeness in Lu	.9614
Likeness in P	.5699
Total relative chance p_{DZ} (blood)	.0470
Total chance $p_{DZ}/(1+p_{DZ})$	.0448
Difference in ridge count	.2288
Total relative chance (blood+ridges)	.0107
Total chance	.0106
Difference in stature	.4671
Total relative chance (blood+stature)	.0219
Total chance	.0214
Total relative chance (all of above)	.0050
Total chance	.0050

basic principle underlying the origin of the tabulated probabilities. Blood typing of a large Caucasian population in Great Britain shows that the frequency of Type B blood is .084509 and the frequency of two sibs being B is .040062. The probability that if one of two sibs is B the other is also becomes .040062/.084509 or .4741. Since DZ same-sex twins are genetically as similar as ordinary sibs, the probability of DZ twins both being Type B is equally .4741. Probabilities derived in this manner are listed in the upper part of Table 2 together with the initial probability that Caucasian, United States twins are DZ and that of a DZ pair being the same sex. Multiplication of all these independent probabilities results in the probability of finding twins who are DZ and alike in all the gene loci involved. A more extensive discussion of the Smith and Penrose method together with the rationale for the derivation of probabilities for the morphological traits used can be found elsewhere (Gottesman, 1960).

As a result of the blood typing, 34 pairs of twins were diagnosed as definitely DZ, that is, they differed on at least one of the independently inherited blood groups. Using only blood, the remaining 34 pairs were diagnosed as MZ with the probability of accuracy no less than 95 times in 100. Table 3 summarizes the results of the accuracy for the various combinations of blood and physical characteristics. It seems paradoxical that while additional information increased the accuracy of some of the diagnoses, it was at the expense of serious errors, e.g., using all the characters resulted in calling 22 pairs MZ at the .01 level or better, but at the expense of 6 pairs failing to meet the criterion of the .05 level. The primary reasons for this are the lack of cross validation and the small samples upon which the fingerprint and height probabilities are calculated, 52 and 50 pairs of MZ, respectively. This resulted in a range of within-pair differences too narrow to allow for those found in the present sample of MZ twins. The probability figure given was too much in the DZ direction to be overcome by any amount of additional information. Eight pairs of MZ twins had differences of ridge count which were tabulated at probabilities greater than 1.0 in favor of the DZ contingency. Similarly, five pairs were "penalized" for differences in height larger than the tabulated ones for the 50 criterion pairs of MZ twins on which they were based. Differential growth rates during adolescence may have been another attenuating factor in the use of the probabilities attached to differences in height.

Let us turn now to two different analyses of the fingerprints: one clinical and the other statistical. Given the 68 pairs of fingerprints and no information as to the base rates, i.e., incidence of DZ and MZ twins in the sample, how accurately can an expert diagnose the two kinds of twins? Douthit undertook this task and was able to correctly identify 30 MZ pairs and 23 DZ pairs. Most important in his clinical decisions were three components (Cummins & Midlo, 1943): differences in pattern slope for the same finger in a twin pair, similarities in slope but different patterns, and differences in the range of dermal ridges for paired fingers. His decisions were not purely clinical in the Meehl (1954) sense of the word in that he subjectively assigned different weights to these components, and used the scores a pair of prints obtained. The statistical method used was for the author to assign what appeared to be the optimum cutting score (Meehl & Rosen, 1955) to the distribution of differences between total ridge count. This cutting score was then cross validated by applying it to the original distribution (Smith & Penrose, 1955) from which the aforementioned probabilities were determined. A cutting score of 30 classified 33 of 34 MZ pairs and 20 DZ pairs correctly. This score correctly classified 51 of the original 52 MZ pairs at the expense of misclassifying 39 of 101 (38.6%) like-sex siblings. The clinical and statistical methods tied in their accuracy for diagnosing the entire present sample with both hitting 78%. Both Newman, Freeman, and Holzinger (1937) and Slater (1953) make use of some aspects of fingerprints in their diagnosis of zygosity.

Judgments of photographs constituted the final method of zygosity determination evaluated in this section. A summary of all the methods attempted is then presented. Three groups of judges were utilized; three geneticists, three child psychologists, and three artists.[3] Although the front and profile pictures

[3] I am grateful for the assistance of Vivian Phillips, Elizabeth Reed, S. C. Reed, J. E. Anderson, Mildred C. Templin, R. D. Wirt, Carol Safer, L. Safer, and Ane Wolfe Graubard.

TABLE 3

Accuracy Summary for Methods of Zygosity Determination

Method	MZ %	DZ %	Total %
Blood ($p \leq .05$)	100	100	100
Blood + Height	85	—	—
Blood + Ridge count	88	—	—
Blood + Height + Ridge count	82	—	—
Height + Ridge count	0	—	—
Fingerprints			
Clinical	88	68	78
Statistical	97	59	78
Photos			
Best geneticist	97	74	85
Best psychologist	59	82	71
Best artist	79	91	85
Pooled judges (6/9 agreement)	68	88	78

of the head were black and white 35 mm. contact prints, expressions of dissatisfaction with their quality were minimal.

It is obvious that previous estimates (e.g., Newman, 1940) of a 10% error in the diagnosis of zygosity by general appearance are subject to doubt. Jackson's (1960) contention that he observed a "striking difference" between photographs of MZ and DZ twins in the literature needs to be re-evaluated. Even allowing for the quality of photographs and the absence of the cues from the twins' physical presence, the median accuracy of 72% for all nine judges was significantly less than an expected 90% ($\chi^2 = 24.33$ $p < .001$). Poor reliability of judgments may be inferred from the fact that for only 13 MZ pairs and 14 DZ pairs were there one or no inaccurate judgments. There was a total of 84 errors in judging the MZ twins and 70 errors in judging the DZ twins. Judging the MZ girls seemed to be the most difficult. To the extent that the data in Table 3 were stable, only the geneticists made sufficient allowance for the variability that existed between MZ twins.

Summary

For the sake of clarity, the accuracy of zygosity determination for all the methods described thus far is presented in Table 3.

It should be noted that the three columns cannot be evaluated independently of one another. A judge of the photographs or fingerprints could maximize his accuracy in one category at the expense of the other. It is the final column which conveys the most meaning. The blanks in the table derive from the fact that the DZ twins were absolutely removed from further consideration by the blood typing methodology in the Smith and Penrose scheme.

Psychometric Devices

Both instruments used to measure personality in this study come under the category of objective as contrasted with projective tests. Within the former group there are two major types of questionnaires: one is derived empirically from its ability to discriminate among behavioral phenotypes and the scales may be said to have functional unity, and another type is derived by factor analysis and the scales may be said to have statistical unity. The MMPI was selected as an example of the first type and the HSPQ of the second.

Widespread usage of the MMPI precludes the necessity for a detailed description (Dahlstrom & Welsh, 1960). The test was constructed to provide, in a single instrument, measures of all the more important phases of personality of interest to the psychiatrist. Items were selected from 26 subject-matter categories, e.g., general health, sensory disturbances, family problems, sexual and social attitudes, masculine and feminine interest patterns, and schizophrenic thinking disturbances. The 550 items were answered *true* or *false*. Use of the MMPI with normal adolescents may be questioned, but this practice is becoming more popular as experience accumulates with this age group (Hathaway & Monachesi, 1961; Wirt & Briggs, 1959). A basic assumption of the present research was that personality scales represent dimensions or continua of behavior, not categories. Such scales thus lend themselves to the assumptions of quantitative inheritance and permit the use of nonpsychiatric subjects. Numbers have been assigned to the MMPI scales instead of their Kraepelinian based named by the developers of the test so as to facilitate fresh associations to the meaning of the scales as personality constructs. There is no simple translation from MMPI data into descriptive terms for normal populations, but adjectives associated with the behavior of subjects with different scale patterns are easily obtained (e.g., Black, 1953).

The group form of the test was used and scored in the usual fashion for the four validity indicators and 10 clinical scales, 1 through 0 (or Hypochondriasis–*Hs*–through Social Introversion–*Si*). In addition, 6 experimental scales (Hathaway & Briggs, 1957), Ego Strength (*Es*), Anxiety (*A*), Repression (*R*), Dominance (*Do*), Dependency (*Dy*), and Social Status (*St*) were scored and analyzed. The 5 scales requiring a *K* correction were analyzed after the correction had been made.

Test-retest reliability coefficients on the 10 standard MMPI scales for a sample of 100 male and female college students after a 1-week interval (Dahlstrom & Welsh, 1960, p. 472) range from .56 to .90. These particular coefficients were chosen because they are most comparable to the circumstances under which the HSPQ reliabilities were computed.

The HSPQ is new to the literature on personality tests and requires more exposition than the MMPI. Cattell, Beloff, and Coan (1958) constructed this instrument by factor analysis especially for adolescents 12 through 17 years in the tradition of the Cattell (1946, 1950) laboratory. It is said to cover all the major dimensions involved in any comprehensive view of individual differences in personality (Cattell et al., 1958).

It consists of 280 forced-choice items, all of which are scores, which form 14 independent, equal length, scales. Although printed in two forms of 140 items each, the authors[4] recommend the use of both to obtain sufficient reliability. It is

[4] R. B. Cattell, personal communication, February 1959.

also suggested that raw scores rather than standard scores be used for research purposes and this suggestion was followed. The scale designations and their titles are given in Table 4.

TABLE 4

HSPQ Symbols and Titles[a] for Test Dimensions

Symbol	Low score	High score
A	Stiff, Aloof	Warm, Sociable
B	Mental Defect	General Intelligence
C	General Neuroticism	Ego Strength
D	Phlegmatic Temperament	Excitability
E	Submissiveness	Dominance
F	Sober, Serious	Enthusiastic
G	Casual, Undependable	Super Ego Strength
H	Shy, Sensitive	Adventurous, Thick-Skinned
I	Tough, Realistic	Esthetically Sensitive
J	Liking Group Action	Fastidiously Individualistic
O	Confident Adequacy	Guilt Proneness
Q_2	Group Dependency	Self-Sufficiency
Q_3	Uncontrolled, Lax	Controlled, Showing Will Power
Q_4	Relaxed Composure	Tense, Excitable

[a] A mixture of technical and popular terms was used here.

Test-retest correlations based on 112 children aged 13 through 15 tested 2 weeks apart with the full test range from .68 to .80.

In the opinion of the constructors, validity for the 14 scales is satisfactorily established. The main technique used to demonstrate this is the computation of a multiple correlation from factor-item correlations. This gives a median *r* of .81. Although no correction for Test-Taking Attitude is used, there are equal numbers of "yes" and "no" keyed answers on each scale.

It should be obvious that the reliabilities of the MMPI and the HSPQ are on much firmer ground than the validities. Unfortunately, the magnitude of the test-retest correlations has no direct bearing on the construct validity of a scale (Loevinger, 1957). An inherent difficulty in measuring personality traits is the observation that they change with the passage of time and with intervention. After the data of the present study are analyzed, there should be more evidence for the validity, or lack thereof, of the various scales. At the least one might expect significant correlations between MZ twin siblings. Another speculation would be the absence of negative correlations between either class of twins unless there were some parsimonious explanation of a within-pair interaction on a trait.

Twin Method

Bacteria, fruit flies, and mice have contributed greatly to the body of genetic knowledge, but the application of this knowledge to the causes of variation in human behavior raises difficulties. Moreover, relatively few direct methods are available to the researcher in human behavior genetics because of such problems as those introduced by uncontrolled mating, small numbers of offspring, heterogeneous environments, and the uniqueness of one individual's heredity. Of the available methods, the twin method approaches the ideal experimental design. Galton (1875) first called attention to the possible usefulness of twins for casting light on the nature-nurture problem. The underlying principle is simple and sound: since MZ twins have identical genotypes, any dissimilarity between pairs must be due to the action of agents in the environment, either postnatally or intrauterine; DZ twins, while differing genetically, have certain environmental similarities in common such as birth rank and maternal age, thereby providing a measure of environmental control not otherwise possible. When both types of twins are studied, a method of evaluating either the effect of different environments on the same genotype or the expression of different genotypes under the same environment is provided. This means, with respect to any given genetically determined trait, that there should be a greater similarity between MZ than between DZ twins. If both members of a twin pair develop the same phenotype in a given environment, they are called *concordant* for the trait under study; *discordant* is the designation for differing phenotypes. When dealing with a single gene difference, such as Huntington's chorea, MZ twins should always be concordant; DZ twins may be either concordant or discordant. The expected difference in concordance can then be used to give a measure of heritability (H) of the trait if the traits are amenable to discrete classification.

Inasmuch as few traits in the normal range of human personality are dichotomous, another approach is needed to estimate H when traits are continuous and the genetic component is of the polygenic variety. In the present research, H will be defined as the proportion of total trait variance associated with genetic factors. Holzinger (1929) suggested that the best comparison to make in evaluating the nature-nurture interaction for a quantitative characteristic is that between the intraclass correlation coefficients (R) for MZ and like-sexed DZ twins. Holzinger's H gives the proportion of variance produced by genetic differences *within families.* The method underestimates the effects of heredity in the general population by a factor of approximately 2 since the genetic variance is estimated from the genetic overlap between DZ twins which is .5. The index of heritability, h^2, computed in animal behavior genetics (e.g., Falconer, 1960; McClearn, 1961) is different from H in that it is an estimate of the proportion of trait variance in a population determined by genotypic variation in that population. Both between- and within-families variance components are used to compute h^2.

Holzinger gave two formulas for his estimate of heritability, one based on R and another, statistically equivalent, based on the within-pair variances.

$$H = \frac{R_{MZ} - R_{DZ}}{1 - R_{DZ}}$$

$$H = \frac{V_{DZ} - V_{MZ}}{V_{DZ}}$$

where,

R_{MZ} = intraclass correlation between MZ twins
R_{DZ} = intraclass correlation between DZ twins
V_{MZ} = within-MZ pairs variance estimate (mean square)
V_{DZ} = within-DZ pairs variance estimate (mean square)

Falconer has suggested that the difference between the MZ and DZ *R*s could be taken as an estimate of half the heritability if there were no nonadditive genetic variance; since the latter assumption is probably not warranted, the difference can only be regarded as setting an upper limit to half the heritability.

Limitations and Criticisms of the Twin Method

After reviewing probable natal and prenatal influences on twin development, Price (1950) was willing to conclude.

> In all probability the net effect of most twin studies has been underestimation of the significance of heredity in the medical and behavior sciences [p. 293].

This appears to be the result of biases of two sorts. Inferences drawn from data on twins are subject to both statistical and biological biases. In the first category, it is basic to the kinds of analyses discussed above that the samples be proportional and therefore representative of the population of MZ and like-sexed DZ twins before *the* concordance or *the* variance used in the formulas can be assumed to be valid enough to support the inferences. This assumption is difficult to meet. Use of the twin method also assumes that the within-pair environmental variance is the same for the two types of twins. This is not necessarily true for the personality traits as measured by the tests, but one can proceed only on the assumption that such variance is not too different for the two types of twins. Loevinger (1943) mentioned some additional difficulties underlying the use of the variance method, chief among which were the assumption that influences combine additively and the assumption that estimates of the error variance are eliminated from the computation of *H*. Again the extent to which these conditions are met is hard to assess. Cattell (1953) concluded that approximations of a solution to the nature-nurture issue, with an awareness of methodological shortcomings, were better than postponing all research in the area. The author is inclined to agree.

Biological biases have been reviewed by Price (1950) who

divided them into natal factors (e.g., position in utero), lateral inversions, and effects of mutual circulation. No attempt to evaluate these factors will be made since data are not available. Postnatal biological biases are often overlooked on the apparent assumption that the general environment for a pair of twins is the same. Once more the assumption is questionable. Should one of a pair, for example, contract some form of encephalitis with its well-known sequelae, the results on personality measurement would be obvious.

The main limitations of twin studies were viewed in somewhat different terms by Kallmann and Baroff (1955) as the following:

> (*a*) twins cannot be separated before they are born, nor can they be provided with two mothers of different age, personality, or health status; (*b*) two-egg twins are no more dissimilar genotypically than brothers and sisters and like them, are rarely raised in different cultures; therefore, even fraternal twins are unlikely to fall into the extremes of theoretically possible genetic and cltural differences; and (*c*) the average difference between one-egg twin partners is no precise measure of environmentally produced variation, nor does an increase over the average difference between two-egg twins represent the exact contribution of genetic influences even in relatively comparable environments [p. 303].

Intraclass Correlation Analysis of Personality Traits

Following the diagnosis of zygosity and the collection of the personality test data, each scale of the two tests and the IQ from the school records were analyzed by means of the intraclass correlation coefficient, first for the two classes of twins and then for the two sexes within each class. A total of 186 coefficients was obtained from 186 simple one-way analyses of variance using *T* scores for the MMPI, raw scores for the HSPQ, and Otis IQs.

Haggard's (1958) book on the intraclass correlation gives a detailed exposition of the method used here. Although the *intraclass* correlation was formerly computed by calculating the *interclass* correlation after constructing a symmetrical table with double entries for a pair of scores and then dividing by 2, it now is recognized as a simple function of variances. Haggard (1958, p. 11) gives this formula for the computation:

$$R = \frac{BCMS - WMS}{BCMS + WMS}$$

where,

$BCMS$ = between-classes (twin pairs) mean square
WMS = within-classes (twin pairs) mean square

This means that the unbiased estimate of R may be obtained in terms of the mean squares (i.e., variance estimates) of the analysis of variance table. This formula is the specific one to use for pairs of scores. The relationship of F, the variance ratio, to R is given by:

$$F = \frac{1+R}{1-R}$$

The level of statistical significance of R is identical with that of the corresponding F (i.e., $BCMS/WMS$). In other words, the hypothesis that an observed R could have come from a population with a true correlation of zero can be tested by the F ratio computed from the same mean squares, with the appropriate degrees of freedom, as were used to obtain R.

In order to test the significance of the differences between two independently obtained Rs, they were converted into Fisher's z (Fisher & Yates, 1949) which has an approximately normal distribution with variance:

$$V = \frac{k}{2(c-2)(k-1)}$$

where,

k is the number of individuals within a class, i.e., 2 (MZ or DZ twins), and c is the number of classes, i.e., 34 (pairs). The distribution of the difference between the corresponding z values is approximately normal with variance:

$$V_d = \frac{k_1}{2(c_1-2)(k_1-1)} + \frac{k_2}{2(c_2-2)(k_2-1)}$$

Dividing the difference between z's by the square root of the above gives a normal deviate, the p value of which is found in the usual manner. A one-tailed test of significance was appropriate and was used.

Recapitulating the objectives of this intraclass correlation analysis of traits are (a) to demonstrate that the traits are significantly and positively correlated in MZ twins and may or may not be in DZ twins and (b) to demonstrate that for any genetically influenced trait the correlation within MZ pairs will be significantly greater than that within DZ pairs.

Subsequent to this analysis, the heritability indexes were computed as described in the previous section using the independently obtained *WMS* or within variances. It should be noted that the two procedures, intraclass correlation analysis and computation of heritability indexes, involve simple and complex assumptions, respectively. The correlation analysis is sufficient to show that heredity has something to do with individual differences. Estimates of the relative importance of nature and nurture as indicated by H must be considered as suggestive rather than definitive in human behavior genetics.

Configural (Holistic) Analyses of Personality Similarity

Following a scale-by-scale analysis, one of the two personality tests was selected for holistic profile analyses. The MMPI was chosen because MMPI configurations have been treated extensively in the literature. Recent emphasis on the study of profiles has resulted from the realization that interpretation of an individual's set of scores must frequently be based on the pattern of scores rather than examination of one scale at a time or the use of a linear sum of the scale deviations. General and specific methodological difficulties arise which weaken any confidence that may be attached to the quantification of profile similarity. Only a few of the difficulties noted by students of the problem (Cronbach & Gleser, 1953; Osgood & Suci, 1952) will be discussed.

Similarity as a general quality of personality is nebulous but necessary for communication. Cronbach and Gleser (1953) say:

> *similarity is not a general quality. It is possible to discuss similarity only with respect to specified dimensions (or complex characteristics).* This means that the investigator who finds that people are similar in some set of scores cannot assume that they are similar in general. He could begin to discuss general similarity only if his original measurement covered all or a large proportion of the significant dimensions of personality [p. 457].

Other general methodological difficulties involve the loss of information by reducing the relationship between two configurations to a single index; lack of comparability between indexes of similarity; and violations of assumptions about ratio scales, uncorrelated measures, and equal reliability among subtests.

There are two aspects of profiles which matching may involve: the shape or configuration of scores and the general elevation from the mean of the norm group. It is logical to distinguish between matching for absolute agreement, in which both shape and elevation are considered, and relative agreement, in which only shape is considered. Three statistical and one clinical indexes of similarity were computed for the two classes of twins. In addition, the profile of each twin was coded according to the methods of both Hathaway and Welsh (Dahlstrom & Welsh, 1960) to facilitate further clinical assessment (Gottesman, 1960).

Statistical Indexes

Rank-Difference Correlation. This well-known measure, Spearman's rho, was the first index computed. It yielded a nonarbitrary number which reflected similarity of shape but disregarded elevation. One of its disadvantages was that a rho of 1.00 did not necessarily indicate perfect similarity and another was that two pairs of profiles with the same coefficient need not be equally similar. Rho's were calculated from the Welsh codes; ties were resolved by using the scales in numerical order.

D *Coefficient.* This index, sometimes known as the generalized distance function, was then computed. Cronbach and Gleser (1953) devote considerable attention to this index which is designed to reflect both shape and elevation. The *D* coefficient is based on the geometric principle that in a space of *N* mutually orthogonal dimensions, the distance between two points is equal to the square root of the sum of the squared differences between the coordinates of the points on each dimension. Since profiles may be considered as points in *N* space, where *N* equals the number of scales (i.e., 10), the distance between them serves as a measure of similarity. Note that orthogonality does not obtain for the MMPI. The *D* coefficient results in an arbitrary number whose value depends on the number of scales.

Concordance of Test Behavior (TT′). In the context of discovery it was decided to compute the absolute percentage of MMPI items answered in the same direction by a pair of twins, i.e., one twin's answer sheet was used to score the other's. Of course the MMPI was not designed to be used this way and in this instance serves primarily as an item pool. The percentage of agreement for the 566 items has been termed TT′, to signify the comparison of one twin with his sibling. No provision was made for the few items which are repeated, but any question omitted by either twin was subtracted from 566 before the percentage was calculated. This process was then repeated using only those items appearing on the 10 clinical scales (337 items).[5]

Clinical Index

Visual Judgment. The only quantifiable clinical index of similarity used was the accuracy of visual judgment in sorting the profiles into four categories: Very Similar, Similar, Dissimilar, and Very Dissimilar. By *accuracy* was meant the number of MZ pair profiles placed in the first two categories and the number of DZ in the last two. Three psychologists[6] skilled in the use of the MMPI were the judges. Another indication of similarity was provided by comparison of the accuracy of visual judgments in the extreme categories with the overall accuracy.

[5] Hathaway's linear statistic *CC′* was also calculated with no improvement over any of the statistical methods reported here. Data are available upon request.

[6] Thanks are due Jan Ducker, H. Gilberstadt, and R. D. Wirt.

Recapitulating, the objective of each of the above four procedures was to search for a greater similarity of personality, as measured by the configural aspects of the MMPI, for the MZ twins than for the DZ twins.

RESULTS

The presentation of the findings is organized around the two instruments of assessment; first the MMPI with its 10 clin-

TABLE 5

MMPI Intraclass Correlations for MZ and DZ Twins for Total Group, Females, and Males

	Total Groups' *R*		Females' *R*		Males' *R*	
Scales	MZ	DZ	MZ	DZ	MZ	DZ
1 *Hs*	39**	21	14	19	44	25
2 *D*	47**	07	44*	25	48*	–19
3 *Hy*	47**	41**	36*	51**	65**	30
4 *Pd*	57***	18	45*	25	66**	–07
5 *Mf*	52***	32*	52**	35*	55*	28
6 *Pa*	44**	18	44*	31	43	–10
7 *Pt*	55***	20	62***	11	40	31
8 *Sc*	59***	19	50**	12	65**	19
9 *Ma*	24	–07	29	–28	11	42
0 *Si*	55***	.08	37*	18	73**	–04
K	32*	–02	30	03	40	–26
Es	25	47**	17	49**	48*	30
A	45**	04	46*	12	42	–35
R	29*	.22	20	15	50*	42
Do	46**	21	22	40*	72**	–44
Dy	52***	25	59**	47**	28	–37
St	47**	53***	28	63***	72**	24

* $p \leq .05$.
** $p \leq .01$.
*** $p \leq .001$.

ical scales, 6 experimental scales, and *K* (a validity scale thought to have personality referents); and second the HSPQ with its 14 factored scales. Indexes of H are presented separately after each correlational analysis. Results are displayed for the total sample of 34 pairs each of MZ and DZ twins, then for the female subsample (22 MZ and 23 DZ pairs) and then for the male subsample (12 MZ and 11 DZ pairs). The results by sex are provocative but any extensive discussion of them is vitiated by the size of the samples. This section concludes with the results of the configural and holistic MMPI profile analyses. Thorough discussion of the results is deferred to the next section.

Minnesota Multiphasic Personality Inventory

The excellent matching of the two classes of twins and their representativeness of adolescents in general were observed on the mean scores and standard deviations of the 3 validity scales and the 10 clinical scales. Intraclass correlation coefficients for the twins are given in Table 5. Nine of the 10 standard MMPI scales were significantly different from zero at the .01 level for the MZ twins. Fifteen of the 17 MMPI MZ scale correlations were larger than the corresponding correlations for DZ twins. It should be noted that for 6 of the 8 total MZ scale correlations for which test-retest reliability data on adolescents were available, the magnitudes are about the same.

Rozeboom (1960) has reminded us that the

> primary aim of a scientific experiment is not to precipitate decisions, but to make an appropriate adjustment in the degree to which one accepts, or believes, the hypothesis or hypotheses being tested [p. 420].

Since the traditional null hypothesis tests in psychological research pay no attention to the utilities of various outcomes, he suggested that the basic statistical report should be in the form of a *confidence interval* whenever possible. It is obvious, in the case of correlation coefficients, for example, that the researcher is concerned with more than the fact that a particular coefficient is not zero—he hopes he is in a position to account for more than a trivial amount of variation. In sympathy with the Rozeboom position, a few of the 90% confidence limits for the data in Table 5 will be mentioned. For the largest nonzero *R* in the total sample of MZ twins, .59 for Schizophrenia (*Sc*), the limits are .35 and .74; for the smallest nonzero *R*, .39 for *Hs*, the limits are .13 and .61. For the corresponding *R*s, in the sample of DZ twins, .19 for *Sc*, the limits are —.10 and .44; for *Hs*, .21, the limits are —.08 and .46.

The results of testing whether or not the correlation between MZ pairs is significantly greater than that between DZ are presented in Table 6. All *R*s were first converted to Fisher's *z*'s. The MZ twins appeared to be significantly higher than

TABLE 6

One-Tailed Test of the Difference between MZ and DZ MMPI Scale Intraclass Correlations for Total Group, Females and Males

	Total group		Females		Males	
Scale	Normal deviate	*p*	Normal deviate	*p*	Normal deviate	*p*
1 *Hs*	.79	.21	–.16		.47	.32
2 *D*	1.76	.04	.69	.24	1.56	.06
3 *Hy*	.30	.38	–.59		1.01	.16
4 *Pd*	1.86	.03	.73	.23	1.88	.03
5 *Mf*	.98	.16	.68	.25	.72	.24
6 *Pa*	1.16	.12	.48	.31	1.22	.11
7 *Pt*	1.66	.05	1.97	.02	.22	.41
8 *Sc*	1.94	.03	1.37	.08	1.27	.10
9 *Ma*	1.26	.10	1.88	.03	–.73	
0 *Si*	2.15	.02	.66	.25	2.11	.02
K	1.40	.08	.89	.19	1.50	.07
Es	–1.02		–1.17		.46	.32
A	1.78	.04	1.20	.11	1.77	.04
R	.30	.38	.16	.43	.22	.41
Do	1.14	.13	–.64		3.00	.001
Dy	1.28	.10	.54	.30	1.47	.07
St	–.32		–1.45		1.44	.08

DZ on 5 of the 10 standard MMPI scales (p less than or equal to the 5% level). The results from the correlational analyses then left 5 of the 10 standard scales, 2—Depression (D), 4—Psychopathic deviate (Pd), 7—Psychasthenia (Pt), 8—Schizophrenia, and 0—Social Introversion, which appeared to have significant genetic (i.e., gene determined) components for the combined group.

The heritability estimates for the MMPI scales, which only utilize within-pair variances, are presented in Table 7. While there is no method yet for the computation of confidence limits for H, the suggested way for testing its significance is a function of the significance of the F ratio formed by the within-

TABLE 7

MMPI Scale Heritability Indexes for Total Group, Females, and Males

	Total group		Females		Males	
Scale	H	F[a]	H	F[a]	H	F[a]
1 *Hs*	.16	1.19	.25	1.33	.01	1.01
2 *D*	.45	1.81	.22	1.28	.65	2.83
3 *Hy*	.00	.86	.00	.56	.43	1.74
4 *Pd*	.50	2.01	.37	1.60	.77	4.35
5 *Mf*	.15	1.18	.00	.99	.45	1.83
6 *Pa*	.05	1.05	.00	.70	.52	2.09
7 *Pt*	.37	1.58	.47	1.89	.24	1.31
8 *Sc*	.42	1.71	.36	1.56	.50	2.00
9 *Ma*	.24	1.32	.33	1.50	.00	.81
0 *Si*	.71	3.42	.60	2.49	.84	6.14
K	.06	1.06	.00	.95	.26	1.35
Es	.00	.73	.00	.69	.00	.84
A	.21	1.26	.22	1.28	.18	1.22
R	.00	.82	.00	.79	.00	.86
Do	.00	.95	.00	.76	.33	1.50
Dy	.24	1.32	.03	1.03	.45	1.82
St	.34	1.52	.14	1.16	.63	2.69

[a] The three values of F required for significance at the .05 level are 1.78, 2.04, and 2.72.

DZ pair variance divided by the within-MZ pair variance. Even when the computation of *H* in the present study showed that 42% of the observed within-family variance for the total group could be accounted for by genetic factors, the associated *F* was not statistically significant at the .05 level. The infrequent reporting of the significance of *H* in the literature together with the paucity of positive results for even intellectual and psychomotor tasks (Vandenberg, 1962) suggests a need for mathematical clarification of *H*. Within the limits of the assumptions for this kind of analysis, this attempt at quantification of the proportion of scale variance accounted for by heredity gave positive results for the same five scales identified by the correlational analysis. Scales *Pt* and *Sc* showed appreciable variance accounted for by heredity but with environment predominating. Scales *D* and *Pd* showed about equal contributions of heredity and environment. Scale 0, Social Introversion, showed a predominance of variance (.71) accounted for by heredity. The value of *H* for the *Si* scale is of the same magnitude as that found in this study and others for intelligence as measured by standard IQ tests.

Cattell's Factored Test

Once again the excellent matching of the two classes of twins and their representativeness of adolescents in general may be inferred from the comparison of mean scores on the 14 scales with the normative data. Intraclass correlation coefficients for the MZ and DZ twins are given in Table 8. Eight of the 14 scales had correlations between MZ twins greater than zero. Two scales, A and J, were not significantly different from zero for both MZ and DZ twins.

Six of the 14 factors resulted in correlation coefficients which were not significantly different from zero for the MZ twins. That DZ should obtain significant correlations on 4 of these 6 was paradoxical. It is difficult to reconcile claims of construct validity for these 6 scales with these results. Unless there were some logical a priori grounds for identical twins to be opposed on some trait, their identical heredity and/or their very similar environment would lead us to expect other than

TABLE 8

HSPQ Intraclass Correlations for MZ and DZ Twins for Total Group, Females, and Males

	Total groups' *R*		Females' *R*		Males' *R*	
Factor	MZ	DZ	MZ	DZ	MZ	DZ
A	19	27	26	40*	—10	05
B	60**	61***	65***	66***	56*	57*
C	28	38*	05	12	33	76**
D	21	47**	23	62***	19	23
E	16	41**	—06	53**	33	—18
F	47**	12	29	16	64**	04
G	49**	42**	23	33	76***	56*
H	38*	20	42*	34	34	—15
I	55***	47**	26	00	37	14
J	26	—04	29	—08	24	—01
O	45**	37*	50**	51**	20	—04
Q_2	60***	15	54**	—02	51*	42
Q_3	30*	12	56**	34	—01	—22
Q_4	27	32*	35*	16	12	73**

* $p \leq .05$.
** $p \leq .01$.
*** $p \leq .001$.

a zero correlation between them for a personality trait. The factor derivation of all the scales and their low intercorrelations permitted acceptance and interpretation of the remaining eight scales on their own merit. Factors B, F, G, H, I, O, Q_2 and Q_3 at this point in the analysis have the potential for showing a predominance of genetic variance.

For the largest nonzero *R* in the total sample of MZ twins (other than the intelligence factor), .60 for Q_2, the 90% confidence limits are .38 and .75; for the smallest nonzero *R*, .30 for Q_3, the limits are .02 and .54. For the corresponding *R*s in the sample of DZ twins, .15 for Q_2, the limits are —.13 and .42; for Q_3, .12, the limits are —.16 and .39.

In Table 9 are presented the results of testing whether or not the correlation between MZ pairs is significantly greater than that between DZ. All *R*s were first converted to Fisher's *z*'s.

The MZ twins were significantly higher than DZ on only one HSPQ factor, Q_2. For 8 of the 14 factors the differences were in the predicted direction. The results from both the correlation analyses then left only 1 factor, Q_2—Group Dependency versus Self-Sufficiency, which appeared to have significant genetic (i.e., gene determined) components.

The heritability indexes for the HSPQ scales, computed only from the within-pair variances, are presented in Table 10. Within the limits of the assumptions for this analysis, this attempt at quantification of the proportion of scale variance accounted for by heredity gives positive results for 6 of the 14 factors. Factors E, Submissiveness versus Dominance; H, Shy,

TABLE 9

One-Tailed Test of the Difference between MZ and DZ HSPQ Scale Intraclass Correlations for Total Group, Females and Males

	Total group		Females		Males	
Factor	Normal deviate	*p*	Normal deviate	*p*	Normal deviate	*p*
A	−.34		−.50		−.32	
B	−.03		−.05		−.02	
C	−.46		−.23		−1.45	
D	−1.21		−1.56		−.07	
E	−1.10		−2.07		1.13	.13
F	1.56	.06	.44	.33	1.56	.06
G	.36	.36	−.33		.77	.22
H	.78	.22	.33	.37	1.09	.14
I	.40	.34	.86	.19	.54	.29
J	1.20	.11	1.23	.11	.56	.29
O	.40	.34	−.04		.53	.30
Q_2	2.13	.02	2.00	.02	.24	.41
Q_3	.74	.23	.88	.19	.46	.32
Q_4	−.25		.63	.26	−1.78	

Sensitive versus Adventurous; and J, Liking Group Action versus Fastidiously Individualistic showed appreciable variance accounted for by heredity but with environment predominating. Factors F, Sober, Serious versus Happy-Go-Lucky; Q_2; and O, Confident Adequacy versus Guilt Proneness showed about equal contributions of hereditary and environmental variance (.56, .56, and .46).

TABLE 10

HSPQ Factor Heritability Indexes for Total Group, Females, and Males

	Total group		Females		Males	
Factor	*H*	*F*[a]	*H*	*F*[a]	*H*	*F*[a]
A	.10	1.11	.00	.97	.24	1.32
B	.05	1.05	.00	.96	.18	1.22
C	.03	1.03	.25	1.34	.00	.43
D	.00	.62	.00	.37	.42	1.72
E	.31	1.44	.00	.80	.74	3.84
F	.56	2.29	.45	1.81	.74	3.83
G	.00	.97	.01	1.01	.00	.79
H	.38	1.62	.42	1.73	.34	1.52
I	.06	1.07	.05	1.05	.10	1.11
J	.29	1.41	.27	1.37	.34	1.51
O	.46	1.85	.22	1.29	.69	3.18
Q_2	.56	2.28	.60	2.52	.47	1.89
Q_3	.12	1.13	.31	1.44	.00	.99
Q_4	.00	.53	.18	1.22	.00	.09

[a] The three values of *F* required for significance at the .05 level are 1.78, 2.04, and 2.72.

Results of the Otis IQ analysis

The results of the school administered intelligence test are given at this point because Factor B of the HSPQ is a brief 20-item measure of intelligence. Intraclass correlations for the MZ and DZ twins were .83 and .59, respectively, both significant at the .001 level with the first significantly greater than the second at the .02 level. The *H* value computed from the Otis within variances was .62. This means that 62% of the within-family intelligence variance measured by the Otis is accounted for by hereditary factors in this sample.

Configural and Holistic Analyses

Rank-Difference Correlations for the Coded MMPI Profiles

Table 11 shows the distribution of the Spearman rho's for the 68 pairs of profiles by twin type. It is obvious that the overlap is too great to permit other than chance discrimination ($\chi^2 = .47$, $p > .25$) between the two kinds of twins on the basis of their MMPI profile rank-difference correlation coefficients. A cutting score for a rho of .40 and above would correctly classify 53% of all the twin pairs.

Table 11

Distribution of MZ and DZ MMPI Profile Code Rho's

Rho	MZ	DZ
.80-.89	2	1
.70-.79	2	1
.60-.69	3	3
.50-.59	6	6
.40-.49	4	4
.30-.39	3	2
.20-.29	3	2
.10-.19	3	5
0-.09	2	1
−.50 to −.01	6	9

D *Coefficient*

Table 12 shows the distribution of the generalized distance function computed from the 68 pairs of profiles. This index abstracts information about the shape and elevation of the profile. There was a tendency for the identical twins to have a lower *D*, i.e., be less dissimilar; 19 of the MZ pairs were

TABLE 12

Distribution of MZ and DZ MMPI Profile *D* Coefficients

D	MZ	DZ
15-19	1	0
20-24	4	3
25-29	10	8
30-34	8	5
35-39	5	4
40-44	3	4
45-49	1	2
50-54	1	3
55-59	0	3
60-64	1	0
65-69	0	1
70-74	0	1

below the median of the combined group as contrasted with 15 of the DZ pairs ($\chi^2 = .94$, $p < .17$). Using the median *D* as a cutting score resulted in correct classification of 56% of the profiles.

Concordance of Test-Behavior (TT′)

The percentage of all MMPI test items (566) answered in the same direction for a pair of twins is given in Table 13. There was a tendency for the identical twins to have a greater overlap in their responses to all the items; 21 of the MZ pairs were above the median of the combined group as contrasted with 15 of the DZ pairs ($\chi^2 = 2.35$, $p < .06$). Using the median *TT′* of 72% as a cutting score resulted in correct classification of 59% of the profiles. There was some improvement in the use of this index when only the items in the 10 clinical scales were used as the denominator (337); 21 of the MZ pairs ex-

TABLE 13

Distribution of MZ and DZ Total MMPI Item Agreement Percentages (*TT'*)

TT'	MZ	DZ
85-89	1	1
80-84	6	4
75-79	4	3
70-74	13	11
65-69	7	10
60-64	2	2
55-59	0	2
50-54	1	1

ceeded the median compared to 14 DZ pairs ($\chi^2 = 2.94$ $p <$ .05). The cutting score was the same and correctly classified 60% of the profiles.

Clinical Judgment of Profile Similarity

The extent to which the judges' clinical assessment of personality similarity agreed with the zygosity of the twin pairs is given in Table 14. Computing the combined *p* levels (Mosteller & Bush, 1954, p. 329) for the accuracy of the three judges on their sorting of all profiles resulted in a *p* equal to .003. The

TABLE 14

Agreement of Clinical Judgments of MMPI Profile Similarity with Twin Zygosity

Judge	Total sort	*z*	*p*	Extreme pile sort	*z*	*p*
A	64.7%	2.43	.01	67.6%	2.06	0.3
B	61.8%	1.95	.03	73.5%	2.74	.006
C	58.8%	1.45	.08	58.8%	1.03	.16
Combined *p*			.003			.004

comparable figure for accuracy of judging the two extreme groups of Very Similar and Very Dissimilar was .004. By pooling the ratings for each pair (i.e., 2 of 3, or 3 of 3 votes) in an effort to correct for the various sources of attenuation in the "configural powers" of the clinician, the accuracy of the total sort increased to 67.6% ($z = 2.90$, $p = .005$).

DISCUSSION

In the introduction the purpose of the present research was said to be to answer this question—How much of the variability observed within a group of individuals in a specified environment on a specific measure of a specific personality trait is attributable to genetic factors? Implicit in the posing of this question was the assumption, subsequently confirmed by the data, that there were measurable genetic influences for at least some of the aspects of human personality tapped by the selected personality tests. Since the quantification of genetic variability derived from the computation of heritabilities by the classical twin method, and since this method rests upon some unproved assumptions, the results are recognized as suggestive and heuristic rather than definitive. While the data language for this research consists of scores on scales of personality tests, the discussion which follows is in terms of the underlying biophysical traits which the scales (read constructs) reflect.

The first section of the discussion deals with the specific traits and their configurations demonstrated to have been influenced by hereditary factors together with some of the implications of these data for personality theory and the etiology of mental illness. This is followed by an attempt to explain the apparent failure of some of the holistic analyses of personality to support strongly the findings of the trait analyses.

To what extent can the results of the present study be applied to human behavior in general? The representativeness of the twin sample to Minnesota adolescents in general suggests that this kind of extrapolation is fairly safe. There is the possibility that the twins, preponderantly of Scandinavian extraction, sample a unique gene pool. Whether the further extrapolation to adults can be safely made is dubious and difficult to assess. When *rate* of development enters the picture, biological influences on a trait might be emphasized compared to a final adult stage. The present results could be very different if derived from adult twins as suggested by studies of morphology (Asborne & DeGeorge, 1959). Another important question is the extent to which these data from normal nonhospitalized individuals can be applied to identifiable psychiatrically ill. A basic assumption throughout this project was that the measured aspects of personality varied continuously, and that any underlying genetic mechanisms were polygenic in nature. The possibility that the extremes of distributions for some personality characteristics constitute discrete series exists, but this phenomenon would be masked by the strategy used. The net effect would be the underestimation of the importance of heredity since concordances in MZ twins for low incidence Mendelian or polygenic characteristics would pass unnoticed in the analysis of quantitative variability.

Genetic Aspects of Personality

A total of 6 traits out of the standard 24 in the two tests met a criterion which classified them as significantly influenced by genetic factors; that is, correlations between the scores of identical twins were significantly higher than those between the scores of fraternal twins. Beyond this, for 23 of the total 31 scales analyzed in the two personality tests, the differences were in the predicted direction. Trait Q_2, Self-Sufficiency, was the only survivor of the 14 HSPQ measures. Although Factor Q_2 is not thought to be clearly established, the item content suggests that a person who is resolute and accustomed to making his own decisions will obtain a high score, while low scorers would be described as followers and conformists. A synthesis of this trait and F, Surgency (significant at the .06 level), is provided by Cattell's large second-order factor, Extraversion versus Introversion, which is composed of four factors. Two of the four are F and Q_2. Tying in neatly with the

MMPI findings which are discussed next is a study on the construct validity of the adult form of the HSPQ by Karson and Pool (1957). These authors found the highest MMPI scale correlate of Factor F to be Scale *Si,* Social Introversion, and the highest correlate of Q_2 to be *Si* also with correlations of − .48 and .32, respectively. Together such results appear to identify a general dimension closely related to introversion-extraversion as one which is heavily influenced by genetic factors. Such a conclusion has also been reached by Eysenck (1947, 1956) who isolated introversion-extraversion as one of the two (now three) dimensions of personality by a factor analysis of ratings and personal data on 700 neurotic soldiers. He considers his findings to represent a confirmation of the theoretical ideas of Jung (1933). Genetic factors are given a prominent place in Eysenck's typology; his twin study (Eysenck, 1956) using statistics similar to those in the present study, yielded a tentative value for *H* on a factored measure of introversion-extraversion of .62.

The trait of introversion as measured by the MMPI may also have implications for one of the genetic theories of schizophrenia. Patients with very high scores on Scale *Si* are clinically described as "schizoid" and Kallmann (1953) and others have suggested that the schizoid individual may represent the genetic "carrier state" of the recessive schizophrenic gene. In other words, the schizoid individual may represent the heterozygote and the schizophrenic may represent the homozygote. If the schizoid carrier can be identified, Kallmann's hypothesis about recessivity is no longer tenable. The mode of inheritance must then be that of incomplete *dominance* or polygenic, thus better accounting for the high familial incidence of schizophrenia (cf. Gregory, 1960). The magnitude of the heritability for Scale *Si* was the largest found in the present study. The belief in the genetic contribution to intelligence has come to have a fairly secure status in contemporary psychology (Gottesman, 1963); the results of this investigation indicate that a similar status is appropriate for the more pure personality trait of introversion.

The results concerning the four remaining MMPI scales, *D, Pd, Pt,* and *Sc,* lend support to the general idea that in human beings psychopathology, especially psychosis, has a substantial genetic component. The scoring keys for these scales were developed for the purpose of locating patients with respect to psychiatrically diagnosed states of depression, psychopathic personality, psychasthenia, and schizophrenia. Patients of each type were compared, item by item, with a normal group. Those items which statistically differentiated a diagnostic group from normals were then included in the appropriate scale. Taken *singly*, these MMPI scales are able to discriminate about 60% of the patients corresponding to their label at a cost of 5% false positives among the normals; however, the single scale approach has been largely supplanted by the interpretation of the entire profile with particular attention to the two or three highest scores. The use of the pattern types formed by the various combinations of high scores has led to their establishment as constructs which can be used in lieu of diagnostic classes. Many correlations have been observed between other variables and the personality construct patterns. For the *D-Sc* pattern, for example, a majority of the diagnoses among psychiatric patients obtaining such a configuration were psychotic ones, either depression or schizophrenia.

> Heredity, defined here rather crudely simply as psychosis in siblings or parents, tended to be unfavorable in these individuals [Hathaway & Meehl, 1956, p. 143].

Recent research on the *D-Pt* type (Gilberstadt & Duker, 1960) showed that it could be analyzed into three subtypes. Among other important findings, the authors observed that the *D-Pt-Sc* type of MMPI profile was characterized by a diagnosis of chronic undifferentiated schizophrenia and that such patients had much in common with descriptions in the literature of pseudoneurotic schizophrenia (Peterson, 1954). Surveys of the significance of major configural patterns are available in Hathaway and Meehl (1951) as well as in Dahlstrom and Welsh (1960).

A discussion of the results of the attempted quantification of hereditary influence adds some new information but it is not on the same firm footing as the correlational results. Eleven of the 24 traits measured by the two tests showed at least an appreciable genetic component. By *appreciable* is meant one third or more of the trait variance accounted for relative to the contribution of environmental factors (this required an *H* of .25 or more so that *H* divided by one minus *H* equalled one third). HSPQ factors, F, Q_2, and O showed about equal roles for heredity and environment. Factor O in the Karson and Pool (1957) study correlated most highly, .77, with MMPI Scale 7 (Psychasthenia) and .54 with Scale 0 (Social Introversion). Scale 7 was also found to have an appreciable genetic component in the present study. Three more HSPQ factors survived the criterion, E, H, and J. H along with F and Q_2 formed three of the four factors in Cattell's second-order factor Extraversion versus Introversion. In a study using an early form of the HSPQ (Cattell et al., 1955) both E and J were found to have appreciable genetic components. All five of the MMPI scales surviving the correlation criterion appeared in the quantitative analysis as having an appreciable genetic component. Only Scale *Si*, as noted above, was predominantly genetically determined.

Heritability values given in Tables 7 and 10 as well as those in other twin studies give the proportion of variance produced by genetic differences *within* families and underestimate the effects of heredity in the general population. *If* we could assume random mating for the traits measured, the values in Tables 7 and 10 would have to be multiplied by a factor of approximately two to indicate the true degree of genetic determination. The values thus obtained are remarkable even when compared with the results from animal behavior genetics. Mc Clearn (1961) reported a value of .69 for activity in strains of mice. Fuller and Thompson (1960) reported a heritability of .59 for exploratory behavior in two strains of mice. In animal breeding intense selection pressure permits the value of heritability to become large for such characters as egg weight and wool length (Falconer, 1960). An overview of the relative contributions of heredity and environment to within-family variance on the personality traits measured leads to the observation that environment is the preponderant influence in a majority of the traits. Cattell, Stice, and Kristy (1957) reached a similar conclusion for both within- and between-family variances.

Sex Differences

It is difficult to evaluate the meaning to be attached to the observed sex differences in heritability of the various traits since the sample sizes are reduced. It is tempting to speculate about the possible evolutionary origin of sex differences. Confidence in, and extension of, the interpretations which follow must await the completion of an ongoing study using 180 pairs of twins. To suggest that variation in a personality trait is more under genetic control in one sex than the other, it is only necessary to be reminded of the range of secondary sex differences already observed in physical and behavioral characters in both man and animal. The differentiation of such traits is brought about by hormones; the latter are one of the links in the gene to behavior pathway.

Changing from an expressive environment to a suppressive one will lower the heritability of a trait. Suppose, for example, that early experience in fighting is essential for inducing aggressive behavior. Genetic differences in variability on the *Pd* scale will not be detected in boys reared without this experience. The process of sex typing in our culture restricts certain types of behavior in the two sexes. Fighting is not tolerated and is suppressed in little girls so that we might expect the value of H for females on the *Pd* scale to be less than that for males. The results for Factor E, Submissiveness versus Dominance, and for Surgency reveal the same pattern of greater heritability for males than females. This attempt at an environmental explanation for the sex difference observed in trait heritability should not be thought of as appropriate for all the patterns observed.

Factorially Derived Scales versus Empirically Derived Scales

The positive correlational results for the 1 of 14 HSPQ factors could have been attributed to chance. In comparison with the positive results for 5 of the 10 MMPI scales, the harvest from the factorially derived personality test looks poor. The validity of 6 of the 14 HSPQ scales was cast into doubt by finding a zero correlation between identical twins. Many psychologists (e.g., Hall & Lindzey, 1957) have noted that factors derived by factor analysis are often not psychologically meaningful and do not agree with reality. Perhaps the empirical derivation of the MMPI scales was such as to allow Nature to be carved, albeit imperfectly, at the joints. The ease with which MMPI scales can be factored into subscales may mean that the original scales and those derived in similar fashion are equivalent to the large second-order factors of the factor analysts. The *Si* scale, for example, correlates with 10 of Cattell's factors (Karson, 1958). Demonstration of behavioral correlates for factors, such as Eysenck's (1950) criterion analysis, could result in a formidable merger of the ideas of Cattell, Eysenck, Hathaway, and Meehl.

Fate of Attempts at a Holistic View of Personality

Although all three statistical measures of MMPI profile similarity, rho, *D*, and *TT′*, tended to support the hypothesis of greater personality similarity between isogenic individuals, only the clinical judgments of similarity gave substantial support. Inasmuch as the statistical method usually surpasses the clinical in psychology (Meehl, 1954), these findings require close attention. Factors favoring successful clinical prediction in a profile sorting task have been suggested by Meehl (1959). Perhaps the most directly relevant factor he mentioned was the clinician's ability to analyze a configural relationship existing between predictor variables and a criterion, when the function is not derivable on rational grounds.

Even casual inspection of the results of the clinical judgments of personality similarity are conducive to accepting the general hypothesis of this research—the greater the gene similarity, the greater the overall personality similarity. The judging task required a discrimination between sisters, for example, of the same age, with the same parents, sharing more or less the same environment for their entire lives, but who differed in the amount of genetic overlap by a factor of only two. We might speculate that the task would have been easier had the pairs consisted of unrelated individuals or first cousins of the same age, sex, and social class background contrasted with identical twins. Misjudgments from the pooled ratings of the judges afford some insight to the range of heredity-environment interactions. The amount of variability available to the same genotypes is shown by the fact that 10 of the 34 pairs of identical twins' profiles were classified as dissimilar. Conversely, the lack of variability in personality available to genotypes with only half their genes in common is shown by the fact that 12 of the 34 fraternal twins' profiles were classified as similar.

SUMMARY AND CONCLUSIONS

The present study was carried out in the context of behavior genetics, the interdisciplinary science combining the knowledge and procedures of modern genetics with those of psychology. By means of the classical twin method, as it has been recently improved, and objective personality tests, the purpose of the research was to answer the question, how much of the variability observed within a group of individuals in a specified environment on specific measures of specific personality traits is attributable to genetic factors? It was recognized that the genetic variance, heritability (H), could vary according to such things as age, sex, culture, trait, and method of trait assessment.

Among the key constructs from genetics, *reaction range* and *polygenic inheritance* are central to the methodology used and interpretation of results. Heredity fixes a reaction range; within this framework a genotype determines an indefinite but circumscribed assortment of phenotypes, each of which corresponds to one of the possible environments to which the genotype may be exposed. The classical Mendelian model of dominant and recessive gene inheritance will not handle the data on continuous variation, the kind observed with human behavior. Polygenic systems are posited to account for quantitative inheritance, the phenotypic effects being simply a function of the number of genes present. Both the twin method and its limitations were discussed.

Thirty-four pairs each of identical (MZ) and fraternal (DZ) same-sex adolescent twins from the public high schools of Minneapolis, St. Paul, and Robbinsdale, Minnesota, served as the sample. The entire population of same-sex twins among the 31,000 children in the schools was first enumerated. Forty-five pairs of girls and 23 pairs of boys volunteered which represented 75% and 43%, respectively, of the total possible twin pairs available in the selected sample of schools. Disregarding sex, the twins comprised 60% of the possible 113 pairs.

At the time of testing, the twins filled out a personal history data sheet, the Minnesota Multiphasic Personality Inventory (MMPI), and Cattell's High School Personality Questionnaire (HSPQ); they were weighed, measured for height, fingerprinted, and photographed. The diagnosis of zygosity was based upon extensive blood grouping with respect to nine blood groups. This resulted in 100% accuracy in the diagnosis of DZ twins and at least 95% for MZ twins. A methodological contribution to twin diagnosis was made by the comparison of the accuracies of various methods and their combinations. Blood typing is necessary and sufficient for the accuracy required in human behavior genetics research.

Each scale of the two personality tests and the school recorded Otis IQ were first analyzed by means of the intraclass correlation coefficient for the two classes of twins and for sex differences. Subsequent to this analysis, the heritability indexes were computed; *H* is defined as the proportion of personality scale variance attributable to genetic factors. The correlation analysis of the 14 HSPQ scales suggested that two factors (F, Sober, Serious versus Enthusiastic, Happy-Go-Lucky, and Q_2, Group Dependency versus Self-Sufficiency) had significant genetic (i.e., gene determined) components. The correlation analysis of the 10 MMPI scales resulted in five, Scale 2 (Depression), Scale 4 (Psychopathic Deviate), Scale 7 (Psychasthenia), Scale 8 (Schizophrenia), and Scale O (Social Introversion), which appeared to have significant genetic components.

Within the limits of the assumptions, the attempt at quantification of the proportion of scale variance accounted for by heredity gave positive results for six of the HSPQ factors. Factors E, Submissiveness versus Dominance; H, Shy, Sensitive versus Adventurous; and J, Liking Group Action versus Fastidiously Individualistic showed appreciable variance accounted for by heredity but with environment predominating. Factors F, Q_2, and O, Confident Adequacy versus Guilt Proneness, showed about equal contributions of heredity and environment. The same kind of analysis of the MMPI gave positive results for 5 of the 10 scales. Scales 7 (Psychasthenia) and 8 (Schizophrenia) showed appreciable variance accounted for by heredity but with environment predominating. Scales 2 (Depression) and 4 (Psychopathic Deviate) showed about equal contributions of heredity and environment. Scale 0 (Social Introversion) showed a predominance of variance ($H =$ .71) accounted for by heredity. The value of H for the Otis IQ in this study was .62.

Following the scale-by-scale analysis, three holistic statistical analyses and one clinical holistic analysis of the MMPI profiles were done. The rank-difference correlations (rho) for the coded MMPI profiles, the generalized distance function (D), and a measure of test item verbal behavior concordance (TT') all showed a tendency for the MZ profile pairs to be more similar. The tendency was not strong enough to discriminate between the two classes of twins on the basis of any of the three measures. Clinical judgments of profile similarity by three experts supported the general hypothesis that the greater the gene similarity, the greater the personality similarity. The pooled accuracy of the agreement of clinical judgments of MMPI profile similarity with twin zygosity was 68% ($p = .005$).

Elaboration of the various biochemical or structural differences in the gene to behavior pathway which correspond to the results reported here might be intellectually satisfying, but progress in personality genetics need not await this step. A useful taxonomy of the aspects of behavior termed personality can be facilitated by the use of relatively invariant psychometric configurations to describe behavioral phenotypes. An important by-product of the study of twins and a concern with human behavior genetics is the emphasis given to the need for a multidisciplinary analysis of behavior ranging from biochemistry through evolution. Granting that the difficulties in accurately assessing the contribution of heredity to variation in socially important behavior are great, such efforts will not have been in vain if they contribute to a greater understanding of the sources of individual differences. The provision of an optimum environment for the optimum development of the various aspects of human behavior should follow such increased understanding.

References

Allen, G. Comments on the analysis of twin samples. *Acta genet. med. Gemellolog.*, 1955, *4*, 143-160.

Allen, G. Patterns of discovery in the genetics of mental deficiency. *Amer. J. ment. Defic.*, 1958, *62*, 840-849.

Allen, G. Intellectual potential and heredity. *Science*, 1961, *133*, 378-379.

Anastasi, Anne. Heredity, environment, and the question "How?" *Psychol. Rev.*, 1958, *65*, 197-208.

Black, J. D. The interpretation of MMPI profiles of college women. Unpublished doctoral dissertation, University of Minnesota, 1953.

Cattell, R. B. *The description and measurement of personality.* New York: World Book, 1946.

Cattell, R. B. *Personality.* New York: McGraw-Hill, 1950.

Cattell, R. B. Research designs in psychological genetics with special reference to the multiple variance method. *Amer. J. hum. Genet.*, 1953, *5*, 76-91.

Cattell, R. B., Beloff, H., & Coan, R. W. *Handbok for the IPAT High School Personality Questionnaire.* Champaign, Ill.: Institute of Personality Ability Testing, 1958.

Cattell, R. B., Blewett, D. B., & Beloff J. R. The inheritance of personality. *Amer. J. hum. Genet.*, 1955, *7*, 122-146.

Cattell, R. B., Stice, G. F., & Kristy, N. F. A first approximation to nature-nurture ratios for eleven primary personality factors in objective tests. *J. abnorm. soc. Psychol.*, 1957, *54*, 143-160.

Cronbach, L. J. *Essentials of psychological testing.* New York: Harper, 1960.

Cronbach, L. J., & Gleser, Goldine. Assessing similarity between profiles. *Psychol. Bull.*, 1953, *50*, 456-473.

Cronbach, L. J., & Meehl, P. E. Construct validity in psychological tests. *Psychol. Bull.*, 1955, 52, 281-302.

Cummins, H., & Midlo, C. *Fingerprints, palms, and soles.* Philadelphia: Blackiston, 1943.

Dahlstrom, W. G., & Welsh, G. S. (Eds.) *An MMPI handbook.* Minneapolis: Univer. Minnesota Press, 1960.

Dobzhansky, T. *Evolution, genetics, and man.* New York: Wiley, 1955.

Eysenck, H. J. *Dimensions of personality.* London: Paul Kegan, 1947.

Eysenck, H. J. Criterion analysis. *Psychol. Rev.*, 1950, *57*, 38-53.

Eysenck, H. J. The inheritance of extroversion-introversion. *Acta psychol.*, Amsterdam, 1956, *12*, 95-110.

Falconer D. S. *An introduction to quantitative genetics.* New York: Ronald, 1960.

Fisher, R. A., & Yates, F. *Statistical tables for biological, agricultural and medical research.* (3rd ed.) New York: Hafner, 1949.

Fuller, J. L. The genetic base: Pathways between genes and behavioral characteristics. In, *The nature and transmission of the genetic and cultural characteristics of human populations.* New York: Milbank Foundation, 1957. Pp. 101-111.

Fuller, J. L. Behavior genetics. *Annu. Rev. Psychol.*, 1960, *11*, 41-70.

Fuller, J. L., & Thompson, W. R. *Behavior genetics.* New York: Wiley, 1960.

Galton, F. The history of twins as a criterion of the relative powers of nature and nurture. *Fraser's Mag.*, 1875, *12*, 566-576.

Gilberstadt, H. G., & Duker, Jan. Case history correlates of three MMPI profile types. *J. consult. Psychol.*, 1960, *24*, 361-369.

Gottesman, I. I. The psychogenetics of personality. Unpublished doctoral dissertation, University of Minnesota, 1960.

Gottesman, I. I. Genetic aspects of intelligent behavior. In N. Ellis (Ed.), *The handbook in mental deficiency.* New York: McGraw-Hill, 1963. Ch. 7.

Gregory, I. Genetic factors in schizophrenia. *Amer. J. Psychiat.*, 1960, *116*, 961-972.

Haggard, E. A. *The intraclass correlation coefficient.* New York: Dryden, 1958.

Hall, C. S., & Lindzey, G. *Theories of personality.* New York: Wiley, 1957.

Hathaway, S. R., & Briggs, P. F. Some normative data on new MMPI scales. *J. clin. Psychol.*, 1957, *13*, 364-369.

Hathaway, S. R., & Meehl, P. E. *An atlas for the clinical use of the MMPI.* Minneapolis: Univer. Minnesota Press, 1951.

Hathaway, S. R., & Meehl, P. E. Psychiatric implications of code types. In G. S. Welsh & W. G. Dahlstrom (Eds.), *Basic readings on the MMPI.* Minneapolis: Univer. Minnesota Press, 1956. Pp. 136-144.

Hathaway, S. R., & Monachesi, E. D. *An atlas of juvenile MMPI profiles.* Minneapolis: Univer. Minnesota Press, 1961.

Holzinger, K. J. The relative effect of nature and nurture influences on twin differences. *J. educ. Psychol.*, 1929, *20*, 241-248.

Jackson, D. D. A critique of the literature on the genetics of schizophrenia. In D. D. Jackson (Ed.), *The etiology of schizophrenia.* New York: Basic Books, 1960. Pp. 37-87.

Jung, C. G. *Psychological types.* New York: Harcourt, 1933.

Kallmann, F. J. The genetic theory of schizophrenia. *Amer. J. Psychiat.*, 1946, *103*, 309-322.

Kallmann, F. J. *Heredity in health and mental disorder.* New York: Norton, 1953.

Kallmann, F. J. Psychogenetic studies of twins. In S. Koch (Ed.), *Psychology: A study of a science.* Vol. 3. *Formulations of the person and the social context.* New York: McGraw-Hill, 1959. Pp. 328-362.

Kallmann, F. J., & Baroff, G. S. Abnormalities of behavior (in the light of psychogenetic studies). *Annu. Rev. Psychol.*, 1955, *6*, 297-326.

Karson, S. Second order personality factors and the MMPI. *J. clin. Psychol.*, 1958, *14*, 313-315.

Karson, S., & Pool, K. B. The construct validity of the Sixteen Personality Factors Test. *J. clin. Psychol.*, 1957, *13*, 245-252.

Ljeune, J., & Turpin, R. Chromosomal aberrations in man. *Amer. J. hum. Genet.*, 1961, *13*, 175-184.

Loevinger, Jane. On the proportional contribution of differences in nature and in nurture to differences in intelligence. *Psychol. Bull.*, 1943, *40*, 725-756.

Loevinger, Jane. Objective tests as instruments of psychological theory. *Psychol. Rep.*, 1957, *3*, 635-694.

McClearn, G. E. Genotype and mouse activity. *J. comp. physiol. Psychol.*, 1961, *54*, 674-676.

Meehl, P. E. *Clinical versus statistical prediction.* Minneapolis: Univer. Minnesota Press, 1954.

Meehl, P. E. A comparison of clinicians with five statistical methods of identifying psychotic MMPI profiles. *J. counsel. Psychol.*, *6*, 1959, 102-109.

Meehl, P. E., & Rosen, A. Antecedent probability and the efficiency of psychometric signs, patterns, or cutting scores. *Psychol. Bull.*, 1955, *52*, 194-216.

Mosteller, F., & Bush, R. R. Selected quantitative techniques. In G. Lindzey (Ed.), *Handbook of social psychology.* Boston: Addison-Wesley, 1954. Pp. 289-334.

Newman, H. H. *Multiple human births.* New York: Doubleday, 1940.

Newman, H. H., Freeman, F. N., & Holzinger, K. J. *Twins: A study of heredity and environment.* Chicago: Univer. Chicago Press, 1937.

Osborne, R. H., & DeGeorge, F. V. *Genetic basis of morphological variation.* Cambridge: Harvard Univer. Press, 1959.

Osgood, C. E., & Suci, G. J. A measure of relation determined by both mean difference and profile information. *Psychol. Bull.*, 1952, *49*, 251-262.

Peterson, D. R. The diagnosis of subclinical schizophrenia. *J. consult. Psychol.*, 1954, *18*, 198-200.

Price, B. Primary biases in twin studies. *Amer. J. hum. Genet.*, 1950, *2*, 293-352.

Race, R. R., & Sanger, R. *Blood groups in man.* (3rd. ed.) Springfield: Charles C Thomas, 1958.

Rozeboom, W. W. The fallacy of the null-hypothesis significance test. *Psychol. Bull.*, 1960, *57*, 416-428.

Slater, E. *Psychotic and neurotic illnesses in twins.* London: Her Majesty's Stationery Office, 1953.

Smith, Sheila M., & Penrose, L. S. Monozygotic and dizygotic twin diagnosis. *Ann. hum. Genet.*, 1955, *19*, 273-289.

Stern, C. Principles of human genetics. (2nd ed.) San Francisco: Freeman, 1960.

Strandskov, H. H., & Edelen, E. W. Monozygotic and dizygotic birth frequencies in the total, in the "white" and the "colored" U.S. populations. *Genetics*, 1946, *31*, 438-446.

Tryon, R. C. Behavior genetics in social psychology. *Amer. Psychologist*, 1957, *12*, 453 (Abstract)

Vandenberg, S. G. The hereditary abilities study: Hereditary components in a psychological test battery. *Amer. J. hum. Genet.*, 1962, *14*, 220-237.

Wirt, R. D., & Briggs, P. F. Personality and environmental factors in the development of delinquency. *Psychol. Monogr.*, 1959, *73* (15, Whole No. 485).

SECTION **III**

Intelligence, Ability, and Race

IN THE WINTER EDITION of the 1969 *Harvard Educational Review,* the psychologist Arthur Jensen presented a lengthy treatise on differences in intelligence between "blacks" and "whites" in America. Professor Jensen's thesis rests upon the following findings: numerous studies reliably report ten- to fifteen-point differences between Negro and Caucasian individuals on a variety of I.Q. tests. Moreover, even when variables such as parental education and socioeconomic status are "controlled" (i.e., when "environmental" factors are "balanced"), the differences persist. He concluded from such data that intelligence *is* highly heritable, that the Negro has "less" of *it,* and that any program of educational advantage designed for Negroes must take these factors into consideration. By and large, he answers in the negative to the question raised in the title of his article; *How Much Can We Boost I.Q. and Scholastic Achievement?*

The responses to the Jensen paper were immediate and varied. At his own institution, a number of students demanded his dismissal and certain rabble threatened his family with unsigned notes and terrifying telephone calls. He was attacked as a "racist" and "elitist." Responsible or at least well-meaning scientists asked in earnest whether it is advisable to make such findings public, even if they are true. Not until the rejoinders offered in the spring edition of the same publication was there more light than heat shed upon the issue.

It is not clear how such problems can continue to emerge at as late a date as 1969. One would hope that the century since Mendel would allow the cultivation of a certain perspective on the matter of heredity and intelligence. One would also wish to see a greater respect for the difference between *intelligence* and *I.Q.* Some two years before his article in the *Harvard Educational Review,* Jensen had presented similar data and views at Stanford University. Hirsch has provided a one-paragraph qualification of the facts and views:

> A most unfortunate speech in 1967 by A. R. Jensen (who came under Shockley's influence while at Stanford during 1966-67) illustrates the dangers of inappropriate use of both the concept of heritability and that of race by the biometrically unsophisticated. Following Shockley, Jensen suggests that, since intelligence is largely genetically determined (sic!), in order properly to carry out Project Head Start for enrichment of the lives of the culturally disadvantaged, we need to know (=the government should spend large amounts of money to learn) the proportion of difference (of mean difference, naturally!) in intelligence between Negroes and whites that is genetically determined, i.e., its heritability. As though intelligence were some "thing" which different (pure?) races shared in unequal proportions. As though "it" had *a* (determinate) heritability. As though such "knowledge" would enable a teacher to do a better job in training either a Negro or white or any other "type" of child![1]

Each of the points made by Hirsch relates to material discussed at the beginning of this book and in the introduction to Section I. First, genes do not transmit "traits" (i.e., phenotypes). They control the reaction range of enzyme physiology. The overall physiology of the organism interacts constantly with the environment. The phenotype (hair color, height, "intelligent" behavior) is a product of the interaction. The nature

[1] Hirsch, *Behavior Genetic Analysis,* p. 434.

and quality of commerce with the environment depends upon the individual genotype. Treating populations as "black" or "white" may offer a categorical convenience; but to treat *a* "black" as one "type" and *a* "white" as another "type," i.e., to invoke typological thinking, is genetically naïve. To be sure, and as has already been indicated, the concept of *race* is genetically supported. That is, there are populations of individuals whose genotypes contain gene frequencies different from those found in other populations. However, a given *individual* may fall within one population as regards one phenotypic character and a different population as regards a different "trait." Even using Jensen's data and a limited view of intelligence (i.e., scores on I.Q. tests), 25 per cent of the "black" population exceeds the mean I.Q. of the "white" population. Other studies indicate that "Negroes" in the northeastern United States score higher than southern "whites." Does this suggest that genotype changes with latitude? Moreover, the I.Q. of children increases with an increase in dollars spent per child in a school system. Does genotype vary with educational expenditures? To demonstrate a difference between two *means* and to infer from this the existence of two "types" of individuals is, at best, careless. It should surprise no one that members of one race would differ from members of another race in many respects. So, indeed, will any two individuals who are not monozygotic twins. Even the latter, reared in very different environments, will reveal *average* I.Q. differences of fourteen points.[2] Furthermore, in some narrowly defined (cultural) sense, observed differences may be treated evaluatively; i.e., one may be judged as "better" or "preferred" or "more adaptable." In American culture (the prevailing attitudes, opinions, and beliefs entertained by the majority of American adults) a high premium is placed on performance in school as measured by grades. The more popular I.Q. tests predict "success" measured in these terms. The predictive efficiency of these tests is far from ideal but is better than existing alternatives. If one defines intelligence as marks in school, then I.Q. tests would appear to

[2] I. I. Gottesman. Biogenetics of Race and Class. In M. Deutsch, I. Katz, and A. Jensen (eds.), *Social Class, Race, and Psychological Development* (New York, 1968).

be measuring intelligence. Interestingly, neither I.Q. (above some criterion level such as 110-115) nor academic record appears to be predictive of excellence after graduation.[3] To other cultures and to certain subcultures within America, such considerations may be irrelevant, unknown, or assumed to be unattainable. Attempts to fashion the so-called "culture-free" and "culture-fair" measures of intelligence have been of marginal success. Even when items are eliminated which have obvious connections with language biases, cultural implications, parental educational levels, etc., subcultures may have radically different attitudes about achievement, about tests, and about "self."

This section contains four papers addressed to the foregoing. The first is the provocative contribution of Sir Cyril Burt summarizing his research on twins. He concludes that intelligence is highly heritable and that the "genetic component" is far more significant than the "environmental." It is to be remembered, however, that the environmental controls reported by Burt are difficult to assess. This problem interacts peculiarly with the nature of twins themselves. First of all, identical twins are in no way statistically representative of the general population. Thus, it is always possible that the high heritability of trait-X observed in monozygotes may not obtain for the same trait in nonidentical pairs. Secondly, even in separated and apparently very different settings, "look-alikes" may be reacted to by others in a more similar way than non-look-alikes. Since body-type, height, and weight are much more similar in identical twins, their general involvements (social, athletic, etc.) may be much more similar even in environments which differ radically. High inter-twin correlations in measured I.Q. may, therefore, be spurious and, even if not spurious, they may not be safely generalized to non-twin populations.

The second paper is an exhaustive review of research comparing American Negroes and whites with respect to a large number of characteristics. Dreger and Miller offer an extensive

[3] L. Hudson. Selection and the problem of conformity. In J. Meade and A. Parkes (eds.), *Genetic and Environmental Factors in Human Ability* (London, 1966).

bibliography on "interracial" comparisons of ability. The article also discloses the enormous diversity of psychometric tests available for the study of human abilities. A more recent review (1969) by the same authors[4] supports the conclusions of the present paper and, since the more recent is considerably longer, the 1960 paper was chosen. Their review indicates that "Negro" and "white" populations attain different scores, *on the average,* in a substantial number of test situations. The extent to which any of the observed differences is the consequence of high trait-heritability is, of course, a matter for conjecture.

The third article is a brief but chronologically comprehensive summary of research addressed to familial similarities in I.Q. The conclusion warranted by the review is that I.Q. scores are more similar among individuals with more common *biosocial* roots.

Finally Wesman offers a concise statement of what an I.Q. test measures. He dismisses a number of myths and in the process sets forth the guidelines for a more wholesome utilization of psychometric instruments.

Suggested Readings

Deutsch, M., I. Katz, and A. Jensen (eds.). *Social Class, Race, and Psychological Development.* Holt, Rinehart & Winston, New York (1968).

Harvard Educational Review, Winter Edition, 1969.

Harvard Educational Review, Spring Edition, 1969.

[4] R. M. Dreger and K. S. Miller. Comparative psychological studies of Negroes and whites in the United States: 1959-65. *Psychol. Bull. Monog. Suppl.,* 1969, *70.*

The Inheritance of Mental Ability[1]

CYRIL BURT

May I first say how deeply I feel the honour of being invited by the American Psychological Association, through its special committee, to deliver this lecture, and how grateful we are to Dr. and Mrs. Walter V. Bingham for their generosity in establishing such a lectureship. The problem which the founders had in mind happens to be one of special interest to us here in Britain; and it was with particular pleasure that we learnt that Mrs. Bingham and the committee had agreed that, on this occasion, the lecture should be given in this country and at this college. British psychologists, many of whom are gathered in this hall, are thus enabled to express their sincere and heartfelt admiration of the work carried out by Bingham himself, not only during two world wars, but also in time of peace, and not only in his own country, but also during his many visits over here. The valuable contributions that he made by his books, his numerous articles, and most of all perhaps by his tactful and untiring activities behind the scenes, I have already attempted to review elsewhere. Here therefore may I content myself with endorsing on behalf of British psychologists the well deserved tribute paid by Professor Egon Pearson on behalf of British statisticians.

In his own autobiographical chapter Bingham has himself observed that one of his "chief functions in life seems to have been to act as an emissary for psychology to the heathen"; and it was during one of these apostolic missions that, when he came to London a year or two before the first world war, he so kindly sought me out to discuss the paper I had just published while at Oxford on the measurement and inheritability of mental capacity (3). It was my first effort at research, and the first paper published in this country on mental tests. I still treasure the letter I received from Ward of Cambridge, who was in those

1 This paper is based on the Walter Van Dyke Bingham Memorial Lecture given at University College, London, on May 21, 1957.

From *American Psychologist,* 1958, *13,* 1-15. Reprinted by permission of the author and the American Psychological Association.

days the acknowledged leader of psychology in this country and editor of our new and only journal of psychology. While accepting my paper, he added that I was myself an outstanding example of wasted capacity, in that I had "spent so much time and industry on a transient problem like mental testing, which held so little promise for the future." "Evidently," said Bingham, "Oxford must begin by converting Cambridge." And it was the encouragement I then received from him that emboldened me to persevere. As it turned out, Bingham's precognition of the future of mental testing was far more accurate than Ward's; and soon afterwards McDougall at Oxford and Myers, then a young lecturer at Cambridge, succeeded in transforming British psychology from a branch of philosophy into a branch of science. In so doing both insisted that in future "psychology must devote itself to the study of the individual and not merely to the study of mind in general."[2]

The Problem of Individual Differences

In Britain as in America, the earliest applications of the new techniques of mental testing were concerned chiefly with the lower end of the intellectual scale. They were readily accepted as practical aids to the doctor in diagnosing the mentally deficient, and to the teacher in discriminating the temporarily retarded from the irremediably dull. It was Bingham's firm belief that: "in the long run it would be even more profitable to discover and aid the bright and the supernormal than to ascertain and provide for the dull and defective."

In this country the traditional method of dealing with the problem was by means of what was popularly known as "the scholarship system." British teachers and parents were apt in those days to think of scholarships rather as a means for rewarding hard work than as a device for detecting and assisting impecunious talent. But considerable misgiving was felt over the erratic local distribution not only of scholarship winners

[2] The change that thus took place can be seen by comparing the treatment of individual psychology in the writings of Ward (e.g., *Psychological Principles,* pp. 433ff.) and in those of Myers (*Experimental Psychology,* 1911, Chap. VI & VII and McDougall (*Psychology: The Study of Behaviour,* 1912, Chap. VI-VIII).

but also of defectives. It was indeed widely affirmed that these variations in the number were "largely the result of the injustice of the ordinary methods of examination." Accordingly, in 1913 the London County Council took the unprecedented step of appointing an official psychologist. His chief duties were to act as referee for what were sometimes known as "problem children"—particularly in disputed cases of feeble-mindedness and of scholarship ability.

In the surveys that I carried out in this capacity I found that, in the poorer districts of the East End, there were three times as many certified defectives, eight times as many backward pupils, but only one-tenth the number of scholarship winners as were reported in the well-to-do areas of London. Who or what was to blame: the teachers, the homes, the traditional examination in arithmetic and English? Or the fact that the varying modicum of ability bestowed on each child was irretrievably fixed at birth? Environment or heredity, those seemed to be the alternative explanations; and each had its own enthusiastic advocates. As a rule the social reformers gave one answer and the psychologists the opposite.

During the nineteenth century the main champions of social reform were the philosophical radicals: Bentham, James and John Mill, and their various followers. Their motto was the maxim of Helvétius: *l'éducation peut tout;* or, as James Mill put it, "if education cannot do everything, there is hardly anything it cannot do." In psychology, their one basic law was the law of association; and the wide differences between one individual and another they attributed solely to the cumulative effects of association, through the operation of "habit, custom, training, and environmental opportunity or deprivation."

By the beginning of the twentieth century, however, most British psychologists had abandoned this simple creed. Impressed by the novel doctrines of the evolutionary school, they began to follow Darwin, Spencer, Huxley, and Galton in holding that the laws of heredity and individual variation applied to men as well as to animals, to mind as well as to body. "It is not so much," said Karl Pearson, "that the slums create the dullards, but rather that the duller stocks gravitate automatically to the slums." It followed that the only cure would be to start breeding citizens as we bred race horses, prize rabbits, or pedigree pups. Huxley's celebrated essay on the "Natural Inequality of Man" won over the few who still wavered. Most teachers were already converted by firsthand experience under the new scheme of universal education.

However, during the last few decades psychologists have been gradually discovering that the problem is far more complicated than either side had originally assumed. And, largely as a result, there had been yet another swing of the polemical pendulum. In part this is due to a reluctance to support what G. K. Chesterton once described as "the modernized dogma of predestination—a dogma calculated to paralyse all progress towards a welfare state"; in part it has been due to a fuller appreciation of the possibilities of experience and learning, which in turn has involved a reversion to something like a rejuvenated belief in the fundamental law of association.

To the British psychologist the experimental work on "conditioning" seemed to offer a more precise interpretation, on a more adequate basis, of the traditional doctrine of association as preached by Hartley and Mill; and the behaviourist school was widely welcomed as substituting a more profitable line of practical research for inconclusive speculations about heredity and innate endowment. As a result, we find contemporary educationists reminding each other of Bagley's well known dictum that, "except for a few cases of pathological deficiency, the factor of heredity plays a very small part in human life, as compared with the factor of environment" (1); while several of their psychological colleagues quote with approval the pronouncements of Watson: "there are inheritable differences in structure, but we no longer believe in inherited capacities." "Give me," he adds, "a dozen healthy infants, and my own world to bring them up in, and I'll guarantee to train any one of them to become any type of specialist I might select—doctor, lawyer, artist, merchant-chief, and even beggar-man or thief" (43, p. 104).

A glance through the current literature will show that, for the most part, the hypotheses which our younger psychologists desire to challenge are the two propositions which, at the beginning of the century, were accepted almost without question as justifying the organization of our system of education: first, the hypothesis of general ability, i.e., the view that "up to adolescence the chief differences distinguishing individual pupils are differences in 'general intelligence' "—a view which led the Chief Inspector of Schools to declare that, "with a few exceptions, each pupil may be taught in the same class for every subject, or, if mentally defective, sent to a special school on the ground that he is defective all round"; secondly, the hypothesis of mental inheritance, i.e., the view that the wide differences in the degree of general ability exhibitd by various pupils were due mainly to genetic endowment.

The Hypothesis of General Ability

Of these two hypotheses the first was firmly rejected by the report of the "Norwood Committee"—a report which strongly influenced the reorganization of our national education as envisaged in the Education Act of 1944. The committee maintained that children differed far more in equality of ability than in amount; and recommended a tripartite classification of "secondary" schools, based on the doctrine that pupils fall into three main types: a literary or abstract type, to be educated at "grammar schools"; a mechanical or technical type, to be educated at "technical schools"; and a concrete or practical type to be educated at "modern schools." This conception was supported by several leading psychologists who argued that "the statistical analyses of Thurstone disclose no evidence for Spearman's factor of general ability, but only a number of 'primary abilities,' each more or less specialized" (8 and refs.).

The scheme has not worked out in actual practice so successfully as was hoped; and the proposal itself at once stimulated a large number of new inquiries. Those who still defend the tripartite theory seems to overlook the vast and varied character of the evidence that has since become available on the issue thus raised. As a rule they content themselves with citing one particular line of reasoning—usually the statistical. If, however, we take a wider view, we shall find that there are three or four independent chains of converging arguments, none perhaps irrefragable by itself, but all of them tending to rehabilitate the older view. The evidence on which these inferences rest is drawn from many different fields: introspection, observation, biology and neurology, as well as statistical research.

In point of fact, the distinction between "general ability" and "special abilities" was due, not (as is so widely supposed) to Spearman but rather to Francis Galton, from whom Spearman, as he expressly states in his earlier papers, originally borrowed it. Spearman, however, went on to deny the existence of "special abilities" which Galton had supported, on the ground that they were "just relics or revivals of the obsolete theory of faculties" (35). It is this aspect of the Spearman doctrine that the modern educationist repudiates, and with it he is a little too ready to discard "general intelligence" as well.

However, even Galton had been partly anticipated by previous writers who, in a vague and inchoate fashion, had inculcated a somewhat similar doctrine on the basis of speculative arguments, partly introspective, partly biological. Introspection, so the faculty school had claimed, revealed a sharp distinction between intellectual or cognitive qualities on the one hand and emotional or orectic qualities on the other. Spencer and his followers accepted this distinction, and supplemented it with biological arguments, which in those days were almost equally conjectural. During the evolution of the animal kingdom, and again during the development of each individual, the fundamental capacity which Spencer termed intelligence

> . . . becomes [so he tells us] progressively differentiated into a hierarchy of more specialized cognitive capacities—sensory, perceptual, associative, and ratiocinative much as the trunk of a growing tree ramifies into boughs and twigs (36, Part iv).

Later writers were not so sure that the term cognition really furnished the best differentia: for them the real distinction was between directive or adaptive processes on the one hand and valuative or dynamic processes on the other; "intelligence," said Sully, "steers like a rudder; emotion and interest supply the steam." A similar reinterpretation has more recently been proposed by Piaget in his book on *The Psychology of Intelligence* (30, pp. 4ff.).

Far more convincing to my mind is the neurological evidence. The alternative doctrine of distinct and independent abilities drew much of its support from early experiments on cerebral localization. These, it was claimed,

> . . . appear to indicate that distinct functions—motor, sensory, perceptual, associative, linguistic, and the like—can be assigned to distinct and definite centres or areas within the brain.

And this conclusion was apparently corroborated by the maps of cell structure subsequently produced by histologists like Campbell, Brodmann, and von Economo. Their inferences, however, have been severely criticized by more recent workers, like Bok, Lorente de Nò, and, at University College, by D. A. Sholl (34). Few of the older map makers (he says) realized "the enormous amount of variability that exists between individual human brains." In their studies of comparable cortical areas from different specimens of *Ateles* Lashley and Clark have shown that the differences between different individuals are at least as large as the differences upon which the distinction of "architectonic areas" was based (26). Hence it now seems clear that the charts that used to figure in popular textbooks "greatly exaggerated the amount and the definiteness of the alleged localization, both as regards structure and as regards function."

After all, as Sherrington so frequently pointed out, in any one indvidual each anatomical tissue—skin, bone, hair, or muscle—tends to be of the same general character all over the body: minor local variations are often discernible; but in the same individual the variations are much slighter than those between one individual and another; and we should naturally expect the same to hold good of nerve tissue. Bolton's studies of the cerebral cortex, both in normal and defective persons, suggest that the quality of the nervous tissue in any given individual tends to be predominantly the same throughout: in low-grade defectives, for example, the nerve cells tend to be "visibly deficient in number, in branching, and in regularity of arrangement in every part of the cortex." The clinical work of Hughlings Jackson, the experimental researches of Sherrington, and the microscopical studies of the infant brain carried out more recently by Conel and de Crinis, seem in many ways to provide increasing confirmation for Spencer's theory of a "hierarchy of neural functions," developing stage by stage out of a simpler basic activity into higher and more specialized forms. Even if we cannot wholly accept Lashley's doctrine of the "mass action" of the brain, nevertheless the actual facts that he has reported are far more in keeping with a theory of general ability (and minor special abilities) than with a theory of major "special abilities" and no general factor. And Sholl himself concludes that "intelligence," like retentivity, is a "general attribute of all cortical tissue" (34, p. 111).

Each of the foregoing arguments—from introspection, from biology, and from neurology—is admittedly inconclusive. The most we could so far claim is that, when taken in combination, the evidence from each of these separate fields sets up a fairly strong a priori probability in favour of Galton's twofold theory of general abilities plus special. It is at this point, therefore, that the psychologist is tempted to invoke the aid of statistical analysis. In my view, the function of statistics, at any rate in psychological research, is simply to provide a more rigorous method of testing alternative hypotheses. How then do the various hypotheses that confront us stand up to this ordeal?

Each of the three main suggestions leads to distinctive corollaries which are open to statistical verification. First of all, if, as Thorndike held, there was no such thing as a general factor, but only a set of independent abilities, then the cross-correlation between the tests or assessments for those abilities should be zero, or at least nonsignificant. They are not: they are always both significant and positive. Secondly, if, as Spearman maintained, there were no such thing as special abilities but only a single general factor, the intercorrelations for every test should diminish proportionately in the same descending order (except, of course, for nonsignificant deviations due to sampling and the like): in technical language, the table of intercorrelations should constitute a matrix of rank one. It does not.

The third hypothesis assumes the presence of both general and special abilities. In the early article (3) to which I have already referred,[3] I endeavoured to apply to the correlations between mental tests the procedure which Karl Pearson had already elaborated for use with measurements of bodily characteristics. In this way I sought to compare each of my various tables of observed coefficients with the best fitting set of theoretical values deduced from the hypothesis of a single "general ability" only; this hypothetical ability was redefined as the "highest common factor" entering into all the processes measured in any given table.[4] The fit was moderately good, but by no means perfect. The differences between the observed and the theoretical values were accordingly tested for statistical significance; and the result was to disclose small but significant clusters of residual correlations, common to certain limited groups of tests only, thus plainly indicating the presence of certain supplementary abilities. This seemed definitely to clinch the double hypothesis put forward by Galton. Precisely the same conclusion was reached in many later analyses carried out both by myself and by other workers with larger numbers of pupils and a greater variety of tests (5, 6, and refs.)

[3] Those to whom this earlier (1909) paper is inaccessible will find the essential figures from one of the tables reproduced in books by Freeman (22) and Thorndike (41); Garnett (24) reprints not only the observed values, but, beneath them, the theoretical values, the discrepancies, and the probable errors, as I myself had done. It should be noted that the procedure subsequently developed by Spearman (35) was rather different. He preferred to secure independent assessments for intelligence, based on teachers' ratings, examination results, or, in later inquiries, the results of one or two accredited tests, and use them as "reference values." In other words, he used an *external* criterion; whereas I have used an *internal* criterion or "factor" derived from the intercorrelations themselves. My reasons were that both examination results and teachers' assessments were commonly influenced by the child's capacity to learn, i.e., by his memory, rather than by his sheer intelligence, and thus often reflected what the child had actually acquired rather than his innate capacity.

[4] The simplified formula used for this purpose in most of these earlier researches (6, p. 53) was the same as that subsequently adopted by Thurstone under the name of the centroid formula (42, p. 94, Equation 13). When in addition "special abilities" were studied as well as "general ability," they were regarded as "group factors" obtainable by an arithmetical rotation of the "bipolar" factor matrix. I may add that we encountered a certain amount of evidence indicating that several of these special abilities were also largely dependent on heredity, but space does not permit me to deal with them here.

The factorization of correlations between persons and between successive occasions (with the same persons and tests) revealed, or at least very plainly confirmed, the constant presence of a random factor, which appeared to be due, not merely to sampling as ordinarily understood, but to something essentially characteristic of the working of the mind or brain. A somewhat similar conclusion has been independently reached by neurologists like Beurle, Cragg and Temperley, and Uttley and Sholl. Thus the type of "neuronal connectivity" suggested by the histological study of dendritic fibres strongly suggests "the semi-random arrangement of a machine working largely in accordance with the principles of probability" (34 and refs.).[5] This partly agrees with, though it is by no means identical with, the doctrine of "equipotentiality" put forward by Lashley; and thus further supports the value of factorial procedures as a means of investigating variations in mental ability.

However, what is far more important from the practical standpoint of educational guidance is this: in nearly every factorial study of cognitive ability, the general factor commonly accounts for quite 50% of the variance (rather more in the case of the young child, rather less with older age groups) while each of the minor factors accounts for only 10% or less. With increasing age, the "group factors," which represent special abilities, contribute increasing amounts to the total vari-

[5] This is not the same as the "sampling theory" of Thomson, which maintains that the apparent emergence of "factors" is "a result of the laws of chance and not of any psychological laws": like all simplified theories, it seems to amplify just one aspect of mental process, at the expense of denying all the rest. What is needed is a model which will combine the "integrative action" of the nervous system with its "stochastic action." As I have suggested elsewhere (*Psychometrika*, 3, p. 160), the mathematical model which most effectively does this is a classificatory model derived from Pearson's method of principal axes. The mechanical model would resemble, not the homeostat of Ashby or the mechanical tortoise of Grey Walter, but rather a classification machine like that proposed by Uttley and Sholl. Similarly a stochastical induction machine would seem to reproduce the results of human learning far better than the model adopted by Hull and his followers. It is therefore tempting to suggest that what has been called the A/S ratio—i.e., the proportionate amount of "association cortex" with which either a species or an organism is endowed at birth—would form an index primarily of the amount of its intelligence and only incidentally of its capacity to learn.

ance; and this fact lends further support to the view that, as a result of maturation, cognitive ability progressively differentiates and tends to become more and more specialized (10 and refs.).

Be that as it may, whatever be our views on the various biological or neurological issues, we may, I think, safely say this: for purposes of prediction—forecasting what this or that individual child is likely to do in school or in after-life—the general factor is by far the most important, though admittedly not our only, guide. For all practical purposes, almost every psychologist—even former opponents of the concept of general intelligence, like Thorndike, Brown, Thomson, and Thurstone—seems in the end to have come round to much the same conclusion, even though, for theoretical purposes, each tends to reword it in a modified terminology of his own. And thus today, save for one or two occasional attacks, current psychological criticism is not so much concerned with the problem of "general intelligence," but is directed rather against the second of our two initial propositions: the assumption, namely, that individual differences in intelligence are hereditary or innate.

The Genetic Component

Here three distinct questions seem to be involved: (*a*) what evidence is there for the *fact* of inheritance, (*b*) what precisely is the *mode* in which intelligence is inherited, and (*c*) what is the *relative importance* of the genetic factor as compared with the environmental?

The Fact. In controversies about the facts of mental heredity most critics have tended to assume that the two causal agencies commonly discussed—heredity and environment—are not merely antithetical but mutually exclusive. The environmentalists apparently suppose that, once they have shown that intelligence tests are affected by environment, it follows that all differences in intelligence are due to nothing but environment. Similarly the thoroughgoing hereditarian is apt to talk as though he believed that differences in intelligence were due to nothing but genetic constitution. This is the old familiar fallacy which I am tempted to label "nothing-buttery." In point of fact, with

Table 1

Correlations Between Mental and Scholastic Assessments

	Identical twins reared together	Identical twins reared apart	Nonidentical twins reared together	Siblings reared together	Siblings reared apart	Unrelated children reared together
Mental						
"Intelligence"						
Group Test	.944	.771	.542	.515	.441	.281
Individual Test	.921	.843	.526	.491	.463	.252
Final Assessment	.925	.876	.551	.538	.517	.269
Scholastic						
General Attainments	.898	.681	.831	.814	.526	.535
Reading and Spelling	.944	.647	.915	.853	.490	.548
Arithmetic	.862	.723	.748	.769	.563	.476

a few rare exceptions, like eye colour or serological differences in the blood, every observable characteristic that geneticists have studied has proved to be the product of the joint action of both heredity and environment. There are, in short, no such things as hereditary characters: there are only hereditary tendencies.

Now, where two contributory factors, such as heredity and environment, are likely to be involved, the obvious procedure will be to keep first one and then the other as constant as possible, and observe the results in either case.

1. *Uniform Environment.* As psychological consultant to the London County Council, I had free access to orphanages and other residential institutions, and to the private files of case records giving the history of the various inmates. My co-workers and I were thus able to study large numbers of children who had been transferred thither during the earliest weeks of infancy, and had been brought up in an environment that was much the same for all. To our surprise we found that individual differences in intelligence, so far from being diminished, varied over an unusually wide range. In the majority of cases, they appeared to be correlated with differences in the intelligence of one or both of the parents. Some of the most striking instances were those of illegitimate children of high ability: often the father (so the case records showed) had been a casual acquaintance, of a social and intellectual status well above that of the mother, and had taken no further interest in the child. In such cases it is out of the question to attribute the high intelligence of the child to the special cultural opportunities furnished by the home environment, since his only home has been the institution.[6]

[6] Details are given in the various *Annual Reports of the Psychologist to the London County Council* (5) and are summarized in (9). In the recent symposium on *Quantitative Inheritance* (31), Woolf, quoting a later paper of mine, regrets that I have "based such far reaching conclusions on the study by Barbara Burks . . . covering only 214 foster children" (2). But the principal basis for my own conclusion was a series of investigations in residential schools under the LCC covering in the course of years over 600 cases. I cited Burks' inquiry merely to show how an independent investigator in a different country had arrived at much the same figures as my own.

2. *Uniform Heredity.* To secure cases in which the children's genetic endowment is the same, we may turn to assessments obtained from monozygotic or "identical" twins. The mother is not infrequently unable or unwilling to bring up two children at the same time, and one twin is consequently sent to a relative or to a foster home. Owing to the strong popular prejudice against separating twins, she not unnaturally tries, as a rule, to keep these arrangements secret. But patient and tactful inquiries show that cases of twins brought up in different environments almost from birth are in fact much commoner than is usually believed. We have now collected over 30 such cases (9, 11). I reproduce the more important correlations for the twins in Table 1 and have added for comparison corresponding coefficients obtained from other pairs, both related and unrelated. As regards intelligence the outstanding feature is the high correlation between the final assessments for the monozygotic twins, even when reared apart: it is almost as high as the correlation between two successive testings for the same individuals. On the other hand, with school attainments the correlations are much lower for twins reared separately than for twins reared together in the same home.

Several of our critics—Heim and Maddox, for example—have cited the account of analogous cases (described by Newman and others) as proving that intelligence is dependent on environment. Thus, to take an oft-quoted pair, "Helen," who had been trained as a teacher, scored with the Stanford-Binet tests an I.Q. of 116; whereas her twin sister, "Gladys," brought up for much of her childhood in an isolated district of the Canadian Rockies, scored only 92. But, says Newman, her score

> . . . was higher than we might expect considering her scant education; and . . . it seems certain that the great deficiency in education had inhibited the development of the rather high grade of mental ability with which she was *endowed by heredity* (*28*, pp. 136-144).

Thus Newman's interpretation in no way conflicts with ours.

It is sometimes alleged (27) that, since twins are born at the same time, the intrauterine environment must have been the same for both before birth, even if later on their environments differ widely, and that quite conceivably it is the former that is crucial. As it happens, however, this rather gratuitous assumption reverses the actual facts. Embryological and obstetric records show that, particularly with twins developed from split ova, the position of each in the uterus, and the subsequent development, is liable to differ widely (12, pp. 123ff. and refs.).

I think, therefore, that it may be safely said that, apart from the influence of some preconceived theory, few psychologists nowadays would be inclined to contest the mere fact of mental inheritance: the most that can be plausibly alleged is that its influence is comparatively slight and distinctly elusive.

The Mode of Inheritance. The majority of those who still question the importance of mental inheritance, and many of those who support it, seem by preference to adopt entirely antiquated notions of the way in which inheritable characteristics are transmitted. If, as I have maintained, mental capacities are dependent on the physical characteristics of the brain (or, to speak a little more precisely, on the structural and biochemical qualities of the nervous system), then we should expect those capacities to be inherited in accordance with the same principles that govern the inheritance of physical characteristics; and these principles (except for obscure and apparently exceptional instances of extranuclear heredity) are essentially those commonly associated with the name of Mendel. Many British psychologists, however, feel a strong and not unreasonable prejudice against applying "atomistic theories like Mendel's" to explain the facts of mental life, and consequently, so far as they admit the possibility of mental inheritance at all, still cling to the old Darwinian principle of blended inheritance. On this view heredity means "the tendency of like to beget like" (the definition quoted by one of them from the *Oxford English Dictionary*). As a result, they commonly assume that the arguments for inheritance must consist in demonstrating resemblances between the parent and his children by means of correlations. When the two parents differ, then the child is still expected to consist in an intermediate blend of both.

The approach of the modern geneticist is the reverse of all this. As he views it, the real problem is rather to explain why in so many instances "like begets unlike." Both for the environmentalist and for the believer in blended inheritance, one of the most puzzling phenomena is the appearance, not merely of extremely dull children in the families of the well-to-do professional classes, but also of extremely bright children in families where both the cultural and the economic conditions of the parents would, one might imagine, condemn every child to hopeless failure. With the Mendelian hypothesis these anomalies are just what we should anticipate. However, the few critics who are familiar with the Mendelian explanation appear, as a rule, to suppose that it can apply only to discontinuous variations; and point out that intelligence, like stature, exhibits not discontinuous but continuous or graded variation. Hence, so they contend (sometimes citing the experiments of De Vries on "pure lines"), the apparent differences in intelligence between one individual and another must be due almost entirely to differences in environmental conditions.

Mendel himself was the first to indicate how his theory could be extended to account for this particular difficulty. When supplementing his experiments on the hybridization of peas by hybridizing beans, and (as before) crossing white flowered plants with purple, he found, that, whereas with peas the two types always sorted out with no hint of any intermediate color, with beans the offspring displayed "a whole range of hues from white to deep purple." This, he suggested, might be explained by postulating that with beans the color was determined, not by a single pair of alternative factors, but by a *number* of such pairs, each positive factor, when present, contributing a small additional amount of color. And if, as before, the recombinations are the effects of chance unions, then the resulting frequencies would obviously approximate to those of the normal curve.

However, in our early surveys of London children (5, 6), we found that, when complete age groups were tested, the distribution of intelligence departed significantly from that of a perfect normal curve: there was a swollen tail at the lower end, due to an excess of mental defectives, and a smaller enlargement at the upper end. This and other considerations led me to put forward the tentative hypothesis that innate variations in intelligence are due partly to unifactor and partly to multifactor inheritance: i.e., they result from Mendelian factors of two main kinds (no doubt overlapping), viz., (*a*) major genes responsible for comparatively *large* deviations, usually of an abnormal type, and (*b*) multiple genes whose effects are *small, similar,* and *cumulative.*

Karl Pearson (29) endeavoured to test the Mendelian theory in its multifactorial form by comparing its implications with actual figures obtained for height, arm length, and similar physical measurements, collected by himself and Alice Lee, from over 2,000 students and their relatives. The expected correlations which he deduced for various degrees of kinship were in every case far smaller than the observed coefficients. He therefore emphatically rejected the hypothsis of Mendelian inheritance, and fell back on the older theory of blending. However, in deriving his formulae and his expected values, Pearson relied on an oversimplified model. Contrary to what we now know to be the case, he assumed that the effect of assortative mating—the tendency of like to marry like—could be ignored as negligible, and that dominance would in every case be perfect. Ronald Fisher has since undertaken the rather formidable task of deducing more appropriate formulae, which allow for these and other complicating factors (18). And with these refinements the calculated correlations fit Pearson's own figures as well as could be wished.

My colleagues and I have applied Fisher's methods (suitably modified) to assessments for intelligence (12). The data were secured in the course of surveys of the entire school population in a representative London borough, and covered nearly 1,000 pairs of siblings, together with ratings for parents, and (so far as they were accessible) grand-parents, uncles and aunts, and first cousins. The final assessments for the children were obtained by submitting the marks from the group tests to the judgment of the teachers who knew the children best: where the teacher disagreed with the verdict of the marks, the child was interviewed personally, and subjected to further tests, often on several successive occasions. The assessments for the adult members of the family were naturally far less accurate. Nevertheless, in almost every case the correlations computed from the actual data agreed with the theoretical values deduced from the multifactorial hypothesis far better than with the values deduced from any other hypothesis hitherto put forward. The only appreciable discrepancy occurred in the case of first cousins. Here, as for stature, the observed correlation for intelligence is larger than the theoretical; but the difference is not statistically significant, and could readily be explained if (as has been suggested above) variations in intelligence are affected by a few major genes as well as by numerous minor genes. I may add that on sorting out figures for cousins of maternal, paternal, and mixed kinship there is also some slight evidence suggestive of sex linkage.

The Relative Influence of Heredity and Environment. In practical work, however, the question most frequently raised is, not whether differences in intelligence are inherited, nor even how they are inherited, but rather what is the relative influence of heredity as compared with environment. To an omnibus inquiry like this there can be no single answer. We can only try to determine, for this or that type of environment, for this or that population, and for this or that type of assessment, how far the observable results appear to be influenced by each of the two main groups of factors.

As Fisher's analysis has shown, formulae analogous to those used to deduce the expected correlations from the theoretical variances can also be devised for deducing the amount of the constituent variances from the observed correlations. I have ventured to amplify Fisher's methods (mainly on the lines of later work by Mather and Sewall Wright) so as to allow for unreliability and for the systematic effects of environment, i.e., of environmental influences which are correlated with those of heredity, as well as for random effects. The genetic contribution may be regarded as comprising two distinguishable portions: that due to the "fixable" component (or, as Fisher expresses it, to the "essential genotypes") and that due to the "nonfixable" (i.e., deviations due to dominance and similar influences). The data analysed consist of (*a*) the marks obtained from intelligence tests of the ordinary type taken just as they stand and (*b*) adjusted assessments obtained by the supplementary methods already described (12).

From Table 2 it will be seen that, with the crude test results, taken just as they stand, nearly 23% of the total variance appears due to nongenetic influences, i.e., to environment or to unreliability, and about 77% to genetic factors; with the adjusted assessments only about 12% (or slightly more) is apparently due to nongenetic influences and 88% to genetic factors. This of course means that the common practice of relying on tests alone—usually a group test applied once only—is by no means a satisfactory method of assessing a child's innate ability. Better assessments are obtained by submitting the test scores to the teachers for criticism or correction, and where necessary adjusting them by the methods described above. But such intensive inquiries would be too costly for routine use except in borderline cases.

Environment appears to influence the test results chiefly in three ways: (*a*) the cultural amenities of the home and the educational opportunities provided by the school can undoubtedly affect a child's performance in intelligence tests of the ordinary type, since so often they demand an acquired facility with abstract and verbal modes of expression; (*b*)

TABLE 2

Analysis of Variance for Assessments of Intelligence

Source	Unadjusted test scores	Adjusted assessments
Genetic component:		
fixable	40.51	47.92
nonfixable	16.65	21.73
Assortive mating	19.90	17.91
Environment:		
systematic	10.60	1.43
random	5.91	5.77
Unreliability	6.43	5.24
Total	100.00	100.00

quite apart from what the child may learn, the constant presence of an intellectual background may stimulate (or seem to stimulate) his latent powers by inculcating a keener motivation, a stronger interest in intellectual things, and a habit of accurate, speedy, and diligent work; (*c*) in a few rare cases illness or malnutrition during the prenatal or early postnatal stages may, almost from the very start, permanently impair the development of the child's central nervous system. The adjusted assessments may do much towards eliminating the irrelevant effects of the first two conditions; but it is doubtful whether they can adequately allow for the last.

Limitations Involved in These Conclusions

As in almost all scientific investigations, the hypothetical model which has formed the basis of our inquiry involves of necessity certain minor simplifications. In particular we have assumed, for purposes of calculation, a sharp distinction between the "major genes" of unifactor inheritance and the "polygenes" of multifactor inheritance, and have treated the latter as contributing equal and additive doses to the sum total of each child's innate intelligence. We have then supposed that the effects of the former would on the whole be excluded if the few obviously pathological cases (mostly found in special schools or institutions) were omitted from our final calculations. However, these assumptions have led several critics to accuse us of

> . . . disrupting the individual personality into atomic bits and discrete pieces which have subsequently to be joined together like a mosaic. . . . Personality [it is argued] is not a mosaic but a seamless whole, and hence the entire Mendelian hypothesis with its particulate genes, each producing a unit-character or adding another unit to the same character, is quite inapplicable to the facts of conscious behavior, and therefore to the study of mental capacity.

Objections of this kind could of course be used just as well to prove that a neuronic theory of the central nervous system is incompatible with the facts of conscious behaviour or of individual variations in ability, since nerve cells or "neural bonds" are equally "particulate." Nevertheless, even those neurologists who prefer to start from a "field theory" do not wholly reject the neuronic hypothesis (cf. 34). In both cases the difficulties raised owe their force chiefly to the fact that there is a vast series of elusive processes intervening between theory and observable results which the critic is exceedingly apt to forget. Moreover, criticisms like those just cited plainly rest on an obsolete version of the Mendelian doctrine. No geneticist today, I imagine, accepts the hypothesis of the autonomous corpuscular gene; and the genotypic endowment of the individual can only affect the phenotypic resultant through the mediation of innumerable obscure biochemical steps.

In our original papers Howard and I tried carefully to guard against recurrent objections of this type. As we pointed out, the phrase "multiple factors" may be used to cover either (*a*) relatively numerous loci each with only two allelomorphs or (*b*) a single locus (or relatively few loci) each with numerous allelomorphs, or possibly (*c*) some combination of the two. Hypothesis *b* by itself would hardly seem to fit the facts. We are inclined to think that factors of all the various types may be operative in varying degrees, and that the attempt to classify factors or genes should not be too severely pressed.

We further assumed that in all probability the influence of such factors on the individual's observable intelligence was mainly the indirect result of their influence on the development of his central nervous system, and was presumably effected by modifying growth rates. And we expressly stated that the ultimate influence of any one "gene" upon intelligence might be but one of its multifarious consequences, and possibly a comparatively remote consequence at that. Some genes may have a larger share in the final result; others a smaller; and the rough classification of genes into "major" genes and "minor" was adopted primarily with a view to simplifying our general discussion.[7] We ourselves find the "theory of chromosomal

[7] If Mather's view that "the major genes occur only in euchromatin, while heterochromatin contains only polygenes" (15, p. 151 and refs.) is eventually confirmed, there would be more adequate grounds for re-

hierarchy," advanced by Goldschmidt (25), especially attractive as a basis for the ultimate hierarchical differentiation of mental ability.

However, this is not the place to enlarge on these speculative interpretations. We fully admit that the simplifications involved in our hypothetical model mean that the figures finally deduced can be no more than approximations. But we maintain that the error of approximation, however large, will nevertheless be smaller than the amount of "unreliability" inevitably involved in all such measurements.

In any case we must repeat that the conclusions reached are at best only valid in reference to the particular conditions under which they were obtained. They would not necessarily hold good (*a*) of other mental traits, (*b*) of different methods of assessment, (*c*) of a population of a different genetic composition, or (*d*) of a population at a different cultural level: much less would they hold good if there were any subsequent change (*e*) in the present distribution of environmental and genetic characteristics, or (*f*) in the influences affecting their mutual interaction.[8]

The Inheritance of Other Mental Qualities

The mental trait in which we have been chiefly interested is the general factor of "intelligence." This, we readily concede, is but one psychological ingredient in ordinary everyday behavior. Our critics are accustomed to reproach us for apparently ignoring the rest. There are several obvious reasons why we have

---

taining a sharp distinction between the two modes of inheritance. However, this view is not universally accepted. For an alternative interpretation of the experimental results on which Mather largely relies, see the papers by Reeve and Robertson quoted in our article (12, p. 116). These writers incline more to what in the text I have called hypothesis *b*. But, as Howard and I contended, their alternative interpretation would affect only the method, not the results, of our statistical deductions.

[8] A more detailed reply to the criticisms urged by Woolf, Heim, and Maddox will appear in the forthcoming issue of the *Brit. J. statist. Psychol.*, 10, Part i.

dealt less fully with other determinants. The first and the most conclusive is that comparatively little evidence is as yet available in regard to the genetic composition of other mental factors. We have, however, in the course of our surveys, met with an appreciable number of cases which strongly suggest the inheritability of certain specific abilities and disabilities, or rather of certain ill-defined tendencies that presumably underlie such disabilities—particularly in our studies of memory, of visual, auditory, and kinaesthetic imagery, and of verbal, numerical, artistic, and musical aptitudes (5). We have encountered other cases which suggest the inheritability of general emotional instability, of temperamental qualities like introversion and extraversion, and of certain quasi-instinctive tendencies like sex and bad temper (or "pugnacity," as McDougall would have termed it). But emotional tendencies are always much more liable to be influenced and altered by postnatal experiences, and in any case undergo considerable changes during the developmental stages, so that all attempts at assessment are apt to be precarious. The most convincing evidence is afforded by the family histories of children brought up in orphanages or residential institutions (7, especially p. 14 and refs.). In all these instances the evidence, such as it is, appears to indicate that each of the characteristics mentioned, cognitive and temperamental alike, are influenced both by unifactor and by multifactor modes of inheritance—the latter tending on the whole to predominate, but in varying degrees. This after all is what is commonly found in investigations on physical characteristics. Characteristics depending mainly (though not entirely) on the development of the long bones—such as arm length, leg length, stature, and perhaps the longer dimensions of the face and head—yield correlations of much the same magnitude as those obtained for general intelligence. Others, like cephalic index, breadth at hips, and particularly facial characteristics, seem to be rather more subject to the influence of single (or relatively few) genes, which often appear to exhibit almost perfect dominance. Variations in the cartilaginous features, and particularly those depending on flesh and fat, are far more readily affected by environmental influences (for data, see 38 and refs.).

But secondly there is a practical reason for concentrating attention first of all, and chiefly, on the inheritability of general intelligence. General intelligence, as we have seen, accounts for four times as much of the variance as any other indentifiable factor. Galton's pronouncement is thus fully borne out by our own results:

> . . . general ability appears far more important than special gifts; and, where the allowance granted by nature is inadequate, the keenest will and the stoutest industry must strive in vain.

Moreover, the child's innate endowment of intelligence sets an upper limit to the best he can possibly attain. No one would

TABLE 3

Estimates for Grammar School Entries

	General population[a]	Proportions over +0.85 SD	Expected entries	Actual entries
Middle Class	20%	.40	.08 (40%)	55%
Manual Class	80%	.15	.12 (60%)	45%
Total	100%		.20 (100%)	100%

[a] The *Census of Great Britain* (1951) gives an estimate of 81.7 for males over 15, occupied or retired, falling into the classes of "Skilled," "Mixed," and "Unskilled" workers: these I have pooled in a single broad group.

expect a mongolian imbecile, with the most skilful coaching in the world, to achieve the scholastic knowledge of an average child. In the same way, no one should expect a child who is innately dull to gain a scholarship to a grammar school, or one whose inborn ability is merely average to win first class honors at Oxford or Cambridge. No doubt, in any individual case the ascertainment of this upper limit can never be a matter of absolute certainty: an I.Q. derived from tests alone falls far short of a trustworthy indication. Hence education authorities, like life insurance companies, have to follow Butler's maxim, and take probability for their guide. They cannot, however, afford to risk a lavish expenditure on cases where there are fairly heavy odds against success.

Nevertheless, it must be owned that, in Britain at any rate, the existing administrative machinery is far less efficient than it might be. In an inquiry made just before the introduction of the Education Act of 1944 we found that "approximately 40 per cent of those whose innate abilities are of university standard are failing to reach the university" (9). The new arrangements proposed by the act, and the postwar changes in the economic circumstances of the various social classes, have appreciably altered the entries to "grammar schools"; but it is too early to say how far they have affected the composition of the universities. For grammar schools the entries vary considerably from one area to another and even from year to year. The general aim is apparently to provide a grammar school education for the brightest 20% of each age group; this corresponds to a borderline of about 113 IQ. On this basis the round figures in Table 3 represent estimates deduced from the information so far available.

A more intensive survey of two contrasted areas ("industrial" and "prosperous") has just been published by J. E. Floud and her colleagues. Briefly their conclusion is that, in both areas, gross material handicaps have been greatly lessened, though not entirely eliminated. Nevertheless, in their view, the "problem of social wastage" is by no means overcome. The fact that, even today, a large proportion of the children from the working classes fail to reach the higher levels of scholastic instruction Floud herself attributes to the way in which the "educational ladder" is still widely regarded as a "middle class prerogative," to be anxiously watched and jealously preserved (21).

I myself should be inclined to look rather to a difference in the aims and aspirations which are traditional in the different classes. The consequences seemed clearly revealed by the differences in the after-careers of LCC scholarship winners from the lower and the middle classes respectively whom my colleagues and I have been able to follow up (cf. 5, 9, and refs.).[9] The figures show that the abler children from the working classes, even when they have obtained free places or scholarships at secondary schools of the "grammar" type, frequently fail to stay the course: by the time they are sixteen the attractions of high wages and of cheap entertainment during leisure hours prove stronger than their desire for further knowledge and skill, and easily overcome their original resolve to face a long prospect of hard sedentary work *in statu pupillari.* As a headmaster of a secondary school which receives both types of boy has put it:

> The working class parent wants his boy to be "selected" chiefly to prove that "his child is as good as anybody else's"; having done so, he will withdraw him at the earliest possible opportunity: I do not press them, for the presence of those whose chief interest in life is television, Hollywood films, football pools, "the dogs," and "the girls" has not improved the tone of my school. And too often these "ignobler attitudes and crude ambitions" are shared by, and encouraged by, both parents and friends.

On the other hand, if I may quote the report of an experienced university teacher:

> That section of the middle class which seeks, by paying fees that it can ill afford, to assist its children to climb, *via* a "public

[9] Additional evidence is furnished by the recent report of the Central Advisory Council for Education (England) on *Early Leaving.* In 1953, at the age of entry to the grammar school (11 plus), 66% of the top intelligence group were the children of manual workers; at the end of the grammar school period the proportion had fallen to 47%. Of those who drop out, some are "premature leavers," i.e., they leave voluntarily, as soon as they are legally able to do so; others fail to reach the top form and so leave at 16: of these many even fail to secure a school leaving certificate. In such cases, the award of a special place in the grammar school would seem itself to have been a wasted effort. Nevertheless, in the long run there may be certain compensating advantages to the community as a whole: even the early leavers may have gained something from the higher education to which they have been for a while subjected, and it would surely be a misfortune were all the brightest youngsters to forsake the social class into which they were born. Their continued presence there must help, not only to elevate its tone, but also (a point too often overlooked) to prevent its genetic constitution from being wholly depleted of its better elements.

> school" or an independent "grammar school," to a University education and a good honours degree in the humaner subjects, is animated by a traditional morale which is comparatively rare in children and parents from the other classes: those who nowadays come here on grants, at no cost to themselves or their parents, are, on the whole, most irregular attendants, and the least satisfactory students.

Or, to quote a still more recent pronouncement of the High Master of St. Paul's:[10]

> The parents are themselves imbued with four traditional ideals which they hand on to their posterity: self-discipline, a community spirit, the Christian religion, and a readiness to accept social responsibility even at the sacrifice of material enjoyments —a genuine *noblesse oblige*.

Underlying all these differences in outlook I myself am tempted to suspect an innate and transmissible difference in temperamental stability and in character, or in the neurophysiological basis on which such temperamental and moral differences tend to be built up. Tradition may explain much: it can hardly account for all. However, it would be idle to pursue such speculations here in the absence of more adequate data.[11]

[10] A. N. Gilkes, *Independence in Education,* 1957.

[11] Since my lecture was written and delivered, some further information has become available in the data published by the Association of Universities of the British Commonwealth (*Report of an Inquiry into Applications for Admissions to Universities,* 1957). It appears that at Oxford and Cambridge only 12% of the men and 7% of the women had fathers in manual occupations; at other universities the proportions were somewhat higher, namely, 31% and 19%, respectively. Among the general population 72% of the population would fall into the manual category. The report points out that to a large extent the decline must occur during the school period, and the real reason, it is suggested, is more often lack of zeal than lack of the requisite ability. "Differing environmental influences and aspirations seem mainly responsible for the fall from 66% at the point of entry to the non-fee-paying grammar school and 36% at the point of university entry. . . . Discrimination against those of humble social origin can be virtually ruled out."

Most education authorities nowadays tend to use what are misleadingly termed "intelligence tests" as tests of suitability for a grammar

Need for Further Research

To my mind the most pressing need at the moment is for more extensive research. Hitherto the most active investigators have been research-students, with little or no experience of the ways of children or the conditions of the classroom. For the requisite facilities, and for any supplementary information they may want, they have to rely on the goodwill of the busy teacher. And one important factor is still more often overlooked: the motivation of the children themselves. This takes two obvious forms. There is first what may be called direct or short distance motivation. As actual trial has repeatedly shown, neither pupils nor students are likely to exert their full powers in a test conducted merely in the interests of someone else's research; but when the examinees' entrance to a grammar school or university depends on their performances, the average score may rise 5 to 10 IQ points higher. Secondly, there are the effects of indirect or long distance motivation: the influence of parental attitudes, of teachers' exhortations, and (most of all perhaps with British lads) of the social pressure that arises from the opinions and the comments of their school fellows, and from the tone and atmosphere of the school or class to which the pupil belongs. It is therefore hardly surprising if the results obtained in mere academic researches are at times disappointing, and evince a comparatively low reliability.

I myself had the good fortune to be appointed a member of the school inspectorate, and so not only acquired a firsthand knowledge of the schools, teachers, and pupils, but enjoyed full authority to interrupt the timetable, examine private rec-

---

school. (Such tests, as I have argued elsewhere, should rather be termed "tests of scholastic aptitude.") The original intention was that the "intelligence test" should, so far as possible, assess innate ability. Bright children with special verbal or literary abilities and interests were then to be allocated to grammar schools; and those with practical or mechanical abilities and interests, to technical schools. But for the majority the age of 11 plus is too early to make a satisfactory distinction. Hence the brightest still seem likely to be sent to secondary schools of the grammar type rather than to a technical school, even if more suited to the latter.

ords, and requisition whatever information might be wanted. Here therefore I must again record my indebtedness both to the London County Council for facilitating these inquiries and financing their publication, and to the teachers, social workers, and school medical officers, who rendered such wholehearted cooperation in all our investigations. It is to be hoped that other education authorities will in the near future perceive the practical value of systematically organized inquiries of this kind, conducted on an official basis by those who know the schools from inside.

The basic researches must be carried out on children rather than on students or adults. With adults, it is much harder to achieve accurate assessments; innate tendencies are already obscured and over-laid; and detailed family histories difficult to obtain. In allocating adults to appropriate occupations, whether in civil life or in the fighting services, innate capacities are of little consequence, and accordingly tend to be ignored. As a result the severest critics of researches on heredity are generally those whose experience of mental testing has been gained in the adult field. This was manifest during the symposium recently arranged by the Psychological Section of the British Association. Several contributors, for instance, echoed remarks like those of H. E. Jones that "potential investigators should now be advised that the nature-nurture controversy has been shown to be an unproductive field of research" (14); and similar conclusions have more recently been voiced by writers like Maddox, Renshaw, and others, who deplore what they call "the resurrection of the nature-nurture controversy" as "wholly anachronistic" (27, 32).

But whether or not this is now the prevailing view among psychologists, it is not the opinion of leading geneticists. Goldschmidt, Snyder, Calvin Hall, and Calvin Stone all deplore the neglect of genetic inquiries by psychologists of the present day. Goldschmidt in particular has criticized "the extreme belief of contemporary psychologists in the power of environment"—a belief which he ascribes to the "doctrine of universal conditioning, which has, until lately, dominated the behaviourist school" (17). In this country eminent authorities like R. A. Fisher (19, 20) and Fraser Roberts (33) have not only urged the importance of genetic studies for psychology, but have made noteworthy contributions of their own.

Mere statistical inquiries, however, can never suffice. In this country one of the most obvious needs is a series of intensive studies of able boys and girls similar to those already carried out on the backward, the defective, the neurotic, and the delinquent. The brilliant investigation of 1,000 gifted children, started by Terman and his colleagues over 30 years ago, and so assiduously followed up, furnishes an admirable model (37, 38). But nothing like conclusive answers can be given to the questions here raised far more extensive research has been undertaken on the fundamental problems of psychogenetics. An adequate understanding of the basic processes can be secured only if we start with carefully planned experiments on lowlier creatures, where pure strains can be obtained, breeding controlled, and successive generations more speedily raised. When we know more about the genetics of intelligence in animals, we may be able to construct with greater confidence a more exact hypothesis regarding the transmission of intelligence in man.

Social Implications

After all, the practical importance of the issues involved is so profound and so far-reaching that it would be fatal to dismiss the problem as "unproductive" or "anachronistic" without attempting to settle it one way or the other. No democratic state can afford to pass it by. If in the end the views of the early pioneers should turn out to have been approximately correct—if "innate intelligence varies between limits at least as wide as IQ's of 50 and 150," and if too "the average intelligence of the general population at maturity is little if at all above that of an average child of thirteen or thirteen-and-a-half"—then the bearing of genetic variation on the national and social questions of the present day would be all too obvious.

In our laudable eagerness to improve the lot of our own generation, we have been tempted to close our eyes to the effects of present policy on the generations to come. The overall efficiency of the citizens who make up a nation or a state must in the last resort depend on what has been called its "chromosomal pool." Improved environmental amenities can of themselves ensure no lasting results; but the changes in a nation's genetic constitution are likely to prove irreversible. Throughout almost the entire Western world, and to a less extent in other areas as well, the last half-century or so has witnessed radical modifications in the conditions that previously governed marriage and fertility. Increased mobility, enhanced freedom, new means of production, new methods of government, the progressive reduction of the traditional barriers separating different people and different social groups, and, in our own country, the recent extensions of the educational ladder, all these transformations are visibly disturbing the stability alike of classes and of races. Almost inevitably they must alter the genetic constitution of what have hitherto been the dominant nations and the dominant stocks within those nations.

From history and from past experience we know full well how changes in the balance between inbreeding and outbreeding[12] can affect the later destinies of populations, remoulding them sometimes for good, sometimes for ill. Geneticists have repeatedly drawn attention to the processes at work, and indicated the need for studying the possible effects of contemporary trends (cf., e.g., 16, pp. 49ff. and 20, pp. 170ff.). Yet current sociological textbooks seem almost wholly to ignore them. Surely it is high time that the social psychologist should now take up the question and plan a systematic series of investigations on the intricate but highly important issues involved.

[12] British readers will find it instructive to compare current copies of such works as *Who's Who* with those of (say) 60 years ago, and note the changes in the pedigrees of the British clergy (who in the nineteenth century were probably the most inbred profession in the country) and in the ancestry of members of the British cabinet and Civil Service during that period. The recent work of Ginsberg, Glass, and Moss has shown very plainly how the extension of the educational ladder is altering the composition of the different social groups, though none of the writers has considered the possible genetic consequences. The most illuminating discussion is still that contained in the concluding chapters of Fisher's book on *The Genetical Theory of Natural Selection* (1930), especially Chap. X, "Reproduction in Relation to Social Class."

May I therefore conclude by endorsing, in the strongest possible terms, the verdict of Stone (17), quoted by Bingham in the last letter he wrote to me:

> It is a matter of shame and regret that only an amateurish beginning has been made by psychologists in applying modern genetic methods to fundamental study in the nature-nurture area.

And may we hope that in the near future Bingham's "deepest wish" may be fulfilled, and that "a small band of enthusiasts will come forward to explore afresh this urgent and many-sided field of research."

References

1. Bagley, W. C. *The educational process.* New York: Macmillan, 1912.
2. Burks, Barbara. *27th Yearbook National Society for the Study of Education,* 1928, *1,* 219.
3. Burt, C. Experimental tests of general intelligence. *Brit. J. Psychol.,* 1909, *3,* 94-177.
4. Burt, C. The inheritance of mental characteristics. *Eugen. Rev.,* 1912, *4,* 168-200.
5. Burt, C. *Annual reports of the psychologist to the London County Council.* London: London County Council, 1914-1931.
6. Burt, C. *The distribution and relations of educational abilities.* London: P. S. King & Son, 1917.
7. Burt, C. Is the doctrine of instincts dead? *Brit. J. educ. Psychol.,* 1941, *11,* 155-172; 1943, *13,* 1-16.
8. Burt, C. The education of the adolescent: The psychological implications of the Norwood Report. *Brit. J. educ. Psychol.,* 1943, *13,* 126-140.
9. Burt, C. Ability and income. *Brit. J. educ. Psychol.,* 1943, *13,* 83-98.
10. Burt, C. The differentiation of intellectual ability. *Brit. J. educ. Psychol.,* 1954, *24,* 76-90.
11. Burt, C. The evidence for the concept of intelligence. *Brit. J. educ. Psychol.,* 1955, *25,* 158-177.

12. Burt, C., & Howard, M. The multifactorial theory of inheritance and its application to intelligence. *Brit. J. statist. Psychol.*, 1956, *9*, 95-131.
13. Campbell, F. *Eleven plus and all that.* London: Watts, 1956.
14. Carmichael, L. (Ed.) *Manual of child psychology.* New York: Wiley, 1946.
15. Darlington, C. D., & Mather, K. *The elements of genetics.* London: Allen & Unwin, 1949.
16. Darlington, C. D., & Mather, K. *Genes, plants, and people.* London: Allen & Unwin, 1950.
17. Dunn, L. C. (Ed.) *Eugenics in the twentieth century.* New York: Macmillan, 1921.
18. Fisher, R. A. Correlation between relatives on the supposition of Mendelian inheritance. *Trans. roy. Soc., Edin.*, 1918, *52*, 399-433.
19. Fisher, R. A. The causes of human variability. *Eugen. Rev.*, 1919, *11*, 213-220.
20. Fisher, R. A. *The genetical theory of natural selection.* Oxford: Clarendon Press, 1950.
21. Floud, J. E., Halsey, A. H., & Martin, F. M. *Social class and educational opportunity.* London: Heinemann, 1956.
22. Freeman, F. N. *Mental tests.* New York: Houghton Mifflin, 1926.
23. Galton, F. *Hereditary Genius.* London: Macmillan, 1869.
24. Garnett, M. *Education and world citizenship.* Cambridge: Cambridge Univer. Press, 1921.
25. Goldschmidt, R. B. *Theoretical genetics.* Univer. California Press, 1955.
26. Lashley, K. S., & Clark, G. The cytoarchitecture of the cerebral cortex of *Ateles:* A critical examination of architectonic studies. *J. comp. Neurol.*, 1946, *85*, 223-305.
27. Maddox, H. Nature-nurture balance sheets. *Brit. J. educ. Psychol.*, 1957, *27*, 166-175.
28. Newman, H. H. *Twins and super-twins.* London: Hutchinson, 1942.
29. Pearson, K. On a generalized theory of alternative inheritance with special reference to Mendel's laws. *Phil. Trans.*, 1904, *203*, 53-87.
30. Piaget, J. *The psychology of intelligence.* London: Kegan Paul, 1950.
31. Reeve, F. C. R., & Waddington, C. H. (Ed.) *Quantitative inheritance.* London: H. M. Stationery Office, 1952.

32. Renshaw, T. Burt's concept of intelligence. *Brit. Psychol. Soc. Bull.*, 1957.
33. Roberts, J. A. F. *An introduction to medical genetics.* Oxford: Oxford Univer. Press, 1940.
34. Sholl, D. A. *The organization of the cerebral cortex.* London: Methuen, 1956.
35. Spearman, C. *Abilities of man.* London: Macmillan, 1927.
36. Spencer, H. *Principles of psychology.* London: Williams & Norgate, 1870.
37. Stone, C. P. Methodological resources for the experimental study of innate behaviour as related to environmental factors. *Psychol. Rev.*, 1947, *54*, 342-347.
38. Tanner, J. M. The inheritance of morphological and physiological traits. *Ap.* Sorsby, A. *Clinical genetics.* London: Butterworth, 1953.
39. Terman, L. M. (Ed.) *Genetic studies of genius.* Stanford: Stanford Univer. Press, 1925, 1947.
40. Terman, L. M. The discovery and encouragement of exceptional talent. *Amer. Psychologist,* 1954, 9, 221-230.
41. Thorndike, E. L. *Educational psychology.* New York: Teachers College, Columbia University, 1914.
42. Thurstone, L. L. *The vectors of mind.* Chicago: Univer. Chicago Press, 1935.
43. Watson, J. B. *Behaviorism.* London: Paul, Trench Trubner, 1931.

Comparative Psychological Studies of Negroes and Whites in the United States[1]

RALPH MASON DREGER
AND KENT S. MILLER

This report is a review of published psychological studies which involve a comparison of Negroes and whites in the United States during the period in the main from 1943-1958. The period chosen covers work reported following Klineberg's (1944) review. Where called for, for completeness, we have selected studies from previous years. Most of the literature to which reference is made reports direct comparisons between Negroes and white groups. However, in some instances direct comparisons were not made, but inferences could be drawn—e.g., where a test standardized on a white population has been administered to a Negro group. For the purpose of inclusion in this review, research has been considered relevant wherever the individual authors have stated that Negro-white comparisons were made or wherever the stated population in studies utilizing white-standardized tests is Negro. Thus the populations compared have consisted of varying degrees of "racial pureness." We have reached into studies of physical development and sociological research where we have deemed it important for psychological completeness.

Two earlier reviews of racial psychology (primarily concerned with mental differences) have appeared in this Journal (Garth, 1925; Woodworth, 1916). For comprehensive discussions of the Negro and his problems the reader is referred to Myrdal (1944), Klineberg (1944) and Canady (1946).

In addition to these references, racial differences are discussed in many general works, some prior to our period (An-

[1] Prepared with the assistance of a grant from the Human Genetics Fund.

From *Psychological Bulletin,* 1960, *57,* 361-402. Reprinted by permission of the author and the American Psychological Association.

astasi & Foley, 1949; Bendix & Lipset, 1953; Benedict & Weltfish, 1943; Boyd, 1950; Dunn & Dobzhansky, 1946; Frazier, 1957; Garth, 1931; Ginzberg, 1956; Kardiner & Ovesey, 1951; Knox, 1945, 1949, 1952; Lindzey, 1954; Montagu, 1952; Sarason & Gladwin, 1958; Tyler, 1956). We have endeavored in the following review to confine ourselves to the experimental literature; only in the sections on values and attitudes and on emotional disturbances have we departed to some extent from the rule. Over 200 additional references were reviewed, but not included among our references, either because they have been adequately covered in other reviews, or they lacked significance or relevancy. Less than a dozen possibly significant theses were abstracted in various places, but were not studied by the writers.

Physical and Motor Development

Physical Status of Infants and Children

Gestation, birth, and early physical development may have decisive influence upon later psychological functioning but are of importance in themselves. Even later infancy is measured psychologically in terms of fine and gross muscle coordinations. Accordingly, we summarize briefly studies of physical and motor development of children.

Brown, Lyon, and Anderson (1945) report a significantly higher rate of prematurity among Negro infants than among whites if criterion of birth weight is used as a standard. However, if allowance is made for the generally known fact that Negro infants weigh less at birth (for whatever reasons), the difference in prematurity rates disappears.

White infants are on the average longer than Negro infants if the results of a 1943 survey (Meredith, 1943) of North American research still hold. The possibility is that some change occurs as socioeconomic differences are reduced. At least, such an inference could be made from studies conducted in which socioeconomic levels have been held constant (Rhoads, Rapoport, Kennedy, & Stokes, 1941; Scott, Cardozo, Smith, & DeLilly, 1950). Rhoads et al. (1941) followed white and Negro children from four months to four years in the outpatient department of a children's hospital. From the same lower socioeconomic groups, Negro infants actually tended to be taller from about nine months of age. Body weight and head circumference, eruption of teeth and motor development of both groups were about the same. Negro children tended to have longer legs, shorter trunks, smaller body width at chest and pelvis, lower hemoglobin count, and better teeth. No differences in height and weight were observed by Scott et al. when measurements of 654 Negro infants in the first year of life from the lower middle class were compared with those of white infants from comparable socioeconomic levels. Michelson (1943) concluded that the weight patterns are very similar when individuals are placed on the same dietary regime.

Psychomotor Development

Tests of infant intelligence have been suspect (possibly unjustifiably) for many years as predictors of subsequent intellectual performance. Nevertheless, as measures of comparative psychomotor developmental level, they perform a useful task.

The most widely known study is that of McGraw (1931) which was reported prior to the period covered by this review. In a comparison of Negro and white infants in a Southern community on the Buhler Baby Tests, McGraw discovered that white babies exceeded Negro babies in performance. The former were also superior in height and weight.

Pasamanick (1946), contending that McGraw was measuring differences in opportunities for adequate care and feeding rather than native abilities, endeavored to overcome objections to the previous work by utilizing New Haven infants somewhat comparable in weight and height factors. Comparison on an infant development scale suggested that differences are not between groups so much as within racial groups. Negro babies proved to be somewhat accelerated relatively in gross motor behavior.

Inadequacies of sampling, as well as the more general difficulty of estimating skin color subjectively, tend to invalidate Pasamanick's conclusions. His infants may very well have been

TABLE 1

Comparisons of Infants and Children in Psychomotor and Physical Development

Author(s)	Topic	Subjects	Results	Comments
Williams & Scott (1953)	Development (Gesell)	W low S-E[a] status, N high S-E status	W>N	Authors attribute to differential child-rearing, e.g., greater permissiveness in lower class
Scott, Ferguson, Jenkins, & Cutler (1955)	12 neuromuscular steps	Infants	N>W	N from low S-E class >N from higher S-E class up to 35th week, same following that age
Irwin (1949)	Speech	Infants:		Professionals' children have steeper parabolic curves of phonetic development than laborers' children
		1-10 days	W≅N	
		1st 30 mos.	W≅N	
Moore (1942)	Eye-hand coordination	Preschool	W≅N	From 24-35 mos. N child superior to W
Espenschade (1946)	Coordination and agility	10th grade girls	W≅N	
Ramsey (1950)	Pubertal changes	Adolescent boys	W≅N	

[a] Socioeconomic.

comparable for other reasons than the quality of diet during and after pregnancy to which Pasamanick attributes his results.

In a follow-up study utilizing the same Negro infants Pasamanick and Knoblock (1955) determined that depressing influences on the Developmental Quotient of exogenous factors were not important up to at least two and one half years of age. Whatever methodological difficulties are suggested by the original study, the fact that the group of Negro infants maintained their DQs for so long a period of time indicates that they were maintaining their relative developmental standing, including intellectual functioning.

What appears to be better-controlled research by Gilliland (1951) employed the Northwestern Infant Intelligence Scale rather than the Gesell scales (understandably, since Gilliland developed the former). In three separate studies, at least one of which controlled for major variables, Negro infants in Chicago had IQs as high as or higher than white infants. The results may, the author suggests, be interpreted as manifestations of the greater maturity in motor behavior Negro babies are known to have, or as a resultant of crowded quarters in which Negro babies receive greater social stimulation. Inasmuch as the former explanation does not seem to be found when socioeconomic variables are controlled, the latter may be more acceptable.

The remainder of the comparisons of psychomotor and physical development is summarized in Table 1. Generally, the results indicate white and Negro children represent the same populations in respect to each of the variables measured.

In a study obliquely related to white-Negro comparisons, Codwell (1949) separated a Negro group into three groups varying in degrees of "Negroidness." Composite motor functioning did not change from one group to another, implying no differences between those more Negroid and those more white.

Overview of Physical and Motor Development

On some physical measurements of children, especially anatomical measurements of substructures, a racial difference appears to exist between whites and Negroes. After equating for socioeconomic variables, investigators on the whole find that differences in psychomotor functioning tend to disappear.

Psychophysical Functions

A few studies have endeavored to compare white and Negro subjects on variables which may roughly be classed as biological, but which have, we know, psychological components of varying degrees of importance.

Measurements of pain thresholds by Chapman (1944) suggest that Negroes perceive radiant heat at a lower threshold than do whites. In this experiment 18 Negroes and 18 Americans of southern European ancestry were employed; but even though for psychophysical experiments a small *N* would be satisfactory, the use of psychoneurotics as well as normal persons might bias the results for a normal population. Further, only raw data are given without appropriate statistical treatment.

Among selectees and inductees, 2,200 Negro and 21,000 whites, it was revealed that more colored selectees had normal vision in each eye than did whites. No decrease in the differences was noted with age. It was also determined that the poorer eye fared better among Negroes than the better one did among whites. Karpinos (1944) who reports the study gives no reason for the findings. In the Canal Zone, Covell (1950) discovered that Negroes develop presbyopia 5 to 10 years before whites do. Other than a higher incidence of lues and tuberculosis, the usually accepted causes of presbyopia do not seem to offer an explanation for the differences. Personal insecurity and social inequality are suggested by the author to be of fundamental importance in setting off organic changes within the lens. Peripheral absolute visual thresholds are lower under dark adaptation in Negroes and become progressively higher as the pigment of the fundus gets lighter or is reduced, according to a study by Helson and Guilford (1933). As the center of vision is approached, the difference grows less (as it does between white persons with different colored eyes). Woisika (1944) concludes that the Negro is superior to the white in dark adaptation. Correlation of age and dark adaptation (Negro mean age = 39, white = 46) may vitiate the conclusion.

Some racial differences appear in psychophysical functions. Much more research of the type Eysenck, Granger, and Brengelmann (1957) have done on perceptual processes and mental illness will have to be done in order to discover what part genetic factors play in these differences.

Intellectual Functions

Shuey (1958) has reviewed the literature comparing Negro and white intelligences at least as far back as 1913. We shall not endeavor to cover the same ground. In her text are found valuable tabular comparisons of Negroes and whites for various age groups, the armed forces, gifted and retarded, delinquents and criminals, and racial hybrids. (Actually a number of studies reported in the literature are of hybrids even though not recognized as such.) Shuey's bibliography and résumés are a must for serious students, for she has not only gathered together the better known studies, but has ferreted out otherwise obscure and inaccessible articles and theses.

The usefulness of Shuey's otherwise excellent work is limited by what appears to be a polemic attitude. Her book seems to be an attempt to prove a nonegalitarian hypothesis rather than being strictly a review of literature. In this case Shuey does the same rationalizing from an hereditarian standpoint that Klineberg (1944) did in his earlier "review" from an environmental standpoint.

North (1957) likewise surveys the literature and comes to an opposite conclusion from the one Shuey reaches. Whereas the latter concluded that all the evidence points ". . . to the presence of some native differences between Negroes and whites as determined by intelligence tests," North maintains that there is no proof of biological inferiority or that the Negro's potentials for educational and cultural development are more limited than the white person's. North coverage of the literature is considerably less extensive than Shuey's.

The following discussion on comparisons of intelligence

between the two racial groups endeavors to supplement Shuey's work especially and correct it where it is in patent error.

Children and Adolescents

Young Children. Aside from infant studies cited previously, studies of preschool children in which comparisons are made between whites and Negroes are sparse. Shuey (1958) cites only nine altogether and only five reported in 1944 or after. On the whole, young Negro children score lower than whites. But the differences are very much less than in older groups; and in all of the reports in which average IQs are given, Negroes average well within the normal IQ range for whites. Shuey offers several explanations of the discrepancy between the results for preschool and school children, including inadequate sampling of preschool, the relative invalidity and unreliability of tests at younger ages (although some of the lowest Binet standard errors are found in the late preschool years), more verbal and abstract tests in the school years; she suggests that mental growth curves may not be the same for both races, and that IQs may be less affected by environment in the preschool years.

Special comment is called for with respect to two investigations. Brown (1944) compared Minneapolis kindergarten children on the Binet, Form L, discovering that the Negro mean of 100.8 was significantly lower than the white mean of 107.1. The white children in Occupational Classes VI and VII (Minnesota scale) averaged about the same as Negro children. Shuey takes exception to Brown's conclusion that at "nominally similar" socioeconomic status Negro children are not inferior to whites. Her observation that Brown has small *N*s in Levels VI and VII does not constitute an objection from a statistical standpoint. Nevertheless, Brown's conclusion would have been stronger if he had, first, differentiated his Negroes by class without assuming their occupational status and, second, employed an analysis of variance design.

The other study to which special attention must be paid is that of Anastasi and D'Angelo (Anastasi & D'Angelo, 1952;

D'Angelo, 1950). Five-year-old children in mixed and unmixed neighborhoods in New York City were administered the Goodenough Draw-a-Man Test and studied for language development in spontaneous conversation recordings. IQs were 101.8 and 101.5 for Negro and white children, respectively. Language development appeared to be somewhat more advanced for white than for colored children. This important study is dismissed by Shuey as permitting no generalizations, because selection of subjects appeared to be biased. The crux of the matter is whether all qualified subjects were utilized or only certain selected ones. Anastasi[2] points out Shuey's error in misjudging the selection procedure, and provides a satisfactory answer to Shuey's objections. With due recognition of the limitations of the Goodenough as a test of intelligence we may yet regard Anastasi and D'Angelo's results as a challenge to nativist theories of intellectual differences between the races.

Older Children and Adolescents. Little question need be raised at this time about the results of testing school children. Almost all evidence points to inferior performance of Negroes on tests of either the more traditional variety or those tending to be more "culture-free" or "culture-fair." What Shuey (1958) has done in this respect is not to present startlingly new conclusions, but to marshall data which have been more or less familiar to scholars for many years. Our purpose here, then, is to re-examine some of the data and present material not covered by Shuey, rather than to repeat what Shuey has done with a fair degree of thoroughness.

McGurk's studies. Evidencing a long time interest (McGurk, 1943) in Negro-white comparisons of intelligence, McGurk (1951, 1943a, 1953b) has received considerable publicity in both scholarly and lay circles because of research which seems to undermine the thesis that cultural factors rather than native endowment factors account for differences between whites and Negroes in intellectual functioning. Two hundred thirteen Negro and 213 white youths were matched for age and curriculum and for social and economic variables by means of a shortened form of Sims' Record Card. McGurk found that even

[2] Anastasi, Anne. Personal communication, July 19, 1958.

when socioeconomic variables were thus controlled, the mean scores of whites and Negroes differed significantly in the direction usually reported. Also, when the specially devised test items were separated into "cultural" and "noncultural," the differences were greater on the noncultural questions than on the cultural questions, contrary to environmentalist indications. Further, as socioeconomic status increased, the differences between Negroes and whites increased rather than decreased, again contrary to expectations from a cultural theory of differences in intelligence.

One point at which McGurk's analysis might be misleading to the statistically untrained has been pointed out by Long (1957). McGurk states that 25% of Negroes overlap whites, from which it might be concluded that only 25% of the specified Negro population have scores in common with the specified white population. Actually, analysis of McGurk's data shows that 91% of the Negroes have scores in common with the whites. Inasmuch as McGurk has addressed himself to the lay public, he should make abundantly clear that "overlap" is used in the technical sense of exceeding (or for an upper distribution, falling below) the mean or median of another group.

McGurk does not claim, though he seems to imply, that the superiority of whites on intelligence tests results from innate factors. In the final analysis the implication may be correct, but research such as McGurk's cannot establish it. Indeed, we do not see how the issue can be resolved by any number of ingenious methods of equating for social and economic variables. The various indices of socioeconomic status already devised or those at present conceivable on the same principles are intended to distinguish social *classes* from one another. How they can be employed to compare individuals in different *castes,* except very roughly, is difficult to see.

In actuality, not only in the South (Dollard, 1949), but North as well (Brown, 1944; Long, 1957) whites and Negroes comprise separate castes; they are not merely representatives of different classes. In the state of Florida where the writers reside there are a number of Negroes whose social and economic statuses exceed those of most white persons. These Negroes, however, cannot yet sit in the same seats on public transportation (in most places), go to the same hotel, restaurant, club, school, church, social events, or even restrooms. Although some of these strictures do not hold in Northern states, attitudes regarding intermarriage and the more personal forms of social intercourse do not appear greatly different from those held in the South. From the Early Childhood Project (Radke, Trager, & Davis, 1949) we learn that in Pennsylvania, from which state McGurk drew part of his sample, children discern within at least the first four or five years of life their social and ethnic roles, with attendant supervaluations or devaluations of self and performance expectations. Interlocked with caste variables are those which influence performance, such as the color of the investigator (Trent, 1954), which in turn may be related to deterioration of intellectual performance under anxiety-provoking conditions (Beier, 1951; Hammer, 1954).

We wish to emphasize that we are not taking sides at this point in the heredity-environment controversy in relation to intellectual differences; we believe both camps (e.g., Canady, 1943, for environmentalists) have mistakenly assumed that if the two racial groups are equated in terms of social class and economic variables that a definitive answer can be found concerning even the *relative* weighting of innate or acquired factors. Involved here are different dimensions (possibly correlated, of course) not merely different quantities along the same scale. Quibbling with McGurk over minor points of methodology should not obscure the value of the valiant attempt he has made to test hypotheses proposed by the environmentalists. The error lies in the assumptions both he and his opponents make.

Studies on the WISC. Only one investigation (Young & Bright, 1954) is cited by Shuey under WISC studies. Not surprisingly in view of its standardization, the WISC was found inappropriate for testing Southern Negro children from 10 to 13 years of age.

Another more extensive study has come to the writers' attention, that of Caldwell (1954). Four hundred and twenty Negro children were tested ranging from 6 to 12 years of age, with equal numbers of males and females, drawn from towns in five deep South states and randomly selected from school rosters. One examiner tested 342 of the subjects. According to the report, "excellent rapport was obtained," although the means of establishing rapport appear inadequate especially in the light of the conspicuous Southern accent of the chief white examiner. Caldwell's hypothesis that a difference exists between Southern Negro children and the white standardizing group was, as might be expected, borne out. Socioeconomic class influence was probably strong, with 75% of the subjects in the lowest third of SE groupings. Nevertheless, the Full Scale IQ mean of 85.52 is considerably higher than that obtained in the Young and Bright study (mean = 67.74). (Something is wrong with the standard deviations reported on p. 18 of Caldwell's dissertation: *SD*s ranging from 0.63 to 1.4 sound more like standard errors, but do not seem to jibe with either the WISC Manual's, the data of Young and Bright (1954), or the spread one might expect.) In this investigation the suggestion is also made that cultural bias results from using the WISC, standardized as it was on a white population.

"Culture-free" or "Culture-fair" studies. McGurk's attempt to differentiate cultural and noncultural tests from one another is only one in a series which began with the use of the Army Beta in the first World War. Although the first performance tests were intended to overcome literacy handicaps, they became in time useful for estimating intelligence across ethnic, racial, and cultural barriers. In Table 2 are summarized endeavors to compare Negro and white children by means of tests which were either, like the Beta, adapted for cross-barrier use, or specifically constructed to eliminate cultural factors. Some of these studies are cited by Shuey, but are included here as essential for understanding this area of comparison.

As far as we have been able to tell, no one has employed Cattell's "culture-free" test (Cattell, 1940, 1951; Cattell, Feingold, & Sarason, 1941) for comparing Negroes and whites. This test has yet to prove itself, however, as even "culture-fair," to say nothing of "culture-free." Marquart and Bailey (1955) present convincing evidence that the "culture-free" test has sufficient variance associated with the Binet to suggest they are measuring much the same capacities. Scale 1 corre-

TABLE 2

Comparisons on "Culture-Free" or "Culture-Fair" Tests

Author(s)	Test	Subjects	Results	Comments
Woods, Boger, & Holman (1954)	Beta (Lindner-Gurvitz Scale)	N adolescents, delinquents, nondelinquents	Subtest 4 low	Authors conclude aim of subtest equality not fulfilled or cultural factors deprive N on subtest 4
Woods & Toal (1957)	Revised Beta	N & W adolescents matched for IQ	W>N on Paper Formboard, Drawing Completion; N≥W on Digit Symbol, Visual Comparisons	Authors suggest Detection of Errors and Drawing Completion are culturally loaded—probably *ex post facto* theorizing
Love & Beach (1957)	Davis-Eells Games	N & W from three S-E levels	W>N	Differences obscured by N's being in two lower S-E levels only
Fowler (1957)	3 "culture-controlled" tests	N & W, ethnic & nonethnic	W>N	Positive correlation also found between conventional and "culture-controlled" tests and S-E status
Hammer (1954)	CTMM (non-language)	N children and adolescents	N<W norms	Results same with both language and nonlanguage factors
Newland & Lawrence (1953)	Chicago Non-Verbal Exam.	E. Tenn. N. School children	N<W norms	Differences 2, 3, and more years
Coppinger & Ammons (1952)	Full-Range Picture Vocabulary Test	N children from 2 La. Parishes	N<W about two years	At upper age levels urban N children's mean scores≅W norms
Hess (1955)	"Culture-fair" test of 16 items	High & low status W, low status N	W high status>N	No difference between W groups. Smaller differences between N & W on "culture-fair" test than on standard intelligence test

lates with the Binet .74 and Scale 2 correlates .81. Most test constructors who would want to measure what the Binet measures would be very happy to obtain such coefficients. What Goodenough and Harris (1950) say about the Draw-a-Person Test appears relevant to all "culture-free" or "culture-fair" tests: The search for a culture-free test is illusory.

Gifted children. Shuey's review of comparative studies (both directly and by implication) reveals that whites produce a greater proportion of gifted children by far than do Negroes. The percentage reported by Shuey for the latter, about 0.14% or 0.15% is well below the 0.95% for white children testing 140 IQ or above on the Binet or comparable scales. This low proportion among Negroes is only an expression of the general situation, i.e., that the whole curve for Negroes on most intelligence tests is displaced downward. Even taking Jenkins' (1950) figure of 0.3% for the 140 IQ or above places the area under the upper end of the curve below that for whites.

In surveying cases of gifted Negro children, Jenkins at first (1943) maintained that the gifted Negro child has essentially the same characteristics as a comparable white child. But later Jenkins (1950) recognized from more intensive investigation that the most important single fact for any Negro, gifted or not, is his being a Negro. Consequently, the performance he manifests on an intelligence test as well as elsewhere is literally colored by this fact.

That Negro children earn IQs of 160 or above (Jenkins, 1943) or even as high as 200 (Theman & Witty, 1943; Witty & Jenkins, 1935) on tests standardized on whites is a remarkable phenomenon. Anyone who has tested Negro children has probably been impressed with the fact that a Negro child whom the examiner knows to be functioning as a normal, not a retarded child, may receive a score which automatically would classify him in the retarded range if scores only were regarded. Roughly speaking, the Negro child seems to operate in everyday life situations in a way expected of a white child about 10 IQ points above. With the curve of measured Negro intelligence displaced downward it is thus a surprise to find any Negro children scoring among the highest levels on white-standardized tests.

Specific abilities. There have been few studies comparing Negroes and whites on specific abilities. Only two such reports worthy of mention came to the attention of the writers. Prothro and Perry (1950) administered the Meier Art Test to 460 high school and college students in Louisiana and found that mean score comparisons placed the whites above the Negroes. However, the authors felt that the differences might be due to the socioeconomic status of the groups. Negro and white sixth graders were tested on the Kwalwasser-Dykema Music Ability Tests with the Negroes obtaining a higher median score (Woods & Martin, 1943). The authors conclude that the Negroes were superior in a number of specific areas, but the meager statistical data presented do not support their conclusions.

Klugman (1944) in a study not mentioned by Shuey discovered that money incentives and praise had no demonstrable effects on white children (CA 7-14) in taking the alternate form of the Binet after an initial first form. However, comparable age Negro children given money rewards showed better performance than those given praise as an incentive. (Incidentally the two groups averaged 99.11 for whites and 98.00 for Negroes, with 6.08 and 5.66 *SDs*, respectively.)

In an investigation which compares with similar studies on white children, Robinson and Meenes (1947) found the relation between Negro children's test intelligence and parental occupation very low. From 1938-39 to 1945-46, the intelligence average rose from 97.0 to 99.7, a statistically significant change, and the relation between occupation of parents and intelligence increased as well—in this case it might be better to say that the relative lack of relation became less pronounced. A greater number of Negroes in higher status occupations could increase the coefficient attenuated by abbreviated spread of scores.

Davis (1948) and Davis and Havinghurst (1946) in a test of the influence of social class upon learning draw attention to differences in child-rearing practices of lower and middle classes, in general, that less rigid behavior is expected of children in lower classes. Comparing 100 Negro and 100 white families, Davis first indicates that the Negroes tended to be in the lower class. Where Negro families could be regarded as middle class, their child-training practices were conservative in regard to feeding, toilet training, and masturbation. It seemed to Davis that differences in adult behavior were greater within either Negro or white groups than between them. On mental tests, in which our interest lies principally, there were significant differences between the high and low socioeconomic groups. Davis seems to assume that these differences are culturally determined, but gives no satisfactory evidence in this regard. He does go on to urge the development of "culture-fair" tests and the elimination of academic intellectual tasks in the testing of intelligence.

Many years ago Sunne (1917) discovered that on some items of a test Negro children excelled whites and on others whites excelled Negroes. Sunne gave no satisfactory explanation for the finding. We have cited several studies above which have produced the same results. At least part of the differential performance may be attributable to status differences if the lead suggested by Eells, Davis, Havighurst, Herrick, and Tyler (1951) is in the right direction. These investigators found that status differences among children affect some test items and not others.

Adults

In addition to Shuey's review (Shuey, 1958), we report here a few other studies from the military services and other special groups for which comparisons have been made. One massive investigation, part of the standardization of the PAT (Tomkins & Miner, 1957), is not mentioned by Shuey, but by all means should be brought to the reader's attention because of the careful selection of representative subjects. The investigator (Miner, 1957) employed a 20-word vocabulary test adapted from the CAVD. The sample of 1500 individuals was chosen from stratified random clusters of blocks or rural areas with quotas on important variables. Whites averaged 11.06, *SD* 3.41, and Negroes 8.08, *SD* 2.72, a significant difference.

Armed Forces. In the period since 1944 debate over testing of intelligence in both world wars has been carried on. Although it seems to have subsided, it is not so much that the issues have been settled as that the contestants have become exhausted. To Shuey's conclusion, that in both wars whites consistently did better than Negroes on the average, few would take exception. She takes up one by one the reasons advanced by others for the discrepancies and provides reasons on the opposite side which purportedly answer an egalitarian view. The reader will have to study Shuey in order to judge whether the data support her reasoning. Additional discussion of some of the issues involved can be found in a series of articles by Garrett (1945a, 1945b, 1945c, 1947).

Several studies not cited by Shuey are mentioned here (Altus, 1946; Altus & Bell, 1947; Fulk & Harrell, 1952; MacPhee, Wright, & Cummings, 1947), primarily for the sake of completeness inasmuch as they do not differ in their results from others. Fulk and Harrell endeavored to equate groups by the last school grade completed; they compared AGCT scores and discovered that whites were favored at every school grade completed. The authors recognized that equal schooling does not render groups equivalent. Origins of their subjects, a highly important variable, were not controlled. Altus and Bell interpreted their discrepant findings on illiterates as due to cultural factors.

In a survey of "the uneducated" Ginsberg and Bray (1953) cite figures to indicate that the number of Negro illiterates declined from three million to slightly more than one million from 1890 to 1940. In World War II, 391,300 whites and 325,100 Negroes were rejected on the grounds of illiteracy, almost three-fourths of these rejectees being from the Southeast and Southwest.

Ginsberg (1956) draws the conclusion from studies of manpower in the Second World War that there is a great wastage of human potential when one realizes that the absolute numbers of Negroes in the upper classes of the General Classification Test are large, even though percentagewise Negroes have small representation in Classes I and II. The late Walter V. Bingham pointed out the waste of potential of all groups implied by the fact that one-fourth of the truck drivers exceeded one-fourth of the bank executives.

Special Civilian Groups. A large number of comparisons of college students reviewed by Shuey can be supplemented by only one series of studies by Roberts (1946, 1948, 1950). Northern Negro college students did better on the ACE than did their Southern counterparts. In a longitudinal comparison original difference in academic achievement total scores were erased after four years of college, even though ACE scores still reflected differences. During the four-year period the subjects made greater gains than those expected by the national norms for a similar period of time.

Comparing Southern Negro and white venereal disease patients, Scarborough (1956) discovered differences on the Wechsler-Bellevue paralleling those of other investigators, but the differences between white and Negro VD patients did not seem to be as great as those between white and Negro controls. Davis (1957) using the same test found no difference between 33 mental patients and 27 controls, though the overall IQs of 67 and 68 are well below white norms. Also employing the Wechsler, DeStephens (1953) answered his own question, "Are criminals morons?" in the negative when he discovered that 200 white and 100 Negro admissions to an Ohio reformatory yielded the following average scores: FS:W 93.55, N 87.90; V:W 90.13, N 86.70; P:W 98.30, N 91.20. On the basis of the standard deviations reported none of these differences between groups is significant. Findings of Reitzes (1958) on Negro applicants to medical schools yield a different result. Generally it is expected that Negro applicants will rank lower than white applicants on the Medical Colleges Admissions Test. Regional differences appeared among Negro applicants with North and West above border applicants, who in turn ranked higher than Southerners.

Overview of Intellectual Functions

Several grave issues arise in a review of comparisons of intelligence of whites and Negroes. We shall phrase these and make a few comments on each. It will be obvious that some of the problems apply more generally than to intelligence comparisons, but they are set forth here inasmuch as all of them are related to the foregoing comparisons.

What Constitutes a Race? For convenience we have assumed that groups belong to the white "race" or Negro "race" as designated by the investigator. Any research, however, which seriously attempts to make comparisons between or among "races" must sooner or later grapple with the problem of this section. Skin color, hair texture, and other physical characteristics have proven illusive as definitive criteria. We cannot settle a question to which anthropologists appear not yet to have given an adequate answer. Nevertheless, psychologists of the future must make explicit the concept of race to which they adhere in order to make clear what populations they are comparing.

Determination of Racial Composition of Groups. Assuming some acceptable definition of race, the investigator needs to determine in each case whether the groups he is comparing actually are differentiated by his criteria. If, for example, he has adopted heredity, the possession of such-and-such a percentage of accepted white or Negro ancestry, as a criterion, the researcher must make certain his groups fall within the limits of the specific heredity he has accepted. One suspects that in a number of cases so-called racial comparisons are being carried out between one group designated as "white" and another designated as "black" which consists of many who are partly or even largely white.

Confusion of Class and Caste Variables. As we have indicated in commenting on McGurk's studies, both hereditarians and environmentalists have fallen into the trap of assuming that if they can get two groups who are equated in socioeconomic terms there can be an *experimentum crucis* which answers the question of what parts heredity and environment play in intelligence differences. Canady (1943a) discussed the great difficulty in equating environments, and Anastasi and Foley (1949) pointed out that formal education and socioeconomic differences were not enough to account for differences in tested abilities of northern and southern Negroes. Actually, we despair of being able to equate groups until caste differences are removed and only class differences remain.

Interaction of Examiner and Subject Variables. In some studies (Trent, 1954) the color of the examiner has been taken into account in assessing results; in others this factor has been ignored. Repeated testing of the same individual, both white and Negro, by both white and Negro examiners would seem to be called for in order to determine the proportion of variance in intelligence test scores attributable to interaction of examiner and subject.

The Functional Value of Intelligence. What kinds of intelligence do white persons need and what do Negro persons need to survive, adjust, and make progress? It is naive to assume that the academic types of intelligence tests which have traditionally been the instruments of comparison compare in reality Negroes and whites in those areas of intelligence which they are called upon to use in "real life" situations. Intelligence test differences between Negroes and whites cannot mean the same as they mean between two groups of whites.

If we assume that intellectual functions develop adaptively and are not entirely determined by heredity, we may suppose that intelligence tests of the usual variety measure in part that which is developed in order to achieve success in a certain culture. A Negro in a white man's world requires a kind of intelligence enabling him to detect from minimal cues how a white man is going to react to a critical situation. Usually his success and sometimes his life depend upon this kind of intelligence, hardly ever upon whether he can define "ethnology." Tests of intelligence tapping the kinds of intellectual functioning called for in achieving success in the actual world men have to face might reveal different results in comparative studies, not only between Negroes and whites, but other groups as well.

Newer Concepts of Intelligence. Practically no research has been done comparing white and Negro subjects on factors of intellect (Guilford, 1956, 1959; Thurstone, 1938). The only studies of this kind we have found are those by Lee (1951) and Michael (1947). Even studies with performance-type tests, which in a sense go outside of academic intelligence quotient concepts, do not meet the need to reach the many factors revealed in modern factor-analytic research. A world of research is called for to determine how the two groups we are considering compare on the 50 or more factors of intellect. Caution needs to be exercised here in respect to the functional value of even Guilford's many factors. These were derived from tasks set for high-level personnel in the military and might not have anything specifically to do with what a person in a different class and caste needs to do. However, *insofar* as these factors cover the entire range of intellect, comparisons of factors should be much more valid than of gross IQs.

Significance of Overlapping Distributions. A legitimate question arises in statistics in relation to assigning individuals to one of two (or more) distributions (Horst, 1956). In statistical theory individual scores in a distribution may be regarded as errors departing from a mean (Yule & Kendall, 1949). When two distributions are compared, then, it is assumed that individual scores are only errors of observation from their respective means, which in turn are regarded as estimates of either one true population mean (if the null hypothesis is not rejected) or of two true population means (if the null hypothesis is rejected). In the case of intelligence comparisons the null hypothesis may be stated: the obtained means of these two distributions of intelligence scores differ only by chance, i.e., they are really only estimates of the same true intelligence score population mean. On the whole, investigators have rejected the null hypothesis in comparing Negroes and whites and have concluded that the two means represent two different intelligence population means.

In connection with any one individual, however, especially in the area beyond the mean of either distribution, the question may legitimately be asked: To which *intelligence* mean does he belong—or, from which *intelligence* mean is his score a deviate? It seems to be assumed that because an individual is white or black, his score must therefore be a deviate from a *white* intelligence population mean or a *black* intelligence population mean. But this assumption begs the question. The individual may be a deviate from the *lower* intelligence population mean whether he is white or black, or from the *upper* intelligence population mean whatever his color. The individual who is white but in the very low part of the scale may in truth be a deviant from the upper mean but we cannot know this fact merely because he is white. We could only be relatively sure of this fact if there were no overlap in the absolute sense between the two distributions.

This statistical consideration is a variant of the more general one which suggests that other factors than color (i.e., genetically determined intellectual concomitants of color) decide whether an individual makes a high or low score on an intelligence scale. A statement of the statistical situation focuses attention on the overlap found in virtually every study and what it must signify in relation to any one individual.

Social Consequences of Research Findings. In the introduction to Shuey's book (1958), Garrett discusses the need not to interpret Shuey's conclusion as a basis for differential treatment of the two groups, especially for mistreatment of Negroes. A view strictly limited to scientific conclusions would be, "Here are the results of examination of the data. It is not our responsibility to recommend courses of action." If there should be significant differences between groups which can be shown to arise principally from genetic factors, the practical response of the man on the street (white or Negro) would almost inevitably be to justify either his treating others as inferiors or to accept his own position of inferiority as natively determined. We are not convinced that genetic differences have been shown; but even if they were so shown, we believe it is incumbent upon the social scientist to set forth the full picture. The wide overlap between white and Negro distributions of scores should be pointed out so that it is evident that within group differences are far greater than between group differences. It should also be shown that oftimes two groups of white persons differ significantly, and probably in some if not all cases, partly because of genetic factors. Social scientists need to be alert to the implications of their findings.

Educational and Post-Educational Attainment

Table 3 presents mostly direct Negro-white comparisons in school achievement and in the case of Rodgers' study some post-educational attainment. It is generally recognized that as a national group Negroes receive a poorer education than whites both in quantity and quality, a condition reflected in scores on achievement tests. Other factors, however, may have to be adduced for explaining Witty and Theman's results.

TABLE 3

Relative Attainment of Negroes and Whites[a]

Author(s)	Subjects	Results	Comments
Thompson (1956)	College teachers	Pay, morale: W>N, teaching load: N>W	Author points out little research done by N
Ferrell (1949)	Grades 4, 5, 6	W>N on Stanford Achievement Test (all areas)	W more variable
Bullock (1950)	High school graduates	W>N on Iowa High School Content Exam.	Some question about choice of samples but same results regardless of type of analysis
Witty & Theman (1943)	N youth, Binet IQs 120-200 identified 6 years before	W gifted<N> average	Authors state attainment other than education relatively high
Rogers (1957)	Baltimore superior and average N identified in 1933	Gifted superior in achieving middle class, education, etc.	Comparison with W only indirect

[a] Bradley (1949) discusses relative literacy rates among military selectees, and Davenport (1946) educational attainment as it relates to military selection.

TEMPERAMENT[3]

The studies in the following paragraphs are subdivided according to the instruments employed. We sought for a more logical

[3] Some would use "personality" rather than "temperament." We follow what seems to be the trend in recent years, to regard personality as the larger set of functions under which temperament is subsumed. The latter includes traits or reaction systems like cyclothymia-schizothymia, introversion-extraversion, neuroticism, and others which are largely of constitutional origin and not so much affected by environmental changes as other functions of the personality such as attitudes and social interests.

set of divisions, but fell back on this artificial, yet serviceable classification. One major reason for employing the test as a separate category is that the reader may judge the validity of the studies partly in terms of his knowledge of the validity of the test. With the exception of the Murray TAT, almost all of the instruments utilized for comparative research have been gravely questioned by qualified sceptics. If studies using some particular test are scattered throughout a section among other results, it would be more difficult for a reader to judge whether the results from the particular test reflect the validity or invalidity of the test or stem from the experimental conditions. This consideration is not a major one in other parts of this review.

Rorschach

It seems to us that there must be more studies comparing white and Negro on the Rorschach, but we have found only two. Morons from both groups were compared by Abel, Piotrowski, and Stone (1944) in terms of specific responses to the ink blots. Out of the entire set of comparisons only one showed a real difference between the two groups: Negroes gave more *M* than whites. What the meaning of this finding is depends first upon whether this is a chance result in comparison with the probabilities involved, and/or secondly, if it is not chance, upon the meaning of *M*.

The group Rorschach was administered by Stainbrook and Siegel (1944) to high school and college students. No sampling statistics are presented, but responses are handled on the "simple basis of the probable differences of mean frequency of occurrences." Various determinants were studied in isolation. The authors conclude that both high school and college Negroes show less fluidity in association, that high school Negro youth are more emotionally stable and less impulsive than their white counterparts, and possess less anxiety, but that white college students though showing more emotional irritability are more mature, possess more "general personality resources" than Negro college students and have a "more daring intellect." Inasmuch as the group Rorschach has better norms but less interpretability than the individual Rorschach, it is difficult to know how to evaluate this investigation in comparison with most Rorschach studies which use the individual form. Harrower-Erickson and Steiner (1945) warn against using interpretations from individual Rorschach practice for group Rorschachs.

Thematic Apperception Test

Murray TAT. With a white female examiner Abel (1945) found that both white males and females and Negro females were more communicative on the TAT than Negro males, who were of at least equal intelligence to the other groups. Communicativeness was measured by the number of ideas and the number of words, both of which yielded significant differences. It scarcely seems surprising in the light of other research on the TAT and some knowledge of Negro-white sociodynamics that Negro males were inhibited in the presence of a white female examiner; but possibly we are engaging in ad hoc reasoning. At any rate, the author failed to point out a major disclosure of her data, that sex differences were greater than racial differences on the measured variables.

Differences between TAT responses of Negro and white boys were investigated by Mussen (1953). Fifty of each group from lower-class, New York City homes, of at least average intelligence, and between the ages of nine and 14 were age-matched. They were given 13 TAT cards plus one special mother-child card and their responses were analyzed by a modified Stein Schema with 28 need and 22 press categories. Scoring was blind. On most of the items the two groups were similar, a not unimportant finding. Nevertheless, 14 out of 50 chi squares were significant at least at the 5% level of confidence, a more than chance number of significance tests. Negro boys expressed greater hostility in thought processes than did white boys, less need for acting out murderous aggression, but about the same for other types of acting out. The Negroes manifested more need to reflect, think and speculate, but showed less desire for establishing and maintaining friendly relations, expressing admiration and respect for others, or being respected, followed and obeyed. They displayed less n Ach. Although white boys regarded others as being more rejectant than Negro boys, they also looked upon them as being more friendly, while the Negro boys viewed the environment as more hostile. On the whole, it appears from Mussen's data that self-attitudes of the Negro boys and their corresponding attitudes towards others have suffered somewhere along the line, even though in most things their temperaments parallel those of white boys.

More research utilizing the (Murray) TAT is called for to make comparisons of temperament variables. Of all the instruments employed for extra-intelligence investigation herein mentioned the TAT has suffered least at the hands of critics. It has demonstrated its value both clinically and experimentally. Consequently, comparative findings are not subject to the double jeopardy of having both methodology and test instrument attacked. The TAT has its faults in terms of interscorer reliability, but it is a fairly well proven technique for uncovering psychodynamics.

Thompson TAT. In a logically-motivated and commendable effort to make it easier for Negroes to identify with TAT characters, Thompson (1949) published his Negro version (T-TAT) with characters of obviously Negroid features. For some reason Thompson's ill-starred attempt has stirred more research than one might expect, all of which appears to invalidate the assumption that Negroes identify better with pictured characters of their own race than with corresponding white characters. Incidental to the attempts to refute the assumption are direct or implied comparisons of Negroes and whites on the T-TAT or M-TAT.

Korchin, Mitchell, and Meltzoff (1950), utilizing both T-TAT and M-TAT, performed an analysis of variance with race and social class as independent variables. No significant differences were found between two Philadelphia racial samples, but there were significant differences between social classes on length of Murray TAT stories. Korchin et al. believe that the assumptions underlying the T-TAT are not justified, so that a Negro modification is not called for. In a fairly well-designed experiment Riess, Schwartz, and Cottingham (1950) reached the same conclusion concerning the Thompson assumptions and strengthened their case in another study (Schwartz, Riess, & Cottingham, 1951).

Four groups of subjects, two white and two Negro, were given the M-TAT and the T-TAT by Cook (1953), the former to one group from each race and the latter to the alternate group from each race. Confirming previous findings on the lack of necessity for a Negro modification of the TAT, Negroes regarded characters in both sets of pictures as representative of people in general. Whites, on the other hand, looked upon the Thompson characters as Negroes rather than as persons-in-general. As a whole, considering the production on both sets, Negroes offered a significantly larger number of alternatives and of words indicating uncertainty, a higher vagueness score, and a large number of references to the pictures as pictures. Whites gave a higher word count.

Carrying out research on whites only, Light (1955) found no significant differences between his matched subjects in the productions on the M-TAT and the T-TAT presented in balanced order. Light's subjects, like Cook's, responded to the T-TAT characters as Negroes, with 14 of 26 subjects using traditional themes of Negro inferiority. One difficulty in interpreting these results comes from our not being able to know whether the examiner was white or Negro.

Although Thompson's attempt to provide easier identification for Negro subjects on a projective test has not borne its intended fruit, the result of comparing whites and Negroes seems to be a highlighting of the cultural pattern of virtually universal use of white characters as illustrations, at least in publications white people see. Negroes perceive white and black as people; whites tend to see white as people and black as Negroes. Whether the explanation is as simple as that of experience with the cultural pattern remains for future research to disclose.

Picture Arrangement Test

The Tomkins-Horn Picture Arrangement Test developed by Tomkins and Miner (1957) is probably the most extensively standardized projective test in the United States if not in the world. We are not prepared to say that it is the most adequately standardized, for we are not certain Gallup Poll interviewers are qualified to administer even an objectively-scored projective technique. However, the more than 1500 persons on whom the norms are based were chosen by block clusters and quotas in approved opinion sampling style. In addition to administering the 25 plates of the PAT according to rigidly standard procedures, the interviewer obtained an intelligence indication by means of a 20-word Thorndike-Gallup vocabulary test, together with information on major demographic variables.

The test itself consists of a series of plates with three positions of one or more persons arranged in a circular fashion so that the examinee must either stand on his head or turn the plate around in order to get a right-side-up view of all three positions. Scoring is objective—it may be done by a clerk—and is intended to reveal information on 158 temperament variables. Rarity of responses, determined empirically from the normative sample, yields an individual's "variant" score which in turn varies depending upon his age, education, and vocabulary IQ.

With the PAT which he helped to standardize in part, Karon (1958) performed an extensive series of analyses, utilizing a special discriminant function, to compare Negroes and whites with the intent of finding what effect, if any, caste sanctions have on Negroes. Karon's book is very confusing in its description of procedure. As far as we can ascertain, three "validation" samples of Northern and Southern Negroes and Northern whites were selected partly from among the 1500 persons mentioned above and compared. Final samples were chosen from ninth grade Negro students from North and South.

Results of the analyses confirmed in each case, according to Karon, the hypothesis that caste sanctions affect adversely the "personality characteristics" of Negroes. Significant differences between discriminant score means are interpreted as the result of differential caste sanctions. These differences appear to indicate that actual possession of and simultaneous denial of aggression is the principal problem of the individual living under caste restrictions, with a secondary resultant of flattened affect. There are differential "human costs" even between areas of more severe and those of less severe sanctions.

Some methodological umbrage could be taken in view of Karon's use of subjects taken from the standardization group for part of his experiment. Nevertheless, the results of his pilot studies were validated on what seem to be entirely independent samples.

Another possibility exists for interpreting Karon's results. Some form of selective migration explanation could apply to the dispersal of Negroes which could account for differences in affectivity and reactions to aggression. It is possible that those Negroes who have remained under the severest caste sanctions have remained because of their temperament characteristics and those who have stayed under less severe restrictions or moved to relavitely caste-free environments have done so because of their characteristics. In other words, they (and their children who presumably resemble them) are under or out of caste sanctions on account of their personalities rather than having certain personality characteristics on account of sanctions. Those who can swallow their aggressions—with occasional outbreaks against their fellows—stay, those who cannot, go. This explanation seems fully as plausible as the other, although we recognize the two are not mutually exclusive.

Picture-Frustration Study

Rosenzweig's Picture Frustration Study has been employed in several comparative investigations which are summarized in Table 4. There seems to be a tendency for Negroes to project more aggressive responses (E) than do whites. On the whole, similarities of responses are much more pronounced than differences, a situation which may merely reflect the relative insensitivity of the instrument.

Miscellaneous Tests and Other Instruments

In 1942, on the basis of intercorrelations of several scales administered to Fisk University students, Negro college students were regarded by Brunschwig (1942) as appropriate

TABLE 4

Comparisons Made with Original and Modified Rosenzweig Picture-Frustration Study

Author(s)	Test	Subjects	Results	Comments
McCary (1951)	PFS	Northern and Southern N and W	SN women> NW women on E[a] compared with I[a]	
McCary (1956)	PFS	Northern and Southern N and W, 14-22 yrs.	Most scores $\cong$; NW and NN> SW and SN on E compared with M[a] and I	Normative study—samples should not be used for experiment
McCary & Tracktir (1957)	PFS	High, medium, and low IQ W and N	Low IQ: N>W males on E compared with I and GCR[a] Medium IQ: W>N females on N-P[a] High IQ: N> W females on O-D[a]	
Portnoy & Stacey (1954)	Children's PFS	Subnormal, matched N and W children	N>W on M	Matching: length of institutionalization, intelligence, etc.
Winslow & Brainerd (1950)	Modified PFS (W or N frustrating agent)	Matched N and W	Both N & W: E more with N frustrator	Matching: age, sex, education, S-E[b] status

[a] E = Extrapunitive, I = Intropunitive, M = Impunitive, GCR = Group Conformity Rating, N-P = Need-persistence, O-D = Obstacle-dominance.

[b] S-E = Socioeconomic.

subjects with whom to employ tests and ratings devised for whites. Whether scales standardized on one group can be applied with equal validity to another racial or ethnic group is still a question. The TAT studies cited above might suggest that there can be. Whether or not they can be, we report studies from uncritical applications and also from unstandardized instruments. Roughly, this portion of the review progresses from children to adults and chronologically.

Reporting results of the same research in two journals, Hammer (1953a, 1953b) compared children in grades one through eight on the H-T-P Test; no control was exerted over socioeconomic variables. Clinician judges ranked the drawings on a six-point scale of neuroticism and a three-point scale of aggression. Negro children rated higher (at the .01 level) in aggressiveness. And whereas white children ranked between mildly neurotic and neurotic, Negro children scored on the average above severely neurotic. At every grade level, Negroes ranked higher in neuroticism, although at the upper levels white children became relatively more neurotic, thus decreasing the difference. Negroes showed specifically more bleak trees, more concern with the house, and more organic signs.

Disregarding the lack of established validity for the H-T-P, we consider the clinician's ratings as too severe. Possibly, as Willoughby observed many years ago, what we call normality may only be a widespread case of arrested development; nevertheless, the students on the whole in Hammer's study were probably not neurotic according to ordinary standards. Should the H-T-P be valid, an investigation controlling for class and caste variables is called for rather than the type reviewed here.

Among Letchworth Village mental defectives, it was found (Abel, 1943) that white and Negro girls, matched on relative level of IQ, more or less leadership ability, and other relevant variables, differed in that the Negro girls displayed more dominant behavior in imposing their judgments upon the white girls and making the decisions. Boyd's (1952) results, working with children of more or less normal intelligence in a nonsegregated elementary school, may be juxtaposed with the above. On two tests and a questionnaire designed to determine levels of aspiration, matched groups (age, IQ, economic status) revealed on the target test, questionnaire, and on the arithmetic test, higher Negro l-o-a. In selecting "the greatest person in the world" and "the person I would like to be like," 24 of 28 persons chosen by the Negro children were Negroes.

The above two studies are placed together for the possible implications of compensatory behavior elicited in a society dominated by one group to the real or apparent detriment of the dominated group. Of course, the institutional behavior may not be compensatory acting out, but could be explained at least in part on the basis that the Negro girls are not as defective in reality as their IQ counterparts (cf. the discussion elsewhere in this article on actual versus measured intelligence of low IQ Negroes).

Gray (1944b) made a study of the wishes of Negro elementary school children and compared them with those of white children from a previous study. The most striking point is the basic similarities of the two groups, although Negro children expressed more wishes concerned with home, animals, and musical instruments. Doll play fantasies of Negro and white children may be inferred comparatively from an interracial comparison by Graham (1955). Individuals ranged from 73-102 months and were regarded as homogeneous within groups. Although the results are difficult to evaluate interracially, it appears that the 30 Negroes produced fewer total fantasies than the 30 white children. With some exceptions, both groups produced approximately the same proportions of stereotype (dining room, kitchen, etc.) and nonstereotype (affection, aggression, etc.) responses.

One other study of children gives by implication some meager comparative information from a temperament test. Anderson (1947) administered two group intelligence scales the Otis and California Test of Mental Maturity, and a group "personality" test, to 153 Negro pupils in an Oklahoma high school. He found the averages on the intelligence scales at or above the means expected, but the Negro youth low in personal and social maturity on the CTP. The author does not mention the question of standardization of the CTP on a white population. The intelligence test results can be cautiously interpreted as running counter to the general findings elsewhere.

Illiterate soldiers revealed sectional differences on the Altus Adjustment Scale according to Altus and Clark (1949). With a higher mean signifying poorer adjustment, the following means were obtained in this study:

Southern Negro	7.92
Northern white	9.40
Northern Negro	9.56
Southern white	11.26

An analysis of variance suggests that these differences are not the result of chance factors.

Contrary to Altus' findings with his scale, Felton (1949), employing the Cornell Selectee Index with 148 Negroes and approximately 2400 whites at Oak Ridge, discovered that the former were significantly more frequently in the neurotic group. The author does not break down the figures, but a direct relation shows up in the data between lack of education and neuroticism. Although the Negro group is described as not from the deep South, there is a possibility that they were in the less well-educated groups. Touchstone (1957) found that on a sentence completion test neither white nor Negro rated higher than the other in passivity, aggression, hostility, or withdrawal, on all of which the investigator had expected to find Negroes scoring higher.

Overview of Temperament Studies

Living in a white culture the Negro seems to have difficulty with his frustration-induced aggression, although some reversals (which may, to be sure, be test artifacts) suggest that even under severe caste restrictions aggressive drives may not be the most important problem in the Negro's handling interracial interactions. It seems to be a truism that more and better research is called for in any area of investigation, but especially here where norms have scarcely begun to be established, where tests which have even failed to prove themselves with white subjects are applied almost unquestioningly to Negro subjects, and tests with white norms are utilized as if class and caste distinctions have no bearing on temperament or personality as a whole.

So-called "personality tests" may be inappropriate for testing most Negroes who are different from whites in socioeconomic status (Auld, 1952; Hoffman & Albizu-Miranda, 1955) and caste. In this area intensive studies of whites and Negroes need to be performed by scientists who understand both psychodynamics, sociodynamics, and adequate scientific procedure.

Values and Attitudes

Various attempts have been made to measure the value systems and attitudes of groups of Negroes and whites. The studies to be reviewed in this section cover a wide range of topics and in some instances are related only by the above broad heading.

Value Systems

The Allport-Vernon Study of Values has been the most frequently employed single measuring instrument (Eagleson & Bell, 1945; Gray, 1947; Milam & Sumner, 1954; Pugh, 1951). Eagleson and Bell (1945) administered this scale to 164 Southern Negroes and compared the results with the original data presented by Allport and Vernon. Subjects in the latter study were primarily from the Northeast. The authors report that the means were not greatly different, but the data were not sufficient to check the significance. Negroes and white females were similar in giving Religious values a top ranking, but differed in that the Negroes scored lowest on Aesthetic values, while the white females gave this a second place rating. The authors felt that these findings could be explained on the basis of a cultural interpretation relating to the "suffering" of Negroes and the white female's role in society.

Gray (1947) compared these findings with the performance of whites from Peabody College and Florida State University. These subjects also gave Religious values the highest rating with the Social scale being placed second. Both of these scales were significantly higher than all of the others. As compared with the normative group these Southern females also placed a low emphasis upon Aesthetic values. Thus Gray reinforces the conclusion of Eagleson and Bell that low Aesthetic and high Religious scores are a function of culture; this factor seems to apply also to Southern white women. Pugh (1951) administered the Allport-Vernon to several groups (ministers, laymen, nonchurch members) in south Georgia and reported more similarities than differences. Religious values again received top rating by all groups except white men who gave it a rank of five. The ratings of a group of male Negro medical students were similar to those reported above, except that they placed a higher emphasis on the Theoretical values (Milam & Sumner, 1954). These differences seem to be fairly consistent, but once again one is struck by the overall similarities.

Southern Negro and Northern white youth value systems have been analyzed by studying classroom themes (Hughes & Thompson, 1954). For the great majority of values studied a similar amount of interest was reflected in the two groups. Negroes gave a greater emphasis to justice and had a weaker identification with family.

Somner and Killian (1954) investigated value differences by means of an attitude scale wherein the subjects rated the behavior of "a Negro." Negroes want the Negro to be *relatively* forward, passionate, elegant, aggressive, persistent; whereas whites want the Negro to be relatively more witty, jovial, practical, quiet, patient. The emphasis should be on the word *relatively* as the range was restricted by the nature of the scale.

Self-Attitudes

Within recent years there have been a number of studies investigating the young Negro's attitude towards himself. Available evidence indicates that the white or Negro child learns very early about racial differences and that a light skin is to be preferred. Marks (1942) had Negro college students rate each other for attractiveness and found that within their own race there is a preference for light coloring, although extremely light skin was not preferred. A variety of techniques have been employed in evaluating this factor, including the use of white and colored dolls, special puzzles, picture tests, choice of playmates, etc. (Goodman, 1952; Helgerson, 1943; Horowitz, 1947; Koch, 1944; Landreth & Johnson, 1953; Radke & Trager, 1950; Radke, Trager, & Davis, 1949). Goodman (1952) reported finding a marked racial awareness in a group of four-year-old Negroes and whites. Consistent patterns of response to different skin colors has been observed in children of three years of age (Landreth & Johnson, 1953). This study, as well as one by Radke, Sutherland, and Rosenberg (1950), indicates that the development of a negative self-concept has its beginning in the early childhood of the Negro. When compared with whites the Negro is much less positive towards his own race. Whites tend to express a strong preference for their own race throughout a wide age range, and when a variety of techniques have been employed. In at least one study the proportion of Negroes expressing a preference for Negroes increased with grade level in school (Koch, 1946). Negroes have tended to prefer light skin colors, but a study among Fisk University students reveals that for at least that group the reference scale used in rating skin color is in part a function of the relative colors of the judge and the subject rated (Marks, 1942).

In clinical interviews and autobiographies of "some eighty Negro youths" ranging principally from 17 to 25 years of age (plus some juvenile delinquents and adult patients), Dai (1953) endeavored to determine those problems which are shared with white youth and those peculiar to Negroes. Clinically-oriented, Dai's approach was in terms of role-self-concepts; his interest lay in part in tracing how self-attitudes developed in childhood relate to adult problems.

Dai cites illustrative case material elicited from Negro youth, which could be found in counterpart in most texts on adolescent psychology, in relation to reaction patterns within the primary group, the family: antagonism and open revolt, passive-aggressive reactions, extreme conformity, and so forth. Problems peculiar to Negro children, Dai concludes, are basically human problems, but colored by the fact of being Negro. Caste restrictions bring the following influences to bear on personality development: a preponderance of lower class families with their special (and to the middle class individual, immoral) codes of conduct, broken homes, (consequent) maternal dominance, preoccupation with skin color and other physical features, and an extraordinary stress on social status. The factor most different from being a white youth, i.e., being a youth with dark skin, yields differential reactions in a sense of unworthiness, i.e., absorbing the white's evaluation of dark skin, in developing a sense of indifference—"So what?"—or in identifying with whites to the point of becoming white in judgments of blacks.

Controlled investigations of some of Dai's conclusions are necessary, for his study suffers from the general defects of a posteriori reasoning of many clinical studies and from lack of the precision which even good clinical studies can have. His conclusions do, however, coincide with those reached in the American Council on Education summary of studies (Sutherland, 1942) on "What does it mean for developing personality to be born a Negro?"

Social Perceptions and Attitudes

One study based upon choice of playmate in preschool children gives the racial factor a secondary weighting (Helgerson, 1943). The sex factor was considered the most important single determinant. It may be significant that this particular study was conducted in Minneapolis. In contrast to the Koch study mentioned above, older children chose colored playmates less frequently in both Negro and white groups.

Relatively detailed analysis of children's social perceptions and attitudes may be found in Radke and Trager (1950), Trager and Yarrow (1952), Radke, Trager, and Davis (1949), Bird, Monachesi, and Burdick (1952), and Horowitz (1947). For the most part the results are consistent with the points mentioned above. However, this entire series of studies suffers from many of the methodological errors referred to elsewhere in this paper. In many instances conclusions are based on what appear to be nonsignificant group differences; the race of the examiner is controlled infrequently; social class differences are seldom considered.

Concern with patterns of preference and attitudes toward other racial groups has also been reflected in a series of studies with adults. Studies prior to the period covered by this review suggested that patterns of preference of Negro and white college students were similar. Prothro and Jensen (1952) administered the Grice-Remmers Generalized Attitude Scale to Southern white and Negro college students with results indicating that the preferences of the two groups are not similar. Among other findings the authors report that the attitudes of the Negro towards the white were no more favorable than the reverse. Attitudes of the Negro towards Jews were generally favorable as were those of the whites towards Jews. Gray and Thompson (1953) report different findings on this latter point. On a modified Bogardus Social Distance Scale the Negro subjects voted all groups except their own lower than did the white subjects. The scale was administered to college and high school students as well as randomly selected adults, with white groups being more liberal in all instances.

Interest in the problems of segregation has resulted in several papers comparing the attitudes of Negroes and whites on specific aspects of this issue. Unfortunately, the subjects of these studies have been drawn almost exclusively from high school and college populations. In 1940 Boynton and Mayo (1942) administered a questionnaire to high school students in order to measure attitudinal responses towards the Negro and Negro-white relations. Differences between races were most pronounced with regard to social relations. The authors felt that the differences were becoming greater, primarily because of a shift on the part of the Negro. The questionnaire was administered again in the same schools in 1948 (Mayo & Kinzer, 1950), with the finding of a more favorable attitude towards the Negro by both races. However, the two groups were further from agreement on issues involving interracial relations, and the shift towards a more positive attitude towards the Negro came primarily on items not implying social proximity.

Holmes (1943), working with groups from four colleges reported the Negro more liberal in racial attitudes than the white and the not surprising finding that Negro students in the South were more conservative than those in the North, with white students in the South vastly more conservative than any of the other groups tested. Greenberg, Chase, and Cannon

(1957) administered the California F Scale and an integration attitude scale to west Texas high school students. In this instance, the Negro was more authoritarian than the white, but the authoritarian attitudes of either race were not indicative of negative attitudes toward integration.

Scholer (1943) had eighth grade students in Louisiana indicate their degree of approval of racial segregation in hypothetical case situations. Total scores failed to reflect a significant difference between Negroes and whites, although the authors report significant differences in 7 of the 11 situations presented. But the authors conclude that race similarities were more frequent than differences and there were no points of complete contradiction between the two groups. Preferences for segregation exceeded those for nonsegregation among both whites and Negroes, but to a greater degree among whites.

In a doctoral study Banks (1950) used a slightly different approach and had Ohio Negroes rank 96 different situations as to the degree of resentment experienced. The results were compared with Myrdal's (1944) rank order of discriminations: (1) economic, (2) legal, (3) political, (4) access to public services, facilities and funds, (5) courtesies, (6) sex relations and intermarriage. Myrdal feels that these are reversed for whites. However, the subjects in this study gave the same rankings except that economic discrimination dropped to third place. Thus, it seems that for at least the population employed in this study white and Negro concerns are very similar. In a study of the influence of discrimination on minority group members in New York, a larger percentage of both Negroes and Jews reported experiencing discrimination more frequently than did Catholics or Protestants (Saenger & Gordon, 1950). When compared with the other ethnic groups studied the Negroes particularly felt discriminated against in terms of job opportunities.

One final study dealing directly with attitudes towards segregation and schools should be mentioned. Turman and Holtzman (1955) surveyed a group of teachers in Texas on this question and found that only 4% of the whites and 1% of the Negroes held out consistently for segregation at all levels of education. Forty-four percent of the whites and 57% of the

Negroes expressed complete approval of mixed classes in public schools.

Attitudes toward other specific situations have been evaluated in several instances. Clarke and Campbell (1955) had junior high school students estimate their Negro classmates' performance on objective tests. Negroes were significantly more accurate in estimating their fellow Negroes' scores. The estimates of the whites were below those of the Negroes. The authors interpreted the data as reflecting a white stereotype of low Negro ability. One manner in which this stereotype gets reinforced may be shown in an analysis of magazine pictures (Shuey, King, & Griffith, 1953). Pictures of Negroes in leading popular magazines were judged on the basis of the socioeconomic level thought to be reflected. Negroes were portrayed in a less favorable light than in reality and actually appeared in only ½ of 1% of the total pictures.

In Flint, Michigan, students were asked to predict teachers' attitudes toward Negro children (Amos, 1952). In this instance the whites were more accurate than the Negroes who showed more prejudice, stronger feelings of rejection, and more consciousness of race. However, once again when social class was considered there was some evidence that this factor was more important than race in determining pupils' attitudes towards teachers. In a relatively unsophisticated study by Lewis and Biber (1951) Negro children when given an opportunity expressed preference for white teachers; but those who had had a Negro teacher inclined towards choosing a Negro teacher.

Attitudes towards the Negro tend to operate in a number of experimental situations and certainly have influenced the results obtained in the studies reported above. In addition to studies involving the use of psychological tests, the color of the investigator has been shown to be a relevant variable in such diverse things as work with GSR (Rankin & Campbell, 1955) and identification tests with young children (Trent, 1954).

Overview of Values and Attitudes

Insofar as generalizations can be made from sheer weight, the evidence points to similarities in the value systems of whites and Negroes. Differences in self-concepts are marked, however, in that being a white person in a white society appears to mean little in respect to the development of self-concepts, whereas being a Negro in a white society seems to be one of the most important factors in such development.

The quality of the studies reviewed in this section has varied considerably. Values and attitudes are difficult to quantify, subject to regional variation and other factors which make generalization difficult. One obvious step in clarifying group differences requires the study of populations other than those drawn from schools and colleges.

Vocational Interests

This is another area in which limited research has been reported. A PhD thesis by Hartshorn [4] purportedly discovered interest scores on the Strong Vocational Interest Blank to be different for lawyers, physicians, and life insurance agents of the two groups, in that white men of the standardization group have higher mean interest and occupational level scores, and also more masculine interests, than do their Negro counterparts, and fewer specific likes than do Negroes. Strong (1952) who reviews Hartshorn's work at some length, disputes the latter's conclusions in part, stating the two groups compared are not equivalent (though the major reason Strong offers is not sufficient reason for his contention). Supporting Strong to a slight degree in his denial of the contention that interests of Negro and white professional persons really differ is his research on medical school seniors of both races (Strong, 1955); the differences in occupational and specialization scales are in most cases relatively small. Yet out of 14 occupational scales one is significantly different and six are very significantly different for whites and Negroes, while three out of five specialization scales differ at least at the 5% level of confidence; Strong's concluding remark in his analysis of Hartshorn's data could conceivably do without the qualifying conditional clauses: "If Negroes are really different from whites

[4] Known to the writers only through Strong's review.

in their interests, which we are not ready to accept as a fact, then their interests must be measured from a Negro, not a white, point of reference" (Strong, 1955).

Milam and Sumner (1954) have provided further information on the spread and intensity of vocational interests of first year Negro medical students. They report that low vocational interest intensity in Negroes has been a consistent finding with the Strong Vocational Interest Blank and the controversy turns only on why it is lower. After comparing performance on the Strong with academic grades, the authors conclude that there is some evidence to the effect that high scholastic ability is correlated with more intense physician interest as well as more intense and/or diversified nonphysician interests.

Gray (1944a) inquired directly of some 800 Negro children (first to sixth grades) as to what their vocational preferences were. Results were compared with the responses of white children obtained six years earlier by Boynton. The Negro and white females voiced similar preferences, but the Negro male was more interested in professional occupations than the white male. The median occupational level chosen by the Negro on a five-point scale was one point higher than the white child's. Data were not presented to the extent that statistical significance could be evaluated, but it appears that the Negro children were certainly less realistic. Several studies contrasting Negro and white occupational patterns clearly point to the fact that jobs of higher quality generally are reserved for whites (Keenan & Kerr, 1952; Mundy, 1949; Turner, 1954).

Social Structure

Leadership

Research on comparative qualities of leadership has been extremely limited and, for the most part, has been concerned with the evaluation of miscellaneous factors. Dexter and Stein (1955) administered various temperament tests to leaders and nonleaders in a white and Negro college, with the finding that differences within each group were greater than differences between groups. Some indirect information of a comparative nature is available from studies of *Who's Who in Colored America* (Monahan & Monahan, 1956; Valien & Horton, 1954). Relatively speaking it seems that the Negro female leader receives more recognition than the white female and achieves distinction at a younger age than does the white person of either sex.

Family Organization

In this section various recent miscellaneous studies relating to Negro and white family practices will be reviewed. For comprehensive reviews the reader is referred to the general references listed in the beginning of this article.

The dynamics of the Negro family have been discussed in some detail (Davis & Havighurst, 1946; Frazier, 1939, 1957; Myrdal, 1944). For the most part, the emphasis has been placed upon family disorganization. Statistics have been quoted from the 1950 census to point out the extent to which disrupting socioeconomic and cultural factors are operating (Ginzberg, 1956). At that time one-third of the Negro women who had been married were divorced or separated from their husbands as compared with only one-fifth of the white women. Thirty-five percent of the Negro mothers under 45 were employed versus 19% of the white women in this same age group. In a theoretical discussion of the problems and needs of Negro youth, Frazier (1950) states that the male parent is absent in about 20% of the Negro homes. These statistics would appear to be relatively significant and confirm the existence of greater family disorganization among Negroes, as in general is the case with lower socioeconomic groups (Hollingshead, 1953). If there is any validity to currently held theories of personality development, we would expect some distinctive group personality differences as a reflection of the above facts.

Several studies on child-rearing practices among Negroes and whites have been referred to under the section on intelligence. Davis and Havighurst (1946), after studying groups of mothers from lower and middle classes in Chicago, concluded that essentially the same types of differences prevailed between middle and lower class whites as between middle and lower class Negroes. The major exception was that Negro mothers were described as more permissive than whites in feeding and weaning, but more rigorous in toilet training. Aside from this the authors conclude that the most striking thing about the study was that the Negro and white middle and lower classes were so much alike.

There have been several comparative investigations of fertility rates. After studying the 1940 census figures, Lee and Lee (1952) state that the pattern of Negro fertility is remarkably similar to that of whites. Within both races fertility declines as socioeconomic level goes up. As with whites, Negro fertility is lower outside of the South. The authors conclude that the patterns of the Negro most closely approach those of the white in those areas where he shares most freely in the general culture. The rate is approximately the same in urban areas; any differences that do appear occur in rural sections. Valien and Vaughn (1951) and Tietze and Lewit (1953), investigated birth control practices in two Southern communities and came up with expected findings: favorable attitudes were correlated with urban birth, education, working mothers, etc.

One investigation of factors entering into mate selection was reported during this period (Sussman & Yeager, 1950). By means of a questionnaire Negro and white college students were rated on 19 different factors. The ratings for the two groups were similar on 14 of these factors, indicating considerable agreement in the qualities thought desirable in a prospective mate. The amount of agreement was surprising since the Negro and white groups were drawn from different sections of the country with differing religious backgrounds. Differences which did appear to be significant are the whites' high rating of "insight and understanding," and the low rating of "good health." For the Negro group these two factors had standings opposite to those of the whites. It is not too remarkable that the health factor is of concern to Southern Negroes as their general level of health is considerably below that of the whites.

Of incidental interest under this section are several studies of Negro-white marriages (Cash, 1956; Golden, 1953, 1954). The subjects of all three of these investigations lived in Philadelphia, and thus generalizations are limited; but the following findings were reported. In most such marriages the male is the Negro; more than 50% of both parties had been previously married; they have few children and tend to marry late; most of the Negroes were Negroid in appearance.

Summarizing the literature on the Negro family dealing with descriptive, theoretical, and quantifiable comparisons with white families, we are led to believe that many of the differences that have been reported can be accounted for in terms of socioeconomic class differences.

Emotional Disturbances and Mental Illness

Attempts to assess the relative emotional stability of Negroes and whites have ranged from the use of paper and pencil tests of neuroticism to comparisons of first admissions to state hospitals. The test approach has not made a significant contribution to a clearer understanding of differences between the two groups, primarily because there are few tests available which have been standardized on Negro populations. In an experimental study Roberts (1944) has shown that measures of attitudes, adjustment, and personality yield results dependent upon the cultural and racial group in which the measure is made; thus extreme caution is necessary when applying popular American tests to Negroes. It is also recognized that the use of hospital statistics is open to a number of errors.

Attempts to measure Negro-white differences in neuroticism have been limited primarily to college populations. The studies reported have been contradictory. Heyman (1945) has observed that there is a tendency to ignore psychoneurotic behavior in the Negro, even though the observed symptoms are very similar to those in whites. Boykin (1957) administered the Bell Adjustment Inventory to college freshmen over a four-year period and found 25% "poorly adjusted." His findings were compared with the original standardization groups; of his two "poorly adjusted" groups, the Negro group was more maladjusted. It was not possible to tell from the statistics presented if this difference was statistically significant. The Bernreuter Personality Inventory has been administered to Negro groups on at least two occasions. Wheatley and Sumner (1946) found no differences between college student performance and the original norms. They concluded that the most neurotic scores were obtained from the lowest socioeconomic classes. Sumner (1948) on the other hand found no relationship between Bernreuter scores and the socioeconomic status of Negro college women. The California Test of Personality has been administered to second and third grade Negro and white groups without obtaining striking differences, although the author concluded that the minority group tends to feel persecuted (Engle, 1945).

Rowntree (1943) studied discharges from the Service for psychiatric reasons in the period 1941-43 and concluded that a diagnosis of psychoneurosis occurs four times as frequently among Negroes as among whites. The sampling in this study leaves something to be desired; and such factors as geographical representation among the two groups were not controlled. In a study of 105 consecutive discharges from the Navy a diagnosis of psychoneurosis occurred approximately three times as frequently among Negroes as among whites (Hunt, 1947). The author, however, checked these results with four other stations and concluded that the prevalence of psychoneurosis among Negroes is lower than among whites. Gardner and Aaron (1946) reviewed consecutive admissions to the psychiatric ward of a Naval hospital and concluded that whites were more prone to a psychoneurotic reaction than Negroes. In seeming contradiction to these conclusions Ripley and Wolf (1947) reported that minor psychiatric illness occurred two and one-half times more frequently among Negro troops overseas than among comparable white troops. These results were felt to be due in part to lower standards of acceptance for service among Negroes. From these studies we are left wondering whether psychoneurosis actually is more extensive among Negroes or if standards of judging psychoneurosis differ for whites and Negroes.

The picture is clearer when serious mental illness is considered. If first admissions to state hospitals are taken as a measure of relative incidence of psychoses, the occurrence is approximately twice as great among Negroes (Frumkin, 1954; Ivins, 1950; Malzberg, 1940; Wilson & Lantz, 1957). This ratio seems to hold over a number of years and in widely separated states, with the reported range being from one-and-one-half to four-and-one-half times as great among Negroes. The only contradiction found was in a study by McLean (1949) who reported that first admissions to Illinois State Hospitals were no greater in the Negro than in the white. In explaining this finding he made note of the fact that Illinois is well established as an integrated state.

Even when allowances are made for the fact that whites are more likely to be able to afford private hospitalization, it seems apparent that the relative incidence of psychoses among Negroes is significantly higher. There are also some consistent differences in terms of psychiatric diagnosis and symptoms. Negroes have a higher relative incidence of schizophrenia, paresis, and alcoholic psychosis (Frumkin, 1954; Ivins, 1050; Malzberg, 1940, 1953; Ripley & Wolf, 1947). The only report contradicting this general conclusion, at least in part, is that of Wilson and Lantz (1957), who found that the higher rate of first admissions among Negroes in Virginia was not due to alcoholism, epilepsy, or mental deficiency. They concluded that the difference was due to senile psychoses (although there was a greater number of whites over 65), arteriosclerotic dementia, and schizophrenia.

Other differences in symptomatology have been noted. Psychosomatic disorders are found more frequently among whites, and in some instances—e.g., peptic ulcers—the rate is as much as ten times greater (Rowntree, 1943). The suicide rate seems to differ according to geographical location. McLean (1949) reports that Negroes in the South have a rate only one-fourth that of the whites, whereas the ratio is almost equal in the North. She advanced the explanation that the higher rate among Northern Negroes was due to "ambition." There are also differences in "acting out" behavior which will be discussed in more detail below. In his studies of New York State Hospital, Malzberg (1956a) found that the admission rate for

Negroes to hospitals for the criminally insane was four-and-one-half times that of whites.

A number of hypotheses have been advanced to explain the higher incidence of emotional difficulties among Negroes. Various factors associated with socioeconomic status have been of major interest in this regard. Pasamanick, Knoblock, and Lilienfeld (1956) conclude that there is a

> positive and probably etiologic relationship between socioeconomic status and prenatal and paranatal abnormalities which in turn are related to retarded behavioral development and certain NP disorders such as cerebral palsy, mental deficiency, and behavior disorders.

Negroes were reported as having a much greater number of prematurities and complications than white controls with the overall incidence of abnormal conditions being almost twice as great in the Negro. This ratio held up when socioeconomic status was taken into consideration, but since many Negroes were below the lowest white, socioeconomic status was assigned greater weight than any "racial" factors. Hollingshead and Redlich (1953) found in their New Haven study that class is directly related to mental disorder and that schizophrenia occurs most frequently in lower classes. This is one of the psychiatric categories in which the Negro has consistently been reported as having a high incidence.

Other factors felt to contribute to emotional illness in the Negro have centered in family structure, economic problems, and prejudice. Gardner and Aaron (1946) studied childhood and adolescent adjustment and found many similarities between white and Negro psychiatric casualties in a Naval hospital. However, the Negroes were more likely to have an unstable or "broken" home background, and more likely to have been eneuretic in childhood. Ellis and Beechley (1950) compared Negro and white children seen at a child guidance center on a number of variables and concluded that the Negro children were more disturbed, of lower socioeconomic status, came more often from broken homes, and were less responsive to treatment. Racial conflict was found in 32 of 45 consecutive Negro clients of a child guidance clinic and was felt by the investigator to be a contributing factor in the emotional problems of the children (Verin, 1944). An attempt to evaluate cultural factors in mental illness is reported by Cervantes (1954) who studied 30 Northern born Negro patients and 30 "average" whites. Socioeconomic differences between the groups were felt to be important, but the design of the study was such that the differences were not clearly demonstrated. In studying first admissions in state hospitals Malzberg (1956b) reported a higher rate among New York Negro males than among foreign born Negro males. Prejudice was suggested as an explanation for this difference; but the same ratio did not hold for females, a fact which weakens his argument. In another study of first admissions to state hospitals Malzberg (1956a) reports that for the most part Negroes tend to follow the overall pattern in that single, divorced, or widowed individuals have higher rates than married.

Several papers have appeared describing the difficulty in treating psychiatric problems in the Negro as opposed to whites (Adams, 1950; Harms, Kobler, & Sweeney, 1945; Heine, 1950; Kennedy, 1952; St. Clair, 1951). In describing the Negro the following features are frequently mentioned: concern with race consciousness, tendency to act out, hostility as a dominant problem, distrustful, self-hating, strong prestige needs, difficulty in establishing rapport. The references cited are not experimental studies but clinical impressions that are frequently reported and of unknown validity. But there have been several attempts to quantify Negro-white differences in response to treatment. Most of these studies reflect a greater likelihood of improvement in the white. Blassingille (1955) compared the rehabilitation of 70 Negro leucotomy patients at a VA hospital with that of similar patients reported in the literature and considered the results similar. Cultural, social, and economic factors peculiar to the southern Negro were felt not to be significant in rehabilitation of this type patient. In other studies whites have been found to respond more favorably to treatment of varying kinds.

In a review of 455 cases treated with electroshock therapy,

the whites were found to be significantly more responsive when the criteria employed were rate of discharge from the hospital and absence of relapse (Palmer, Sprang, & Hans, 1951). Kahn, Buchmueller, and Gildea (1951) describe a program of group therapy for parents of behavior problem children which failed when employed with Negroes where it had been considered successful with white parents. The adjustment of Negro and white schizophrenics was studied by Hewlett (1946) with a conclusion that the whites were making the better adjustment. She concluded that economic, financial, occupational, and familial problems were the factors most often related to the adjustment made and that these factors were most severe among Negroes. Ellis and Beechley (1950) feel that some of the same factors were operating to cause a poorer response to treatment among Negro children seen in a child guidance clinic.

For the most part these attempts to evaluate objectively the response to treatment have resulted in conclusions in agreement with clinical impressions reported in the literature. More definite evidence could not be expected in terms of the absence of well accepted criteria of improvemnt as well as the many extraneous variables operating simultaneously. In evaluating research pertaining to psychiatric disorders among Negroes, Schermerhorn (1956) comments that investigators have frequently been in error in assuming that the Negro and white populations are alike except for the variable in question and that the experimental meaning of these variables is culturally equivalent in both groups. By and large, also, treatment has been offered by white therapists to Negro patients. What effect this factor has in determining results has not been assessed.

On the basis of the evidence available it does appear that the Negroes more frequently experience psychiatric difficulties, particularly of a severe nature. For the most part the specific forms and course appear to be similar with the differences attributable at least in part to identifiable socioeconomic and cultural factors. Whether genetic endowment accounts for the unknown variance is difficult to tell. However, the trend in recent years to find many factors of previously unquestioned hereditary nature attributable to experience makes it also reasonable to assume environmental causation here. Thus, the facts concerning emotional disturbances are fairly clear, but their interpretation is far from being settled.

Crime and Delinquency

Statistics reporting the incidence of crimes and delinquency have consistently shown a higher prevalence among Negroes. The few studies that have appeared in the literature during this period seem not to have been particularly concerned with the question of incidence per se, but have explored possible causative differences between the two races and differential treatment by police officials and the courts.

On the basis of a questionnaire completed by Philadelphia policemen, Kephart (1954) concluded that both white and Negro patrolmen are more strict with Negro offenders. He also noted that Negro offenders more frequently resisted arrest and therefore reinforced the attitude of the officers. Moses (1947) studied crime rates in Baltimore within four socioeconomically equated areas and found in two of them that the Negro rate was higher. He did not find evidence that the Negro offender is convicted more readily. The pattern of offense was similar except that crimes involving loss of life were concentrated among Negroes. The author had some reservations about the attempt to equate the areas in terms of socioeconomic status and found that as a whole the whites had been settled much longer. A greater percentage owned their homes, etc.

Pollack (1944) reviewed Negro and white admissions to Pennsylvania State prisons in 1941-42, being concerned only with subjects who were aged 50 or over. The Negro was charged more frequently with aggressive assault, criminal homicide, and liquor law violations. Pollack reported that the decrease of incidence of various crimes at age 50 and over was comparable in the two groups. The significance of the difference reported could not be evaluated as the data were not reported in the article.

On the basis of statistics compiled in Philadelphia in 1948, Diggs (1950) found several significant differences in both offenses and disposal of offenders between children of the two races. A smaller number of Negro children were dismissed or discharged and a larger number were institutionalized or referred to a criminal court. The Negro was less likely to be referred to private agencies for treatment as opposed to public agencies. The leading offense for Negro boys was that of taking the property of another as opposed to carelessness or mischief for non-Negro boys. Sexual offenses led for Negro girls, while white girls came to the attention of the court most frequently for running away from home. Diggs reported that only one-fourth of the Negro delinquents had both natural parents in the home, but the comparable figure for whites was not reported.

Axelrad (1952) studied the records of 300 institutionalized delinquents in New York City. He concluded that in comparison with whites, Negro children were committed younger, for less serious offenses, with fewer previous court appearances; they came from more unstable homes and homes with a different kind of pathology than that of white delinquents.

The Negro crime rate seems to be consistently higher when a single overall breakdown is made by race. On the basis of recent comparative studies it is difficult to make any judgment as to the relative weights of heredity and environment in this area. It is the writers' opinion that only environmental factors are responsible for differential criminal or delinquent behavior. But if we were to depend on the studies reported here, we must frankly admit we should have little basis for the opinion.

Summary and Concluding Remarks

In psychophysical and psychomotor functions, differences appear between whites and Negroes which may not be accounted for by differential environmental conditions. However, a tendency is present in the literature to indicate that most differences of this nature may be leveled off when social and economic variables are controlled. Intelligence differences reported in the period covered by this review are in the same direction as those seen previously, although infant and young child comparisons suggest greater similarities between Negroes and whites in the early years. Educational achievements of Negroes relative to whites follow the pattern of intellectual differences.

In temperament ("personality") studies Rorschach, TAT, PAT, and P-F Study differences are found, but once again there is insufficient evidence to determine the relative contributions of genetic constitution and experience. At least in those reactions which indicate responses to a dominant-group culture, experience seems to be the major if not sole determinant. Overall, likenesses in psychodynamics appear more extensive than differences. Some, especially paper-and-pencil, tests imply that Negroes are more neurotic than whites, although these tests may be limited for cross-barrier comparisons.

Religious values are ranked first by Negroes, a condition holding generally for white females also, but not for white males. Self-concepts seem to suffer in the Negro subculture in contrast to those of whites. Social perceptions correspondingly differ from one group to the other, with, however, a number of likenesses which may not fit into the stereotype the social scientist holds, for example, in attitudes toward segregation where some Negroes maintain expressed attitudes very much like those of the white majority, or in rank-order of attitudes toward discrimination. Vocational interests may be somewhat similar though less intense (by Strong's criteria) among Negroes; Negro children may be less realistic.

As might be expected, in the areas of psychological functioning most closely related to the sociological, social class differences show up more clearly as bases of differentiation between the two groups. Leadership, family life, child-rearing practices, fertility, and mate selection all seem to conform to social structure rather than to racial lines per se.

Mental illness is considerably more prevalent among Negroes than whites even though specific symptoms like psychosomatic disorders are more frequent in the white group. For various reasons advanced in the body of this article treatment of psychiatric disturbances is more difficult with Negroes than with whites. Rates of delinquency and crime are reported to be higher for Negroes, especially in regard to violent crimes. Most of the evidence indicates that Negroes are given less adequate treatment at the hands of officers of the law and courts.

It is clear from the foregoing review that (*a*) there are still wide differences between Negro and white in many areas of psychological functioning and (*b*) a number of differences attributed in times past to heredity have been shown to be the result of social class determination. It is not clear whether some differences adumbrated here, specifically in the intelligence and temperament realms, are genetically based or not. We agree with Garrett in his Foreword to Shuey's (1958) book, that there are some wholly well-meaning persons who hold that ". . . racial differences ought not to be found; or if found, should immediately be explained away as being somehow immoral and reprehensible." Nevertheless, we are not satisfied that either those who like Garrett believe that genetic differences exist in psychological functions or those who maintain that no such differences can be found have succeeded in establishing their position. Most students in the period of this review have leaned to environmentalist explanations. But this concurrence of judgment may be the result of a possibly unjustified extrapolation from Point b above. We do not agree with Garrett when he says in the same Foreword: "The honest psychologist, like any true scientist, has no preconceived racial bias." As clinical psychologists we are convinced that scarcely anyone undertakes investigation in this field without preconceived biases. We frankly are environmentalist in our bias; but we also hope that we are "honest psychologists" enough to recognize that many research results can yet be interpreted from an hereditarian viewpoint without doing violence to them.

As a last note, from our survey we have come to the conclusion that research within the United States, to which we have limited this review, must be supplemented by investigations between American and other cultures and within other cultures where caste differences are relatively nonexistent.

Bibliography

Abel, Theodora M. Dominant behavior of institutionalized Negro girls: An experimental study. *Amer. J. ment. Defic.*, 1943, *47*, 429-436.

Abel, Theodora M. Responses of Negro and white morons to the Thematic Apperception Test. *Amer. J. ment. Defic.*, 1945, *49*, 463-468.

Abel, Theodora M., Piotrowski, Z., & Stone, G. Responses of Negro and white morons to the Rorschach test. *Amer. J. ment. Defic.*, 1944, *48*, 253-257.

Adams, W. A. The Negro patient in psychiatric treatment. *Amer. J. Orthopsychiat.*, 1950, *20*, 305-310.

Alper, Thelma G., & Boring, E. G. Intelligence test scores of northern and southern white and Negro recruits in 1918. *J. abnorm. soc. Psychol.*, 1944, *39*, 471-474.

Altus, W. D. The validity of the Terman vocabulary for Army illiterates. *J. consult. Psychol.*, 1946, *10*, 268-276.

Altus, W. D., & Bell, H. M. The validity of a general information test for certain groups of Army illiterates. *J. consult. Psychol.*, 1947, *11*, 120-132.

Altus, W. D., & Clark, J. H. Some sectional differences among Negro and white illiterate soldiers. *J. soc. Psychol.*, 1949, *30*, 97-104.

Amos, R. T. The accuracy of Negro and white children's predicdictions of teachers' attitudes toward Negro students. *J. Negro Educ.*, 1952, *21*, 125-135.

Anastasi, Anne, & D'Angelo, Rita Y. A comparison of Negro and white preschool children in language development and Goodenough Draw-a-Man I.Q. *J. genet. Psychol.*, 1952, *81*, 147-165.

Anastasi, Anne, & Foley, J. P., Jr. *Differential psychology, individual and group differences in behavior.* (Rev. ed.) New York: Macmillan, 1949.

Anderson, W. E. The personality characteristics of 153 Negro pupils, Dunbar High School, Okmulgee, Oklahoma. *J. Negro Educ.*, 1947, *16*, 44-48.

Auld, F., Jr. Influence of social class on personality test responses. *Psychol. Bull.*, 1952, *49*, 318-332.

Axelrad, S. Negro and white male institutionalized delinquents. *Amer. J. Sociol.*, 1952, *57*, 569-574.

Banks, W. S., Jr. The rank order of sensitivity to discrimination of the Negroes in Columbus, Ohio. *Ohio State U., Abstr. doctoral Dissertations. 1948-49*. 1950, No. 60, 7-13.

Beier, E. G. The effect of induced anxiety on flexibility of intellectual functioning. *Psychol. Monogr.*, 1951, *65*(9, Whole No. 326).

Bendix, R., & Lipset, S. M. *Class, status, and power: A reader in social stratification*. Glencoe, Ill.: Free Press, 1953.

Benedict, Ruth, & Weltfish, G. *The races of mankind.* (Public Affairs Pamphlet No. 85) New York: Publ. Affairs Comm., 1943.

Bird, C., Monachesi, E. D., & Burdick, H. Infiltration and the attitudes of white and Negro parents and children. *J. abnorm. soc. Psychol.*, 1952, *47*, 688-699.

Blassingille, B. Rehabilitation of Negro post-leucotomy patients. *J. nerv. ment. Dis.*, 1955, *121*, 527-534.

Boyd, G. F. The levels of aspiration of white and Negro children in a non-segregated elementary school. *J. soc. Psychol.*, 1952, *36*, 191-196.

Boyd, W. C. *Genetics and the races of man: An introduction to modern physical anthropology*. Boston: Little, Brown, 1950.

Boykin, L. L. The reading performance of some Negro college students. *J. Negro Educ.*, 1955, *24*, 435-441.

Boykin, L. L. The adjustment of 2,078 Negro students. *J. Negro Educ.*, 1957, *26*, 75-79.

Boynton, P. L., & Mayo, G. D. A comparison of certain attitudinal responses of white and Negro high school students. *J. Negro Educ.*, 1942, *11*, 487-494.

Bradley, Gladyce H. A review of educational problems based on military selection and classification data in World War II. *J. educ. Res.*, 1949, *43*, 161-174.

Brown, Estelle W., Lyon, R. A., & Anderson, Nena A. Courses of prematurity: Influence of maternal illness on incidence of prematurity; employment of new criterion of prematurity for negro race. *AMA J. Dis. Child.*, 1945, *70*, 314.

Brown, F. An experimental and critical study of the intelligence of Negro and white kindergarten children. *J. genet. Psychol.*, 1944, *65*, 161-175.

Brunschwig, Lilly. Some personality adjustments of Negro college students as indicated by tests and ratings. *Psychol. Bull.*, 1942, *39*, 578. (Abstract)

Bullock, H. A. A comparison of the academic achievements of white and Negro high school graduates. *J. educ. Res.*, 1950 *44*, 179-192.

Caldwell, M. B. An analysis of responses of a southern urban Negro population to items on the Wechsler Intelligence Scale for Children. Unpublished doctoral dissertation, Penn. State College, 1954.

Canady, H. G. The problem of equating the environment of Negro-white groups for intelligence testing in comparative studies. *J. soc. Psychol.*, 1943, *17*, 3-15. (a)

Canady, H. G. A study of sex differences in intelligence test scores among 1306 Negro college freshmen. *J. Negro Educ.*, 1943, *12*, 167-172. (b)

Canady, H. G. The psychology of the Negro. In P. Y. Harriman (Ed.), *The encyclopedia of psychology.* New York: Philosophical Library, 1946. Pp. 407-416.

Cash, E., Jr. A study of Negro-white marriages in the Philadelphia area. *Dissertation Abstr.*, 1956, *16*, 1006-1007.

Cattell, R. B. A culture-free intelligence test: I. *J. educ. Psychol.*, 1940, *31*, 161-180.

Cattell, R. B. Classical and standard score IQ standardization of the I.P.A.T. Culture-Free Intelligence Scale 2. *J. consult. Psychol.*, 1951, *15*, 154-159.

Cattell, R. B., Feingold, S. A., & Sarason, S. B. The Culture-Free Intelligence Test: II. Evaluation of cultural influence on test performance. *J. educ. Psychol.*, 1941, *32*, 81-100.

Cervantes, Esther L. A study of cultural factors affecting mental illness. *Smith Coll. Stud. soc. Work*, 1954, *24*, 62-63.

Chapman, W. P. Measurements of pain sensitivity in normal control subjects and in psychoneurotic patients. *Psychosom. Med.*, 1944, *6*, 252-257.

Clarke, R. B., & Campbell, D. T. A demonstration of bias in estimates of Negro ability. *J. abnorm. soc. Psychol.*, 1955, *51*, 585-588.

Codwell, J. E. Motor function and the hybridity of the American Negro. *J. Negro Educ.*, 1949, *18*, 452-464.

Cook, R. A. Identification and ego defensiveness in thematic apperception. *J. proj. Tech.*, 1953, *17*, 312-319.

Coppinger, N. W., & Ammons, R. B. The Full-Range Picture Vocabulary Test: VIII. A normative study of Negro children. *J. clin. Psychol.*, 1952, *8*, 136-140.

Covell, L. L. Presbyopia: Comparative observations of white and Negro populations. *Amer. J. Ophthal.*, 1950, *33*, 1275-1277.

Dai, B. Some problems of personality development among Negro children. In C. Kluckholm, H. A. Murray, & D. M. Schneider (Eds.), *Personality in nature, society, and culture.* (2nd ed.) New York: Knopf, 1953.

D'Angelo, Rita Y. A comparison of white and Negro pre-school children in Goodenough I.Q. and language development. Unpublished master's thesis, Fordham Univer., 1950.

Davenport, R. K. Implications of military selection and classification in relation to universal military training. *J. Negro Educ.,* 1946, *15*, 585-594.

Davis, A. *Social class influence upon learning.* Cambridge, Mass.: Harvard Univer. Press, 1948.

Davis, A., & Eells, K. Manual for the Davis-Eells Test of General Intelligence on Problem Solving Ability. Yonkers, N. Y.: World Book, 1953.

Davis, A., & Havighurst, R. J. Social class and color differences in child rearing. *Amer. sociol. Rev.,* 1946, *11,* 698-710.

Davis, J. C. The scatter pattern of a southern Negro group on the Wechsler Bellevue Intelligence Scale. *J. clin. Psychol.,* 1957, *13,* 298-300.

DeStephens, W. P. Are criminals morons? *J. soc. Psychol.,* 1953, *38*, 187-189.

Dexter, Emily S., & Stein, Betty. The measurement of leadership in white and Negro students. *J. abnorm. soc. Psychol.,* 1955, *51,* 219-221.

Diggs, Mary H. Some problems and needs of Negro children as revealed by comparative delinquency and crime statistics. *J. Negro Educ.,* 1950, *19*, 290-297.

Dollard, J. *Caste and class in a southern town.* (2nd ed.) New York: Harper, 1949.

Dunn, L. C., & Dobzhansky, T. G. *Heredity, race, and society.* New York: Penguin, 1946.

Eagleson, O. W., & Bell, Eleanor S. The values of Negro women college students. *J. soc. Psychol.,* 1945, *22,* 149-154.

Eells, K., Davis, A., Havighurst, R. J., Herrick, V. E., & Tyler, R. *Intelligence and cultural differences.* Chicago: Univer. Chicago Press, 1951.

Ellis, A., & Beechley, R. Comparisons of Negro and white children seen at a child guidance clinic. *Psychiat. quart. Suppl.,* 1950, *24*, 93-100.

Engle, T. L. Personality adjustments of children belonging to two minority groups. *J. educ. Psychol.,* 1945, *36*, 543-560.

Espenschade, Anna. A note on the comparative motor ability of Negro and white tenth grade girls. *Child Develpm.*, 1946, *17*, 245-248.

Everhart, R. W. The growth and development of Negro and white elementary children with articulatory defects. *Dissertation Abstr.*, 1953, *13*, 335-336.

Eysenck, H. J., Granger, G. W., & Brengelmann, J. C. *Perceptual processes and mental illness.* New York: Basic Books, 1957.

Felton, Jean S. The Cornell index used as an appraisal of personality by an industrial health service: I. The total score. *Industr. Med.*, 1949, *18*, 133-144.

Ferrell, G. V. Comparative study of sex differences in school achievement of white and Negro children. *J. educ. Res.*, 1949, *43*, 116-121.

Fowler, W. L. A comparative analysis of pupil performance on conventional and culture-controlled mental tests. In, *Fourteenth Yearbook of the National Council on Measurements Used in Education.* Princeton, N. J.: Educational Testing Serv., 1957.

Frazier, E. F. *The Negro family in the United States.* Chicago: Univer. Chicago Press, 1939.

Frazier, E. F. Problems and needs of Negro children and youth resulting from family disorganization. *J. Negro Educ.*, 1950, *19*, 269-277.

Frazier, E. F. *The Negro in the United States.* New York: Macmillan, 1957.

Frumkin, R. M. Race and major mental disorders: A research note. *J. Negro Educ.*, 1954, *23*, 97-98.

Fulk, B. E., & Harrell, T. W. Negro-white army test scores and last school grade. *J. appl. Psychol.*, 1952, *36*, 34-35.

Gardner, G. E., & Aaron, Sadie. The childhood and adolescent adjustment of Negro psychiatric casualties. *Amer. J. Orthopsychiat.*, 1946, *16*, 481-495.

Garrett, H. E. Comparison of Negro and white recruits on the Army tests given in 1917-1918. *Amer. J. Psychol.*, 1945, *58*, 480-495. (a)

Garrett, H. E. A note on the intelligence scores of Negroes and whites in 1918. *J. abnorm. soc. Psychol.*, 1945, *40*, 344-346. (b)

Garrett, H. E. Psychological differences as among races. *Science*, 1945, *101*, 16-17. (c)

Garrett, H. E. Negro-white differences in mental ability in the United States. *Scient. Mon.*, 1947, *65*, 329-333.

Garth, T. A. A review of racial psychology. *Psychol. Bull.*, 1925, *22*, 343-364.

Garth, T. A. *Race psychology: A study of racial mental differences.* New York: McGraw-Hill, 1931.

Gilliland, A. R. Socio-economic status and race as factors in infant intelligence test scores. *Child Develpm.*, 1951, *22*, 271-273.

Ginzberg, E. *The Negro potential.* New York: Columbia Univer. Press, 1956.

Ginzberg, E., & Bray, D. W. *The uneducated.* New York: Columbia Univer. Press, 1953.

Golden, J. Characteristics of the Negro-white intermarried in Philadelphia. *Amer. sociol. Rev.*, 1953, *18*, 177-183.

Golden, J. Patterns of Negro-white intermarriage. *Amer. sociol. Rev.*, 1954, *19*, 144-147.

Goodenough, Florence L., & Harris, D. B. Studies in the psychology of children's drawings: II. 1928-1949. *Psychol. Bull.*, 1950, *47*, 369-433.

Goodman, Mary E. *Race awareness in young children.* Cambridge, Mass.: Addison-Wesley, 1952.

Graham, T. F. Doll play fantasies of Negro and white primary school children. *J. clin. Psychol.*, 1955, *11*, 29-33.

Gray, J. S., & Thompson, A. H. The ethnic prejudices of white and Negro college students. *J. abnorm. soc. Psychol.*, 1953, *48*, 311-313.

Gray, Susan W. The vocational preferences of Negro school children. *J. genet. Psychol.*, 1944, *64*, 239-247. (a)

Gray, Susan W. The wishes of Negro school children. *J. genet. Psychol.*, 1944, *64*, 225-237. (b)

Gray, Susan W. A note on the values of southern college women, white and Negro. *J. soc. Psychol.*, 1947, *25*, 239-241.

Greenberg, H., Chase, A. L., & Cannon, T. M., Jr. Attitudes of white and Negro high school students in a west Texas town toward school integration. *J. appl. Psychol.*, 1957, *41*, 27-31.

Guilford, J. P. The structure of intellect. *Psychol. Bull.*, 1956, *53*, 267-293.

Guilford, J. P. Three faces of intellect. *Amer. Psychologist,* 1959, *14*, 469-479.

Hammer, E. F. Frustration-aggression hypothesis extended to socio-racial areas: Comparison of Negro and white children's H-T-P's. *Psychiat. Quart.*, 1953, *27*, 597-607. (a)

Hammer, E. F. Negro and white children's personality adjustment as revealed by a comparison of their drawings (H-T-P). *J. clin. Psychol.,* 1953, *9,* 7-10. (b)

Hammer, E. F. Comparison of the performance of Negro children and adolescents on two tests of intelligence, one an emergency scale. *J. genet. Psychol.,* 1954, *84,* 85-93.

Harms, H. E., Kobler, F. J., & Sweeney, F. J. Negro Army psychiatric casualties. *War Med.,* 1945, *7,* 309-312.

Harrower-Erickson, Molly R., & Steiner, M. E. *Large scale Rorschach techniques.* Springfield, Ill.: Charles C Thomas, 1945.

Heine, R. W. The Negro patient in psychotherapy. *J. clin. Psychol.,* 1950, *6,* 373-376.

Helgerson, Evelyn. The relative significance of race, sex, and facial expression in choice of playmate by the preschool child. *J. Negro Educ.,* 1943, *12,* 617-622.

Helson, H., & Guilford, J. P. The relation of visual sensitivity to the amount of retinal pigmentation. *J. gen. Psychol.,* 1933, *9,* 58-76.

Hess, A. D. Controlling culture influence in mental testing: An experimental test. *J. educ. Res.,* 1955, *49,* 53-58.

Hewlett, Jane D. A comparison of the adjustment of white and Negro schizophrenics. *Smith Coll. Stud. soc. Work,* 1946, *17,* 108-109.

Heyman, Dorothy. Manifestations of psychoneurosis in Negroes. Ment. Hyg., 1945, *29,* 231-235.

Hoffman, M. L., & Albizu-Miranda, C. Middle class bias in personality testing. *J. abnorm. soc. Psychol.,* 1955, *51,* 150-152.

Hollingshead, A. B. Class differences in family stability. In R. Bendix & S. M. Lipset (Eds.), *Class, status and power: A reader in social stratification.* Glencoe, Ill.: Free Press, 1953. Pp. 284-292.

Hollingshead, A. B., & Redlich, F. C. Social stratification and psychiatric disorders. *Amer. sociol. Rev.,* 1953, *18* (2), 163-169.

Holmes, Jean B. A study of racial attitudes found in four colleges, including a scale to measure these attitudes. *J. Negro Educ.,* 1943, *12,* 69-75.

Horowitz, E. L. Development of attitude toward Negroes. In T. M. Newcomb & E. L. Hartley (Eds.), *Readings in social psychology.* New York: Henry Holt, 1947. Pp. 507-517.

Horst, P. Multiple classification by the method of least squares. *J. clin. Psychol.,* 1956, *12,* 3-16.

Hughes, J. H., & Thompson, G. G. A comparison of the value

systems of southern Negro and northern white youth. *J. educ. Psychol.*, 1954, *45*, 300-309.

Hunt, W. A. The relative incidence of psychoneurosis among Negroes. *J. consult. Psychol.*, 1947, *11*, 133-136.

Irwin, O. C. Infant speech. *Scient. Amer.*, 1949, *181* (3), 22-24.

Ivins, S. P. Psychoses in the Negro: A preliminary study. *Del. State med. J.*, 1950, *22*, 212-213.

Jenkins, M. D. Case studies of Negro children of Binet IQ 160 and above. *J. Negro Educ.*, 1943, *12*, 159-166.

Jenkins, M. D. The upper limit of ability among American Negroes. *Scient. Mon.*, 1948, *66*, 399-401.

Jenkins, M. D. Intellectually superior Negro youth: Problems and needs. *J. Negro Educ.*, 1950, *19*, 322-332.

Jenkins, M. D., & Randall, Constance M. Differential characteristics of superior and unselected Negro college students. *J. soc. Psychol.*, 1948, *27*, 187-202.

Kahn, Jane, Buchmueller, A. D., & Gildea, Margaret C. L. Group therapy for parents of behavior problem children in public schools: Failure of the method in a Negro school. *Amer. J. Psychiat.*, 1951, *108*, 351-357.

Kardiner, A., & Ovesey, L. *The mark of oppression: A psychosocial study of the American Negro.* New York: Norton, 1951.

Karon, B. P. *The Negro personality: A rigorous investigation of the effects of culture.* New York: Springer, 1958.

Karpinos, B. D. Visual acuity of selectees and Army inductees. *Hum. Biol.*, 1944, *16*, 1-14.

Keenan, V., & Kerr, W. A. Unfair employment practices as viewed by private employment counselors. *J. appl. Psychol.*, 1952, *36*, 361-364.

Kennedy, Janet A. Problems posed in the analysis of Negro patients. *Psychiatry*, 1952, *15*, 313-327.

Kephart, W. M. The Negro offender: An urban research project. *Amer. J. Sociol.*, 1954, *60*, 46-50.

Klineberg, O. *Race differences.* New York: Harper, 1935.

Klineberg, O. (Ed.), *Characteristics of the American Negro.* New York: Harper, 1944.

Klugman, S. F. The effect of money incentives versus praise upon the reliability and obtained scores of the revised Stanford-Binet test. *J. gen. Psychol.*, 1944, *30*, 255-269.

Knoblock, Hilda, & Pasamanick, B. Further observation on the behavioral development of Negro children. *J. genet. Psychol.*, 1953, *83*, 137-157.

Knox, E. O. The Negro as a subject of university research in 1944. *J. Negro Educ.*, 1945, *14*, 182-196.

Knox, E. O. The Negro as a subject of university research in 1948. *J. Negro Educ.*, 1949, *18*, 489-497.

Knox, E. O. The Negro as a subject of university research in 1951. *J. Negro Educ.*, 1952, *21*, 484-491.

Koch, Helen L. A study of some factors conditioning the social distance between the sexes. *J. soc. Psychol.*, 1944, *20*, 79-107.

Koch, Helen L. The social distance between certain racial, nationality, and skin pigmentation groups in selected populations of American school children. *J. genet. Psychol.*, 1946, *68*, 63-95.

Korchin, S. J., Mitchell, H. E., & Meltzoff, J. A critical evaluation of the Thompson Thematic Apperception Test. *J. proj. Tech.*, 1950, *14*, 445-452.

Landreth, Catherine, & Johnson, Barbara C. Young children's responses to a picture and inset test designed to reveal reactions to persons of different skin color. *Child Develpm.*, 1953, *24*, 63-80.

Lee, E. S. Negro intelligence and selective migration: A Philadelphia test of the Klineberg hypothesis. *Amer. sociol. Rev.*, 1951, *16*, 227-233.

Lee, E. S., & Lee, Anne S. The differential fertility of the American Negro. *Amer. sociol. Rev.*, 1952, *17*, 437-447.

Lewis, Claudia, & Biber, Barbara. Reactions of Negro children toward Negro and white teachers. *J. exp. Educ.*, 1951, *20*, 97-104.

Light, B. H. A further test of the Thompson TAT rationale. *J. abnorm. soc. Psychol.*, 1955, *51*, 148-150.

Lindzey, G. (Ed.) *Handbook of social psychology*. Cambridge, Mass.: Addison-Wesley, 1954.

Long, H. H. The relative learning capacity of Negroes and whites. *J. Negro Educ.*, 1957, *26*, 121-134.

Love, M. I., & Beach, S. Performance of children on the Davis-Eells Games and other measures of ability. *J. consult. Psychol.*, 1957, *21*, 29-32.

McCary, J. L. Reactions to frustrations by some cultural and racial groups. *Personality*, 1951, *1*, 84-102.

McCary, J. L. Picture Frustration Study normative data for some cultural and racial groups. *J. clin. Psychol.*, 1956, *12*, 194-195.

McCary, J. L., & Tracktir, J. Relationship between intelligence and frustration-aggression patterns as shown by two racial groups. *J. clin. Psychol.*, 1957, *13*, 202-204.

McGraw, Myrtle B. A comparative study of a group of southern white and Negro infants. *Genet. psychol. Monogr.*, 1931, *10*, 1-105.

McGurk, F. C. J. Comparative test scores of Negro and white school children in Richmond, Va. *J. educ. Psychol.*, 1943, *34*, 474-484.

McGurk, F. C. J. *Comparison of the performance of Negro and white high school seniors on cultural and non-cultural test questions.* Washington, D. C.: Catholic Univer. Amer. Press, 1951. (Microcard)

McGurk, F. C. J. On white and Negro test performance and socio-economic factors. *J. abnorm. soc. Psychol.*, 1953, *48*, 448-450. (a)

McGurk, F. C. J. Socio-economic status and culturally-weighted test scores of Negro subjects. *J. appl. Psychol.*, 1953, *37*, 276-277. (b)

Machover, S. *Cultural and racial variations in patterns of intellect.* New York: Teachers Coll., Columbia Univer., 1943.

McLean, Helen H. The emotional health of Negroes. *J. Negro Educ.*, 1949, *18*, 283-290.

MacPhee, H. M., Wright, H. F., & Cummings, S. B. The performance of mentally subnormal rural southern Negroes on the verbal scale of the Bellevue Intelligence Examination. *J. soc. Psychol.*, 1947, *25*, 217-229.

Malzberg, B. *Social and biological aspects of mental disease.* New York: N. Y. State Hospital Press, 1940.

Malzberg, B. Mental disease among Negroes in New York State, 1939-41. *Ment. Hyg.*, 1953, *37*, 450-476.

Malzberg, B. Marital status and mental disease among Negroes in New York State. *J. nerv. ment. Dis.*, 1956, *123*, 457-465. (a)

Malzberg, B. Mental disease among native and foreign-born Negroes in New York State. *J. Negro Educ.*, 1956, *25*, 175-181. (b)

Marks, E. S. Factors affecting skin color judgments of Negro college students. *Psychol. Bull.*, 1942, *39*, 577. (Abstract)

Marquart, Dorothy, & Bailey, Lois L. An evaluation of the Culture-Free Test of Intelligence. *J. genet. Psychol.*, 1955, *86*, 353-358.

Mayo, G. D., & Kinzer, J. R. A comparison of the "racial" attitudes of white and Negro high school students in 1940 and 1948. *J. Psychol.*, 1950, *29*, 397-405.

Meredith, H. V. Physical growth from birth to two years: I. Stature. A review and synthesis of North American research for the period 1850-1941. *U. Ia. Stud. child Welf.*, 1943, *19*, No. 407.

Michael, W. B. An investigation of the contribution of factors to tests and to their predictive value in two Army Air Force pilot populations. *Amer. Psychologist*, 1947, *2*, 417-418. (Abstract)

Michelson, N. Studies in the physical development of Negroes: II. Weight. *Amer. J. phys. Anthrop.*, 1943, *1*, 289-300.

Milam, A. T., & Sumner, F. C. Spread and intensity of vocational interests and evaluative attitudes in first year Negro medical students. *J. Psychol.*, 1954, *37*, 31-38.

Miner, J. B. *Intelligence in the United States: A survey—with conclusions for manpower utilization in education and employment.* New York: Springer, 1957.

Monahan, T. P., & Monahan, Elizabeth H. Some characteristics of American Negro leaders. *Amer. sociol. Rev.*, 1956, *21*, 589-596.

Montagu, M. F. A. *Man's most dangerous myth: The fallacy of race.* (3rd ed.) New York: Harper, 1952.

Moore, J. E. A comparison of Negro and white pre-school children on a vocabulary test and eye-hand coordination test. *Child Develpm.*, 1942, *13*, 247-252.

Moses, E. R. Differentials in crime rates between Negroes and whites, based on comparisons of four socio-economically equated areas. *Amer. sociol. Rev.*, 1947, *12*, 411-420.

Mundy, P. The young Negro worker in Washington, D. C. *J. Negro Educ.*, 1949, *18*, 104-113.

Mussen, P. H. Differences between the TAT responses of Negro and white boys. *J. consult. Psychol.*, 1953, *17*, 373-376.

Myrdal, G. *An American dilemma: The Negro problem and modern democracy.* New York: Harper, 1944. 2 vols.

Newland, T. E., & Lawrence, W. C. Chicago non-verbal examination results on an east Tennessee Negro population. *J. clin. Psychol.*, 1953, *9*, 44-47.

North, R. D. The intelligence of American Negroes. In M. M. Turmin (Ed.), *Segregation and desegregation: A digest of recent research.* New York: Anti-defamation League of B'nai B'rith, 1957.

Palmer, D. M., Sprang, H. E., & Hans, C. L. Electroshock therapy in schizophrenia: A statistical survey of 455 cases. *J. nerv. ment. Dis.*, 1951, *114*, 162-171.

Pasamanick, B. A comparative study of the behavioral development of Negro infants. *J. genet. Psychol.*, 1946, 69, 3-44.

Pasamanick, B., & Knobloch, Hilda. Early language behavior in Negro children and the testing of intelligence. *J. abnorm. soc. Psychol.*, 1955, *50*, 401-402.

Pasamanick, B., Knobloch, Hilda, & Lilienfeld, A. M. Socioeconomic status and some precursors of neuropsychiatric disorder. *Amer. J. Orthopsychiat.*, 1956, *26*, 594-601.

Pollack, O. A statistical investigation of the criminality of old age. *J. clin. Psychopath.*, 1944, *5*, 745-767.

Portnoy, B., & Stacey, C. L. A comparative study of Negro and white subnormals on the children's form of the Rosenzweig P-F Test. *Amer. J. ment. Defic.*, 1954, *59*, 272-278.

Prothro, E. T., & Jensen, J. A. Comparison of some ethnic and religious attitudes of Negro and white college students in the deep South, *Soc. Forces,* 1952, *30*, 426-428.

Prothro, E. T., & Perry, H. T. Group differences in performance on the Meier Art Test. *J. appl. Psychol.*, 1950, *34*, 96-97.

Pugh, T. J. A comparative study of the values of a group of ministers and two groups of laymen. *J. soc. Psychol.*, 1951, 33, 225-235.

Radke, Marian J., Sutherland, Jean, & Rosenberg, Pearl. Racial attitudes of children. *Sociometry,* 1950, *13,* 154-171.

Radke, Marian J., & Trager, Helen G. Children's perceptions of the social roles of Negroes and whites. *J. Psychol.*, 1950, *29*, 3-33.

Radke, Marian J., Trager, Helen G., & Davis, Hadassah. Social perceptions and attitudes of children. *Genet. psychol. Monogr.*, 1949, *40*, 327-447.

Ramsey, G. V. Sexual growth of Negro and white boys. *Hum. Biol.*, 1950, *22*, 146-149.

Rankin, R. E., & Campbell, D. T. Galvanic skin response to Negro and white experimenters. *J. abnorm. soc. Psychol.*, 1955, *51*, 30-33.

Reitzes, D. C. *Negroes and medicine.* Cambridge, Mass.: Harvard Univer. Press, 1958.

Rhoads, Teresa F., Rapoport, M., Kennedy, Ruth, & Stokes, J. Studies on the growth and development of male children receiving evaporated milk: II. Physical growth, dentition, and intelligence of white and Negro children through the first four years as influenced by vitamin supplements. *J. Pediat.*, 1941, *26*, 415-454.

Riess, B. F., Schwartz, E. K., & Cottingham, Alice. An experimental critique of assumptions underlying the Negro version of the TAT. *J. abnorm. soc. Psychol.*, 1950, *45*, 700-709.

Ripley, H. S., & Wolf, S. Mental illness among Negro troops overseas. *Amer. J. Psychiat.*, 1947, *103*, 499-512.

Roberts, S. O. The measurement of adjustment of Negro college youth: Personality scales for whites versus criteria intrinsic to Negro groups. Unpublished doctor's dissertation, University of Minn., 1944.

Roberts, S. O. Socio-economic status and performance of Negro college women, north and south, on the ACE. *Amer. Psychologist,* 1946, *1*, 253. (Abstract)

Roberts, S. O. Socio-economic status and performance on the ACE of Negro freshmen college veterans and non-veterans from the north and the south. *Amer. Psychologist,* 1948, *3*, 266. (Abstract)

Roberts, S. O. Socio-economic status and performance over a four-year period on the ACE of Negro college women from the north and south. *Amer. Psychologist,* 1950, *5*, 295. (Abstract)

Robinson, Mary Louise, & Meenes, M. The relationship between test intelligence of third grade Negro children and the occupations of their parents. *J. Negro Educ.*, 1947, *16*, 136-141.

Rodgers, E. G. The relationship of certain measurable factors in the personal and education backgrounds of two groups of Baltimore Negroes, identified as superior and average in intelligence as fourth grade children, to their educational, social, and economic achievement in adulthood. *Dissertation Abstr.*, 1957, *17*, 1709.

Rowntree, L. G. The unfit: How to exclude them. *Psychosom. Med.*, 1943, *5*, 324-330.

Saenger, G., & Gordon, N. S. The influence of discrimination on minority group members in its relation to attempts to combat discrimination. *J. soc. Psychol.*, 1950, *31*, 95-120.

St. Clair, H. R. Psychiatric interview experiences with Negroes. *Amer. J. Psychiat.*, 1951, *108*, 113-119.

Sarason, S. B., & Gladwin, T. Psychological and cultural problems in mental subnormality. *Genet. psychol. Monogr.*, 1958, *57*, 3-290.

Scarborough, B. B. Some mental characteristics of southern colored and white venereal disease patients as measured by the Wechsler-Bellevue Test. *J. soc. Psychol.*, 1956, *43*, 313-321.

Schermerhorn, R. A. Psychiatric disorders among Negroes: A sociological note. *Amer. J. Psychiat.*, 1956, *112*, 878-882.

Scholer, E. A. Attitudes toward racial segregation in Baton Rouge, Louisiana. *J. soc. Psychol.*, 1943, *18*, 33-53.

Schwartz, E. K., Riess, B. F., & Cottingham, Alice. Further critical evaluation of the Negro version of the TAT. *J. proj. Tech.*, 1951, *15*, 394-400.

Scott, R. B., Cardozo, W. W., Smith, A. D., & DeLilly, M. R. Growth and development of Negro infants: III. Growth during the first year of life as observed in private pediatric practice. *J. Pediat.*, 1950, *37*, 885-893.

Scott, R. B., Ferguson, Angella D., Jenkins, M. E., & Cutler, F. E. Growth and development of Negro infants: V. Neuro-muscular patterns of behavior during the first year of life. *Pediatrics*, 1955, *16*, 24-30.

Shuey, Audrey M. *The testing of Negro intelligence.* Lynchburg, Va.: Randolph-Macon Women's College, 1958.

Shuey, Audrey M., King, Nancy, & Griffith, Barbara. Stereotyping of Negroes and whites: An analysis of magazine pictures. *Publ. opin. Quart.*, 1953, *17*, 281-287.

Somner, R., & Killian, L. M. Areas of value differences: I. A method of investigation. II. Negro-white relations. *J. soc. Psychol.*, 1954, *39*, 227-235, 237-244.

Stainbrook, E., & Siegel, P. S. A comparative group Rorschach study of southern Negro and white high school and college students. *J. Psychol.*, 1944, *17*, 107-115.

Strong, E. K., Jr. Interests of Negroes and whites. *J. soc. Psychol.*, 1952, *35*, 139-150.

Strong, E. K., Jr. Are medical specialist interest scales applicable to Negroes? *J. appl. Psychol.*, 1955, *39*, 62-64.

Sumner, F. C. Neurotic tendency and socioeconomic status of Negro college women. *J. soc. Psychol.*, 1948, *28*, 291.

Sunne, D. Comparative study of white and Negro children. *J. appl. Psychol.*, 1917, *1*, 71-84.

Sussman, M. B., & Yeager, H. C., Jr. Mate selection among Negro and white college students. *Sociol. soc. Res.*, 1950, *35*, 46-49.

Sutherland, R. L. *Color, class, and personality.* Washington, D. C.: Amer. Council on Education, 1942.

Theman, Viola, & Witty, P. Case studies and genetic records of two gifted Negroes. *J. Psychol.*, 1943, *15*, 165-181.

Thompson, C. E. The Thompson modification of the Thematic Apperception Test. *J. proj. Tech.*, 1949, *13*, 469-478.

Thompson, D. C. Teachers in Negro colleges: A sociological analysis. *Dissertation Abstr.*, 1956, *16*, 1531-1532.

Thurstone, L. L. Primary mental abilities. *Psychometr. Monogr.*, 1938, No. 1.

Tietze, C., & Lewit, Sarah. Patterns of family limitations in a rural Negro community. *Amer. sociol. Rev.*, 1953, *18*, 563-564.

Tomkins, S. S., & Miner, J. B. *The Picture Arrangement Test.* New York: Springer, 1957.

Touchstone, F. V. A comparative study of Negro and white college students' aggressiveness by means of sentence completion. *Dissertation Abstr.*, 1957, *17*, 1588-1589.

Trager, Helen G., & Yarrow, Marian R. *They learn what they live.* New York: Harper, 1952.

Trent, R. D. The color of the investigator as a variable in experimental research with Negro subjects. *J. soc. Psychol.*, 1954, *40*, 281-287.

Turman, J. A., & Holtzman, W. H. Attitudes of white and Negro teachers toward nonsegregation in the classroom. *J. soc. Psychol.*, 1955, *42*, 61-70.

Turner, R. H. Occupational patterns of inequality. *Amer. J. Sociol.*, 1954, *59*, 437-447.

Tyler, Leona E. *The psychology of human differences.* (2nd ed.) New York: Appleton-Century-Crofts, 1956.

Valien, P., & Horton, Carrell. Some demographic characteristics of outstanding Negro women. *J. Negro Educ.*, 1954, *23*, 406-420.

Valien, P., & Vaughn, Ruth E. Birth control attitudes and practices of Negro mothers. *Sociol. soc. Res.*, 1951, *35*, 415-421.

Verin, Olga. Manifestation of racial conflict in Negro clients of a child guidance clinic. *Smith Coll. Stud. soc. Work*, 1944, *15*, 126-127.

Wheatley, L. A., & Sumner, F. C. Measurement of neurotic tendency in Negro students of music. *J. Psychol.*, 1946, *22*, 247-252.

Williams, Judith, & Scott, R. B. Growth and development of Negro infants: IV. Motor development and its relationship to child rearing practices in two groups of Negro infants. *Child Develpm.*, 1953, *24*, 103-121.

Wilson, D.C., & Lantz, Edna M. The effect of culture change on the Negro race in Virginia, as indicated by a study of state hospital admissions. *Amer. J. Psychiat.*, 1957, *114*, 25-32.

Winslow, C. N., & Brainerd, J. E. A comparison of the reaction of whites and Negroes to frustration as measured by the Rosenzweig Picture-Frustration Test. *Amer. Psychologist,* 1950, *5,* 297. (Abstract)

Witty, P. A., & Jenkins, M. D. The case of "B," a gifted Negro girl. *J. soc. Psychol.,* 1935, *6,* 117-124.

Witty, P., & Theman, Viola. A follow-up study of the educational attainment of gifted Negroes. *J. educ. Psychol.,* 1943, *34,* 35-47.

Woisika, P. H. An evaluation of the dark test. *Amer. intern. Med.,* 1944, *21,* 101-118.

Woods, R. C., & Martin, L. R. Testing in musical education. *Educ. psychol. Measmt.,* 1943, *3,* 29-42.

Woods, W. A., Boger, J., & Holman, G. An investigation of revised beta scores (Lindner Gurvitz Scale) among Negro adolescents. *Va. J. Sci.,* 1954, *4,* 321.

Woods, W. A., & Toal, R. Subtest disparity of Negro and white groups matched for IQs on the Revised Beta Test. *J. consult. Psychol.,* 1957, *21,* 136-138.

Woodworth, R. S. Comparative psychology of races. *Psychol. Bull.,* 1916, *13,* 388-397.

Young, Florene M., & Bright, H. A. Results of testing 81 Negro rural juveniles with the Wechsler Intelligence Scale for Children. *J. soc. Psychol.,* 1954, *39,* 219-226.

Yule, G. U., & Kendall, M. G. *An introduction to the theory of statistics.* (13th ed.) London: Griffin, 1949.

Genetics and Intelligence: A Review

L. ERLENMEYER-KIMLING
LISSY F. JARVIK

Nomothetic psychological theories have been distinguished by the tendency to disregard the individual variability which is characteristic of all behavior. A parallel between genetic individuality and psychologic individuality has rarely been

From *Science,* 1963, *142,* 1477-79. Reprinted by permission of the author and publisher.

drawn because the usual assumption has been, as recently noted in these pages (*1*), that the organisms intervening between stimulus and response are equivalent "black boxes," which react in uniform ways to given stimuli.

While behavior theory and its analytic methods as yet make few provisions for modern genetic concepts, the literature contains more information than is generally realized about the relationship between genotypic similarity and similarity of performance on mental tests. In a search for order among the published data on intellectual ability, we have recently summarized the work of the past half century (*2*). By using the most commonly reported statistical measure, namely, the correlation coefficient, it has been possible to assemble comparative figures from the majority of the investigations.

Certain studies giving correlations had to be excluded from this compilation for one of the following reasons: (i) type of test used (for example, achievement tests, scholastic performance, or subjective rating of intelligence); (ii) type of subject used (for example, mental defectives); (iii) inadequate information about zygosity diagnosis in twin studies (*3*); (iv) reports on too few special twin pairs.

The 52 studies (*2*) remaining after these exclusions yield over 30,000 correlational pairings (*4*) for the genetic relationship categories shown in Fig. 1. The data, in aggregate, provide a broad basis for the comparison of genotypic and phenotypic correlations. Considering only *ranges* of the observed measures, a marked trend is seen toward an increasing degree of intellectual resemblance in direct proportion to an increasing degree of genetic relationship, regardless of environmental communality.

Furthermore, for most relationship categories, the *median* of the empirical correlations closely approaches the theoretical value predicted on the basis of genetic relationship alone. The average genetic correlation between parent and child, as well as that between siblings (including dizygotic twins) is 0.50. The median correlations actually observed on tests of intellectual functioning are: 0.50 for parent-child, 0.49 for siblings reared together, and 0.53 for dizygotic twins, both the opposite-sex and like-sex pairs. Although twins are presumably exposed

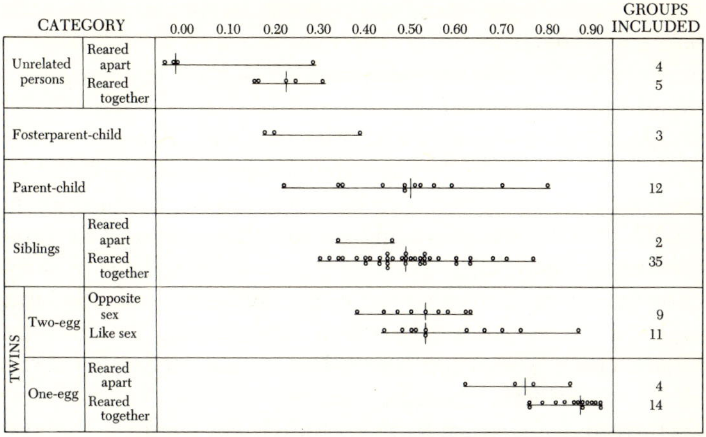

Fig. 1. Correlation coefficients for "intelligence" test scores from 52 studies. Some studies reported data for more than one relationship category; some included more than one sample per category, giving a total of 99 groups. Over two-thirds of the correlation coefficients were derived from I.Q.'s, the remainder from special tests (for example, Primary Mental Abilities). Midparent-child correlation was used when available, otherwise mother-child correlation. Correlation coefficients obtained in each study are indicated by dark circles; medians are shown by vertical lines intersecting the horizontal lines which represent the ranges.

to more similar environmental conditions than are siblings spaced apart in age, the correlations for mental ability do not indicate a sizable difference between the groups. Since only two studies dealt with siblings reared *apart*, it is possible to state only that the reported correlations for that group fall within the range of values obtained for siblings reared together and exceed those for unrelated children living *together*.

For unrelated persons in a large random-mating population, the theoretical genetic correlation is usually considered to be zero; for smaller populations, or those that deviate substantially from panmixia, however, the genetic correlation between presumably unrelated individuals in fact may be considerably higher. The observed median for unrelated persons reared apart is —0.01. Medians for unrelated individuals reared together (children reared in the same orphanage or foster home from an early age) and for the fosterparent-child group are 0.23 and 0.20, respectively. The relative contributions made by environmental similarity and sample selection to these deviations from zero are still to be analyzed.

At the other end of the relationship scale, where monozygotic twins theoretically have 100 percent genetic correlation, medians of the observed correlations in intellectual functioning are 0.87 for the twins brought up together, and 0.75 for those brought up apart (*5*). The correlations obtained for monozygotic twins reared together are generally in line with the intra-individual reliabilities of the tests. The median for the separated twins is somewhat lower, but clearly exceeds those for all other relationship groups.

In further reference to twin studies, our survey (*2*) shows that mean intrapair differences on tests of mental abilities for dizygotic twins generally are between 1½ to 2 times as great as those between monozygotic twins reared together. Such a relationship appears to hold also for the upper age groups, as suggested by a longitudinal study of senescent twins (*6*).

Taken individually, many of the 52 studies reviewed here are subject to various types of criticism (for example, methodological). Nevertheless, the over-all orderliness of the results is particularly impressive if one considers that the investigators had different backgrounds and contrasting views regarding the importance of heredity. Not all of them used the same measures of intelligence (see caption, Fig. 1), and they derived their data from samples which were unequal in size, age structure, ethnic composition, and socioeconomic stratification; the data were collected in eight countries on four continents during a time span covering more than two generations of individuals. Against this pronounced heterogeneity, which should have clouded the picture, and is reflected by the wide range of correlations, a clearly definitive consistency emerges from the data.

The composite data are compatible with the polygenic hypothesis which is generally favored in accounting for inherited differences in mental ability. Sex-linkage is not supported by these data (for example, under a hypothesis of sex-linkage the correlations for like-sex dizygotic twins should be higher than those for opposite-sex twins), although the possible effects of sex-linked genes are not precluded for some specific factors of ability.

We do not imply that environment is without effect upon intellectual functioning; the intellectual level is *not* unalterably fixed by the genetic constitution. Rather, its expression in the phenotype results from the patterns laid down by the genotype under given environment conditions. Two illustrations of the "norm of reaction" concept in relation to intellectual variability are seen in early total deafness and in phenylketonuria. Early deafness makes its stamp upon intellectual development, in that it lowers I.Q. by an estimated 20 score points (7). Phenylketonuria is ordinarily associated with an even greater degree of intellectual impairment. However, early alteration of the nutritional environment of the affected child changes the phenotypic expression of this genetic defect (8). Individual differences in behavioral *potential* reflect genotypic differences: individual differences in behavioral *performance* result from the nonuniform recording of environmental stimuli by intrinsically nonuniform organisms.

References and Notes

1. J. Hirsch, *Science,* this issue.
2. This material was included in a report presented at the XVII International Congress of Psychology, Washington, D.C., 1963 (L. Erlenmeyer-Kimling, L. F. Jarvik, and F. J. Kallmann). Detailed information about the data presented here is available upon request and is in preparation for publication.
3. This survey does include reports on opposite-sex (hence dizygotic) twin pairs from these studies.
4. Correlational pairings refer to the number of individual pairs used in deriving the correlation coefficients. Some investigators constructed a large number of pairings on the basis of a relatively small number of individuals. Altogether, we have been able to identify the following minimum numbers: twins, 3134 pairs (1082 monozygotic and 2052 dizygotic); sibs apart, 125 pairs plus 131 individuals; sibs together, 8288 pairs plus 7225 individuals; parent-child, 371 pairs plus 6812

individuals; fosterparent-child, 537 individuals; unrelated apart, 15,086 pairings; unrelated together, 195 pairings plus 287 individuals.

Correlational data are now available on 107 separated pairs of monozygotic twins from four series: H. H. Newman, F. N. Freeman, K. J. Holzinger, *Twins: A Study of Heredity and Environment* (Univ. of Chicago Press, Chicago, 1937); J. Conway, *Brit. J. Stat. Psychol. 11*, 171 (1958); N. Juel-Nielsen and A. Mogensen, cited by E. Strömgren, in *Expanding Goals of Genetics in Psychiatry*, F. J. Kallmann, Ed. (Grune and Stratton, New York, 1962), p. 231; J. Shields, *Monozygotic Twins Brought Up Apart and Brought Up Together* (Oxford Univ. Press, London, 1962).

L. F. Jarvik and A. Falek, *J. Gerontol. 18*, 173 (1963).

R. M. Salzberger and L. F. Jarvik, in *Family and Mental Health Problems in a Deaf Population*, J. D. Rainer *et al.*, Eds. (N.Y. State Psychiatric Institute, New York, 1963).

F. A. Homer, C. W. Streamer, L. L. Alejandrino, L. H. Reed. F. Ibbott, *New Engl. J. Med. 266*, 79 (1962).

Intelligent Testing[1]

ALEXANDER G. WESMAN

The nature of intelligence has been a favorite subject for contemplation and disputation for centuries—perhaps from the dawn of man as Homo sapiens. The topic is being studied and debated today by educators, sociologists, geneticists, neurophysiologists, and biochemists, and by psychologists specializing in various branches of the discipline. Despite this attention and effort, however—or perhaps *because* of it—there appears to be no more general agreement as to the nature of intelligence or the most valid means of measuring intelligence than

[1] Presidential Address presented to Division 5 at the meeting of the American Psychological Association, Washington, D. C., September 1967.

From *American Psychologist*, 1968, 23, 267-74. Reprinted by permission of the author and the American Psychological Association.

was the case 50 years ago. Concepts of intelligence and the definitions constructed to enunciate these concepts abound by the dozens, if not indeed by the hundreds.

With so many diverse definitions of intelligence, it is perhaps not surprising that we cannot agree on how to measure intelligence. It is my conviction that much of the confusion which plagued us in the past, and continues to plague us today, is attributable to our ignoring two propositions which should be obvious:

1. Intelligence is an attribute, not an entity.
2. Intelligence is the summation of the learning experiences of the individual.

We have all too often behaved as though intelligence is a physical substance, like a house or an egg crate composed of rooms or cells; we might better remember that it is no more to be reified than attributes like beauty, or speed, or honesty. There are objects which are classifiable as beautiful; there are performances which may be characterized as speedy; there are behaviors which display honesty. Each of these is measurable, with greater or lesser objectivity. Because they can be measured, however, does not mean they are substances. We may agree with E. L. Thorndike that if something exists it can be measured; we need not accept the converse notion that if we can measure something it has existence as a substance.

Intelligence as here defined is a summation of learning experiences. The instances in which intelligent behavior is observed may be classified in various ways that appear to be logical or homogeneous, but they are individual instances all the same. Each instance represents a response the organism has learned; each learned response in turn predisposes the organism for learning additional responses which permit the organism to display new acts of intelligent behavior.

For our present purposes, it matters little whether we are more comfortable with stimulus-response bonds, with experience-producing drives, with imprinting, or with neuropsychological explanations of *how* or *why* learning occurs; whatever the learning theory, the fundamental principle is universal. We start with an organism which is subject to modification by interaction with the environment; as a product of that interaction, the organism has been modified. Further interaction involves a changed organism—one which is ready to interact with its environment in a new way.

Organisms may differ from one another in their susceptibility to modification. One organism may need a more potent stimulus to trigger reaction to the environment than does another. A particular organism may respond to a given class of stimuli more readily than it does to other kinds of stimuli. Organisms may differ from one another in their readiness to respond to different classes of stimuli. There may be important differences in the ability of organisms to modify their behavior in effective ways as a result of experience.

We may develop and investigate hypotheses as to whether such differences in response as are displayed arise from variations in neurological endowment or in conducive environment. All that we can be sure of, at least as of now, is that what we are dealing with is a response-capable organism which has been exposed to environmental stimuli, has interacted in some way with those stimuli, and has been modified thereby.

The bits or modules which constitute intelligence may be information or may be skill; i.e., they may be content or process. Furthermore, they are multidimensional, and some modules may have more dimensions than do others. Each module is subject to essential change as the individual is exposed to further learning experiences. Each act of learning serves to create new modules, change the existing ones, or both. Modules are not independent; rather, they may overlap with several or many other modules; thus, they are complex both in their number of dimensions and in their interrelationships. Even early percepts are rarely if ever simple. A toy ball when first seen has at least size, shape, and color; if it is touched, it has texture and hardness as well. Accordingly, few if any modules of learning are truly simple.

The whole of a person's intelligence at any given moment may well be thought of as an amorphous mass—not a regular geometric figure. Within this mass, modules may cluster with greater or lesser permanence, and may be organized along principles of relatedness. Thus, word knowledge may form a cluster—but the words of which one has knowledge will be components of other clusters as well. A pencil is an object one writes with; it has shape in common with other objects, it has function in common with pens and crayons, it produces color of varying intensity, it has a number property, it is usually associated with paper. The learned module "pencil" may thus be part of many clusters.

One need not posit that a learning module is permanent. It could, presumably, disappear entirely, although far more often we would expect it to undergo essential change by taking on a more complex character. This model does assume that higher learning depends so intimately and essentially on certain previous learnings that the more complex modules cannot exist without the antecedent modules from which they grew. For example, if the ability to subtract numbers should disappear, the ability to do long division could not remain unaffected. Thus, retention of learning is integral to the concept here proposed.

The simple-minded conceptualization outlined above may have horrified those of my colleagues who are even moderately sophisticated with respect to modern learning theories. To those colleagues I apologize, but I also beg their indulgence. Oversimplified as the conceptualization undoubtedly is, I believe it does no *essential* violence to any current theory; it has, I hope, the virtue of permitting a view of the organization of intelligence, and of the nature of the testing of intelligence, which may prove illuminating for several issues which confront us.

Issue I: The Classification of Ability Tests into Aptitude, Achievement, and Intelligence Measures

As soon as we have agreed that what we know and what we can do intellectually is learned, the artificiality of the above classification becomes self-evident. Historically, we have recognized that what achievement tests measure is what the examinee has learned. We have been less ready to accord similar recognition to intelligence tests. In their case, we have too often behaved as though what these tests measure is somehow independent of the learning phenomenon. We have played the role of Aladdin seeking a magical lamp, complete with a genie ready to spring forth with full power to accomplish all sorts of wondrous things. We have pondered wistfully on the number of critical issues that would be resolved if we could only somehow measure "intelligence" separately from "achievement."

We have been similarly unrealistic in treating the concept of "aptitude." Our textbooks enunciate the distinction that aptitude tests measure what the individual *can* learn, while achievement tests measure what he *has* learned. Some of our leading theorists aggravate the confusion by ignoring the implications of their special use of the term. "Aptitude" is typically used in laboratory learning experiments as a matching or otherwise controlling variable; it is employed to assure that groups to be compared start the experiment as equal in initial ability. One gets a strong impression that the aptitude instrument is perceived as measuring the innate potential of the individual as distinguished from what is to be achieved (i.e., learned) in the experimental process. If learning theorists recognize that what they are calling "aptitude" (or, for that matter, "intelligence") is "previously learned" (as, clearly, at least some of them do), the artificiality of the distinction between "aptitude" or "intelligence" and "achievement" should be eminently apparent.

I wish that at least a few of my psychometric colleagues would leave off searching for *the* structure of intelligence, and devote their wisdom and energy to learning more about the learning process, and to teaching learning theorists about testing. I am convinced that both specialties would profit immeasurably from the cooperative enterprise. It is my strong impression that the inattention of the psychometrician to the facts of learning is matched fully by the unsophisticated treatment accorded to testing by many learning theorists.

All ability tests—intelligence, aptitude, and achievement—measure what the individual *has* learned—and they often measure with similar content and similar process. Let us take, for example, an item[2] such as this: A square and a rectangle have

[2] This item was proposed by G. K. Bennett in another context as an example of an arithmetic problem which might be correctly answered by any of several methods.

the same perimeter. The square has an area of 10,000 square feet. The rectangle has an area of 9,324 square feet. What are the dimensions of the rectangle?

This item would clearly be deemed appropriate whether it appeared in an achievement test in high school mathematics, a test of aptitude for mathematics, or the numerical portion of an "intelligence" test. I submit that a great many items can equally readily fit any of the three categories.

Such justification as we have for our labeling system resides entirely in the *purpose* for which the test is used, not in the test document itself. If our intent is to discover how much the examinee has learned in a particular area, such as a school course, we may select items which probe for the distinctive learnings the schooling was intended to stimulate. We label the test an "achievement" test. If our intent is to predict what success an individual is likely to attain in learning a new language, or a new job, we seek those specific previous learnings the possession of which bodes favorably for that future learning, and we label the test an "aptitude" test or a "special aptitude test." If our intent is to predict future acquisition of learning over broad areas of environmental exposure, we seek those previous learnings the possession of which will be relevant to as many, and as important, future learning situations as we can anticipate. This test we label an "intelligence" test. The selection of test items or sample tasks for the three purposes may or may not differ; but in each instance what is measured is what was previously learned. We are not measuring different abilities; we are merely attending to different criteria. It is the *relevance* of the learnings we select for investigation that largely determines how we name our test, and whether we will succeed in our purpose.

Issue II: The Utility of Culture-Free and Culture-Fair Tests

The notion of relevance of previous learnings leads naturally to a consideration of some follies we have committed in the search for culture-free or culture-fair instruments. I do not wish to impugn the high social motives which stimulate the search for such devices; I do wish to question that such a search, in its usual setting, is sensible. A culture-free test would presumably probe learnings which had not been affected by environment; this is sheer nonsense. A culture-fair test attempts to select those learnings which are common to many cultures. In the search for experiences which are common to several different cultures or subcultures, the vital matter of relevance of the learning for our purpose is subordinated or ignored.

The implicit intent in the attempt to create culture-free or culture-fair tests is somehow to measure intelligence without permitting the effects of differential exposure to learning to influence scores. This contains the tacit assumption that "native intelligence" lies buried in pure form deep in the individual, and needs only to be uncovered by ingenious mining methods. If we recognize that intelligence comprises learning experiences, it becomes clear that our attempts are not ingenious, but ingenuous.

It is true that we can probe learnings that have occurred in nonverbal, nonnumerical domains. This means only that we can test selected aspects of intelligence. The question immediately arises of the relevance of these special domains to the kinds of learnings we will want to predict. The measurement purpose for which culture-fair tests are ordinarily developed is that of predicting academic or industrial performance. Most academic courses and most industrial jobs involve some use of verbal abilities. Since further learning is conditioned by relevant past learnings, the individual who has developed more of the prerequisite ability inevitably has an advantage over the individual with less of the prerequisite ability. If we wish to predict whether an individual will profit appreciably from additional exposure to learning, our best predictor must be a measure which appraises what prerequisite learning he has acquired heretofore. Appropriate verbal abilities are more relevant to the largely verbal learning we usually wish to predict than other abilities are.

It has on occasion been suggested that tests be developed which sample the verbal skills or factual information which are peculiar to a given subculture. Such tests are proposed as a "fairer" measure of the "intelligence," or readiness to learn, of the members of that subculture. The response to this proposal is "readiness to learn *what*?" If our purpose is to distinguish members of that subculture from their peers with respect to how much of that special culture they have assimilated, such a test might well be useful. If, as is more likely the case, we wish to predict future learnings of the content of the more general culture (e.g., the so-called white, middle-class culture such as typifies what the majority of our schools are organized to transmit), tests designed for the subculture will be less relevant than those which sample from the general culture. This is not intended to imply that the members of the subculture *could* not learn what the schools as constituted are offering. It does emphasize that, at the moment at which we make our appraisal, what the individual has already learned from the general culture domain is the most significant information as to what he is then ready to learn. The less relevant the previous learnings we appraise, the more hazardous must be our predictions of future learnings.

As long as our educational system and our general culture are dependent on conventional verbal abilities, those who aspire to progress in that system and that culture will need to command those abilities. In a verbal society, verbal competence cannot sensibly be ignored.

Issue III: "Verbal Ability" Synonymous with "Intelligence"?

To say that we cannot afford to ignore learnings in relevant verbal areas when we are appraising "intelligence" does not imply that *only* the verbal domain is important. The development of tests of "general mental ability" which sample only the verbal domain implies that since verbal tests predict school criteria best, it is unnecessary to attend to other cognitive abilities the student has developed; in other words, that, in effect, "verbal ability" is synonymous with "intelligence." It would be most unfortunate if, consciously or unconsciously, we adopted this too narrow perspective.

That verbal tests are typically good predictors of grades in many academic courses is undeniable. *Why* this is the case warrants some thought. Is it because all, or even most, of what constitutes "intelligence" is represented by verbal ability? Certainly the chief symbol system of our society is verbal. Even when we deal with numerical, spatial, or figural problems we often transform them to verbal expressions. It is one thing, however, to recognize the involvement of verbal abilities in all kinds of learning experiences and quite another to grant them exclusive sovereignty over learning domains. Many domains require the possession of other abilities as well, but our appraisal methods are often inadequate to reveal that need. Because it is easier to employ verbal criteria, or more convenient—or because we have given insufficient thought to criterion validity—we predetermine the finding that verbal abilities dominate the scene.

A particularly revealing demonstration of this phenomenon came to the attention of the authors of the Differential Aptitude Tests some years ago. Grades in an auto mechanics course were found to be better predicted by the Verbal Reasoning test of the DAT than by the Mechanical Reasoning test. We had the unusual good fortune of having access to further information about the course. We discovered that early in the course the teacher had been called from the room for almost a half-hour. In his absence, the students had disobeyed his instructions not to fool around with the automobile motors. To let the punishment fit the crime, he conducted the rest of the course almost entirely by lecturing, giving the students minimum opportunity for actually working with the engines. That grades in a course taught by lecture and evaluated by a written test should be best predicted by a verbal test is not too surprising!

An illustration such as the above should force us to stop and think. As we study tables replete with validity coefficients, how many of those coefficients represent similar instances? As we develop hypotheses as to the importance of particular aspects of intelligence, how well do we understand the *criteria* which gave rise to the coefficients on which our hypotheses are based? Would the use of more valid criteria in courses for which curricular goals transcend verbal skills, have produced similar data, or different? Would the admittedly broad pervasiveness of verbal skills seem quite so broad if more appropriate measures of learning were employed? If we remain complacent about criteria composed largely of behaviors from the verbal domain, we are unlikely to see the relevance of other abilities.

In his APA presidential address in 1964, McNemar paid flattering attention to the Differential Aptitude Tests; he quite accurately reported that the verbal tests were most frequently the best predictors of course grades. The data he cited certainly supported the point he was making: Verbal tests predict grades in many academic courses. What might well have been added was recognition that the nature of our educational criteria exaggerates the real importance of verbal skills. If (and it is hoped *when*) grades or other criterion statements become more content valid, the relevance of a number of other skills will be more demonstrable.

Industry has perforce learned this lesson. Few mechanical apprentices are selected solely, or even primarily, because they can describe a process, rather than perform it. The military has learned that the ability to diagnose a malfunctioning torpedo is poorly demonstrated by verbal exposition, but well demonstrated by a work sample requiring actual mechanical repairs. It is to be hoped that education will increasingly become more realistic with respect to what *its* criteria *should* be.

Issue IV: The Growth and Decline of "Intelligence"

So preoccupied have we been with reifying intelligence as some mystical substance that we have too often neglected to take a common-sense look at what intelligence tests measure. We find ourselves distressed at our failure to predict with satisfactory accuracy the intelligence test scores of a teen-ager from his intelligence test scores as an infant. Why should this occasion surprise, let alone distress? If we look inside the tests, it should be obvious that the kinds of learnings we typically appraise at the earlier ages bear little resemblance, and may have little relevance, to the kinds of learnings we appraise later.

At the earlier age levels, we have typically tested for such characteristics as motor dexterity, perception, and similar features of physical development. When intellectual concepts become available for testing as baby grows to infant, to child, to teen-ager, we change the focus of our testing from the physical domains to the cognitive—we appraise knowledge, concept formation, and reasoning.

It is possible that future research will disclose that possession of certain physical abilities or tendencies is prerequisite to the development of concept formation, and that earlier possession of these characteristics will foretell the future intellectual development of the individual. Interesting and promising research now being conducted is directed toward this goal. It is my opinion that, because learning experiences vary so from one child to another, there is some practical limit beyond which we will be unable to predict, however penetrating our research. In any event, we would do well at this moment to recognize that since we are measuring in different ability domains at infant and school-age levels, we should not expect good prediction from one level to the other—and we should certainly not behave as though the data permitted confident prediction.

At the other end of the age spectrum we have, with similar lack of insight, proceeded to corollary errors. We have accepted the gloomy dictum that once we have passed the age of 18, or 25, or 35, we have passed our peak; from that age, our ability to learn declines. Our texts are peppered with charts showing that depressing downhill slide. What is the basis for this widely accepted conclusion? The principal basis is that when we apply our conventional measures of intelligence to older people, we find that average scores decrease. We have implicitly accepted the idea that intelligence is defined by what is measured by these particular intelligence tests. If, however, we return to our previous formulation of intelligence as what we know in a wide variety of domains, and hence as a base for what we can learn at a given moment, our perspective changes. We then proceed to compare what the intelligence tests measure with the kinds of learning individuals have engaged in from age 30 or 40 on. The relevance of the tests, except perhaps as measures of retention, is seen as increasingly remote with each passing year. Most individuals have not failed to learn more with added years of life. Their learnings have occurred in areas (science, business, politics, psychology, psychometrics), often relatively specialized, which are not measured by conventional intelligence tests.

It is true that new learnings of adults occur in such a variety of endeavors that it would be virtually impossible to find a common core on which all could be examined. We should not, however, pretend we do not suffer this psychometric disability; we should not continue to use less relevant measures to support deceptive graphs and curves of the decline of "intelligence." We might better recognize the limitations of our measure, until such time as we can devise relevant measures of the significant learnings which do occur. For the present, we can conclude only that with each passing decade older people do less well on tests designed for younger adults.

Issue V: The Search for Purity

A discussion of the nature of intelligence, and of intelligent testing, should not ignore the topic of factor analysis. It is a method which has influenced test construction and test selection. It is a technique which has stimulated the promulgation of theories of the structure of intellect.

The history of psychometrics gives evidence that each new major technique has attained a heyday of popularity, during which unrealistic hopes led to unbridled use. In the 1920s and 1930s, Pearson product-moment coefficients held the stage; everybody seemed to be correlating everything with everything else with wild abandon. We appear, in more recent times, to have been engaging in factor analyses with almost equal frenzy. With so much activity going on, it is perhaps to be expected that some studies, and some conclusions, would be characterized more by enthusiasm than by wisdom.

To criticize factor analysis as a procedure because individuals have misled themselves through its use would be very silly indeed. Among the benefits it has provided are the ability to summarize vast masses of data, and to facilitate the organization of information in a way that inspires, and then leads to investigation of interesting and often fruitful research hypotheses. At the same time, we need not believe that the power of the tool assures the validity of the product. Some of the conclusions which have been drawn, some attitudes which have been adopted, and some theories which have occasionally been treated as though they were established fact might well be exposed to scrutiny.

There have been instances in which a test battery was chosen for practical use *because* it had its origins in a program of factorial research. Presumably, the rationale was that such a battery consists of relatively "pure" tests, and would show near-zero intercorrelation among the tests; it would therefore be more efficient than a battery of similar measures not derived from factorial studies. If this rationale survived empirical study, it would still not of itself be adequate justification for selecting one set of tests rather than another. Efficiency is certainly desirable—but *validity* is *crucial.* How tests were constructed is interesting and even germane; how they *work* is the critical issue.

Let us return, however, to the rationale. Is the leap from "factorial origin" to "purity" defensible? The "pure" tests developed in psychometric laboratories often do correlate very little with one another. To some degree, at least, this low correlation is frequently ascribable to the unreliability of short, experimental tests, or to restriction in range of the various abilities of the subject, or both. For exploratory and research purposes, these conditions represent a reasonable situation. Practical test use situations are something else again.

When batteries of reliable tests with factorial ancestry, and batteries testing in similar domains but not factor oriented, are given to the same students, the within-battery intercorrelation of scores is ordinarily of about the same order. For example, with one ninth-grade group of boys, the average inter-r among the Differential Aptitude Tests was .37; for the same group, the average inter-r of the Primary Mental Abilities Tests was

.36. Similar results were obtained in a comparison of the DAT and the General Aptitude Test Battery scores for a twelfth-grade group. Thus, there was little evidence of greater "purity" in the factorially derived batteries than in the DAT, which were not so derived. (In the everyday world, it appears, "purity" is all too likely to be an illusion.) Accordingly, we would be well advised when choosing tests for practical use to concentrate on how they work, not on how they were built.

Let us now turn briefly to the role of factor analysis as a stimulator of hypotheses concerning the structure of intellect. Its influence has often seemed to be not so much mathematicodeductive as mathematico*se*ductive! The power of the method as a way of manipulating great masses of data appears all too often to have led us astray. Even our more eminent protagonists of the technique have not always appeared immune. When expounding on the theory of factor analysis, experts almost invariably agree that factors are merely descriptive categories; they are not functional entities. But when engaged in interpreting the factors which have emerged from their studies, some analysts apparently succumb to the mystic charm of rotating axes and perceive entities which, they have told us, do not exist. The lure of the temptation to discover a psychological structure analogous to the periodic table of the elements is too powerful to resist. We then hear of "primary mental abilities" or are shown "the three faces of intellect." Though the authors of such creations have sometimes demonstrated in other writings that they well understand the difference between the reality of descriptive categories and the illusion of underlying entities, some of their disciples and many of their readers seem less clear in their perception.

If we accept the thesis that the modules or bits which constitute intelligence are themselves complex, a combination of such modules can hardly be expected to be simple or "pure." A 6-year-old who assembles three alphabet blocks to spell out "cat" has employed, at a minimum, verbal and spatial skills; if he is aware that there are three blocks or letters, he has engaged in numerical perception as well. The ability to peform the task has required cognition, memory, convergent thinking, and evaluation. The product is figural, symbolic, and semantic.

All this, and we have not yet taken into account such considerations as the motor-manipulative activity, the perception of color, the earlier learning experiences which enabled him to perform the task successfully, or the imagery which the concept "cat" induces in him. We, as analysts, may choose to attend to only a single aspect of the behavior—but the behavior itself remains multifaceted and complex. To assume that we can abstract from a host of such activities a pure and simple entity is to ignore the psychological meaning of intelligent behavior.

Let us continue to explore, by all means available to us (including factor analysis) the nature of man's abilities. Let us *not* assume that the results of research obtained under closely managed conditions in the laboratory will hold true as realities in day-to-day situations. Let us not unwittingly forget that the descriptive categories we adopt for convenience in communication do not have real existence as ultimate psychological entities.

Conclusion

To what view of a structure of intellect am I led by the ideas I have enunciated here? Essentially, I believe intelligence is *un*structured. I believe that it is differently comprised in every individual—the sum total of all the learning experiences he has uniquely had up to any moment in time. Such structure as we perceive is structure which we have imposed. We can so select samples of previous learnings to examine as to reveal a general factor, or group factors, or specifics. We can sample from domains that are relatively homogeneous and apply labels such as verbal, numerical, spatial; we can sample from a wider variety of learnings, and apply labels such as "general mental ability" or, simply, "intelligence."

There are many bases on which we may choose which kinds of learnings we will sample. The most reasonable basis, I believe, is that of predictive purpose. Those previous learnings should be probed which are most relevant to the particular future learnings we wish to predict. In addition to criterion—

or, more likely, *criteria*—relevance, the principles of band width and fidelity (as enunciated by Cronbach and Gleser) might well serve as guides. If we are interested in forecasting narrow-band criteria, selection of highly homogeneous, directly relevant behaviors is indicated. If we are interested in a wide range of criteria, we have at least two options: we may choose to select small samples from widely scattered domains—as in a Binet, a Wechsler, or a broader gauge instrument still to be devised—or examine more intensively with several narrower gauge tests, as in the Differential Aptitude Tests. The broader gauge instruments will offer economy, but lesser fidelity for selected criteria. The narrower gauge instruments will be longer and more time consuming—but the possibility of more direct relevance to one or more particular criteria should permit higher fidelity.

The critical issue, then, is not which approach measures intelligence—each of them does, in its own fashion. No approach save sampling from every domain in which learnings have occurred—an impossible task—fully measures intelligence. The question is rather which approach provides the most useful information for the various purposes we wish the test to serve. Recognition that what we are measuring is what the individual has learned, and composing our tests to appraise *relevant* previous learnings, will yield the most useful information. We, and those who utilize the results of our work—educators, personnel men, social planners—face problems for which intelligence test data are relevant, and sometimes crucial. We must remember, and we must teach, what our test scores really reflect. The measurement of intelligence is not, and has not been, a matter of concern only to psychology. It has always been, and continues to be, an influence on educational and social programs. If we are to avert uninformed pressures from government agencies, from school administrators, from the courts, and indeed from psychologist colleagues, we must understand and we must broadly communicate what these scores truly represent. Only then can we who build tests and they who use them properly claim that we are indeed engaged in intelligent testing.

SECTION IV

Directions for the Future

FOUR PAPERS ARE INCLUDED in this final section. In each case, the author points sharply to a subject matter which psychology has either ignored or examined improperly. The impetus of this collection is as follows:

a. Behavior-genetics is a research discipline in its own right and one whose cultivation is essential if both psychologists and geneticists are to be spared a repetition of former errors (Hirsch).

b. Because psychology's "Snark" turned out to be a "Boojum," today's behavioral psychologist possesses a wealth of information about an organism which may be representative of itself alone! (Beach)

c. Multi-species comparisons offer sufficient evidence in support of the view that the laws governing ability, performance, capacity, etc., may depend heavily upon the organism studied and the developmental level attained by the organism at the time of study (Scott).

d. Individual differences, like it or not, remain one of psychology's central issues. Even when the environment has been homogenized to all practical limits, different people behave differently. They do so for reasons that are not simply reducible to genetic *or* environmental factors (Dobzhansky).

Collectively, these articles outline a new discipline and the modification of several older ones. Psychology can no longer rest securely within a typological conceptual framework. Learning "in rats," emotionality "in man," and intelligence "in Negroes" are examples of what Hirsch decries as the *uniformity* hypothesis. This hypothesis assumes the presence of processes whose characteristics are essentially identical within all members of a species. Departures from a given "norm" are treated either as "errors" or as "environmental" (e.g., experimental) manipulations. Thus, the billions of albino rats which have lived and which will live—most of them with unique genotypes—become The White Rat. By analogous logic, society and some scientists arrive at The Negro and The Greek. Extending typology from the domain of individuals to that of biological processes allows concepts such as The Learning Process or The Personality to emerge. The concepts are treated as if they were independent of the individuals within whom they are presumed to occur.

Through an even greater inductive leap, the uniformity hypothesis hurdles over successive species such that The Learning Process which is studied in The White Rat is applied to concepts of The Personality in The Human Being. It is little wonder that Boojums and Snarks are in such violent disagreement!

Of course, the foregoing is not intended to imply that a different scientific psychology must be authored for each individual member of each species of interest. Processes, after all, do go on in generally similar fashion among different individuals. By the same token, departures from some empirically determined norm are not "errors" nor do they necessarily require "correction." Biological individuality conspires with environmental pressures at various times in the development of an organism to create behavioral phenomena of a most complex character. Homogenization of phenotypes through environmental manipulation may obscure genetypic heterogeneity —*for one generation.* Society must not be surprised to find among its membership widely varying reactions to what society considers to be "the good life," "success," and "character." As long as societies homogenize environments, teaching practices, standards of excellence, and concepts of ethics and morality, and as long as they do so in the behalf of genotypically unique members, human achievement will continue to be more of a gift from without than an expression of what is within.

Behavior Genetics and Individuality Understood

JERRY HIRSCH

Individual differences are no accident. They are generated by properties of organisms as fundamental to behavioral science and biology as thermodynamic properties are to physical science. Much research, however, fails to take them into account. The behavioral sciences have attempted to erect a superstructure without paying sufficient attention to its foundation. A uniformity of expression over individuals, and even across species, has too often been assumed for behaviors under study. The uniformity assumption is explicitly incorporated into a spate of mathematical models that have been developed to formalize the study of behavior: Bush and Mosteller (*1*), for example, built theirs for "organisms that can be considered 'identical' at the start of an experiment. . . ." Rosner (*2*) speaks of "a fundamental attitude" which keeps psychophysics (*3*) "oriented toward the sources of uniformity in behavior." In this article I consider some effects that such assumptions about heredity, individuality, and behavior have had on the behavioral sciences.

Three Approaches to Behavior

In the study of behavior, three points of view can be distinguished. (i) Only common properties of behavior are studied among individuals and species. (ii) Only common properties

From *Science*, 1963, *142*, 1436-42. Reprinted by permission of the author and publisher.

of behavior are studied among individuals, while both similarities and characteristic differences are studied among species. (iii) Similarities and differences are studied among individuals, populations, and species.

The first view prevails when an organism is used as a tool for studying behavioral correlates of stimulus conditions, reinforcement schedules, deprivation regimens, pharmacological agents, or physiological mechanisms. It is hopefully assumed that the form of any relation observed—for example, that between stimulus and response—will have universal generality. The organism's role is essentially that of an analyzer, like the role of the Geissler tube in physics. In their illuminating discussion "The misbehavior of organisms," The Brelands (*4*), drawing on over 14 years of faithful application of the methods and assumptions of behaviorism, show that behaviorism also assumes "that the animal comes to the laboratory as a virtual *tabula rasa,* that species differences are insignificant, and that all responses are about equally conditionable to all stimuli." They relate (*4*) a history of "egregious failures" which they feel "represent a clear and utter failure of conditioning theory."

From the second viewpoint the behavior of animals is as characteristic of their species as is their form. This view prevails in ethologically oriented studies—for example, studies of such instincts as reproductive, parental, or territorial behavior. All members of a species are assumed to manifest a given behavior pattern, in some typical way. In Mayr's cogent analysis (*5*) this represents typological thinking whose replacement "by population thinking is perhaps the greatest conceptual revolution . . . in biology."

The third approach characterizes behavior genetics: the study of the relations between the genetic architecture of a taxon and the distributions of its behavioral phenotypes. It employs the methods of both the behavioral sciences and genetics. The growth of this field can be attributed to protest against the counterfactual uniformity postulate, combined with the realization that we can now have a description and analysis of behavior based on a deeper understanding of the materials on which the behavioral sciences make their observations.

The key to our present understanding of the structure of life came during the first half of this century, from investigations of transmission cytogenetics (6) and population cytogenetics (7). Through study of cell division and reproduction (mitosis, meiosis, and fertilization), together with statistical analysis of variations in the expression of traits among offspring of specified matings, transmission cytogenetics gave us our first picture of the fundamental units of life (genes and chromosomes) and of the variation-generating probability mechanisms (meiosis) by which lawfully combined random samples of these units are passed on from parents to offspring. Through study of (i) the distributions of genes in populations, (ii) the mechanisms responsible for both stability and change in gene frequencies, and (iii) the role of such mechanisms in evolution, population cytogenetics has given us some understanding of ensembles of these units that comprise the gene pools of populations and species—the taxa that are natural units of evolution.

Understanding Individuality

The phenotype (appearance, structure, physiology, and behavior) of any organism is determined by the interaction of environment with its genotype (the complete genetic endowment), Each genotype is the end product of many mechanisms which promote genotypic diversity in populations.

Ordinarily members of a cross-fertilizing, sexually reproducing species possess a diploid, or paired, set of chromosomes. Most species whose behavior we study are sexually dimorphic. The genetic basis of this dimorphism resides in the distribution of the heterosomes, a homologous pair of sex chromosomes (XX) being present in mammalian female and an unequal pair (XY), in the mammalian male. Sexual dimorphism guarantees that any population will be variable to the extent of at least two classes. Whether sex-chromosome or other genotypic differences are involved in any particular behavior remains an empirical question to be investigated separately for every population. It can no longer be settled by dogmatic attitudes and assumptions about uniformity.

Chromosomes other than sex chromosomes are called autosomes. Every autosome is normally represented by a homologous pair whose members have identical genetic loci. Alternative forms of a gene any of which may occupy a given locus are termed alleles. If an individual receives identical alleles from both parents at homologous loci, he is said to be homozygous for that gene. If he receives two alleles that differ, however, he is said to be heterozygous for that gene. The process by which a gene changes from one ellelic form to another is called mutation.

When a gene is represented in the population gene pool by two allelic forms, the population will be genotypically polymorphic to the extent of at least three classes. That is, individuals may be homozygous for either of two alleles or heterozygous for their combination.

Study of populations has revealed that often extensive series of alleles exist for a locus. Well-known examples are the three (actually more) alleles at the ABO-blood locus in man and a dozen or more alleles at the white-eye locus in *Drosophila*. Benzer (8), in his study of the internal genetic architecture of *one* "gene" with a corresponding physical structure of probably less than 2000 nucleotide pairs, the *r*II region of the T_4 bacteriophage, found 339 distinguishable mutational sites, and he expects to eventually find some 428. There is no reason to believe that we shall find less complexity in cellular organisms as further refinement increases the resolving power of our techniques for analyzing them. In general, for each locus having n alleles in the gene pool, a population will contain $n(n + 1)/2$ genetypic classes. Mutation insures variety in the gene itself.

Sexual reproduction involves meiosis—a complex cellular process resulting in a meristic division of the nucleus and formation of gametes (reproductive cells) having single genomes (a haploid chromosome set). One homolog in every chromosome pair in our diploid complement is of paternal origin and the other is of maternal origin. In meiosis, the homologs of a pair segregate and a gamete receives one from each pair. The assortment to gametes of the segregating homologs occurs independently for each pair. This process insures diversity because it maximizes the likelihood that gametes will receive unique genomes. For example, gametogenesis in *Drosophila willistoni* produces eight alternative gametic genomes, which, if we represent the three chromosome pairs of this species by Aa, Bb, and Cc, we designate ABC, ABc, AbC, aBC, Abc, aBc, abC abc. In general, n pairs of chromosomes produce 2^n genomes (if we ignore the recombination of gene linkages that actually occurs in crossover exchanges between chromosomes). Man, with 23 chromosome pairs, produces gametes with any of 2^{23} alternative genomes. This makes vanishingly small the chances that even siblings (other than monozygotes) will be genetically identical. Since the gamete contributed by *each* parent is chosen from 2^{23} alternatives, the probability that the second offspring born to parents will have exactly the same genotype as their firstborn is $(\frac{1}{2^{23}})^2$, or less than 1 chance in over 70 trillion! The probability that two unrelated individuals will have the same genotype, then, is effectively zero (9).

The argument for the genotypic uniqueness of members of populations is even more compelling, since other conditions also contribute to diversity. So, it is clear, the organisms which the behavioral sciences study are intrinsically variable before they undergo differentiating experiences. The mechanisms responsible for this variety are mutation, recombination, and meiosis. Add to these individual experience, and it becomes evident why individuals differ in behavior. In fact, the more reliable our methods of observation become, the more evident will this variety be.

The Abnormality of the Normal

For Watson, its founder, behaviorism was "a natural science . . . [whose] closest scientific companion is physiology. . . . It is different from physiology only in the groupings of its problems, not in fundamentals or in central viewpoint" (*10*). Assumptions about the uniformity and normality of material under investigation are often made in physiology, the science after which, more than any other, experimental psychology has attempted to pattern itself. We may, therefore, get a better grasp of the individuality-uniformity distinction by examining the differences between organisms whose behavior is studied by behavioral scientists and systems whose functioning is studied by physiologists.

Since the two disciplines are working at distinctly different levels of biological organization, the meaning of "normality" as operationally determined by them is quite different. Physiologists choose a normal organism to work with—one that looks healthy and does not appear unusual—and study one or more of its systems, such as the adrenals, gonads, or other endocrines, or regions of the nervous system. Either pre- or post-experimentally, anatomical, histological, or biochemical verification is made of the normality of the material under study, and sometimes of related or adjacent functions to boot. In the behavioral sciences we choose normal-appearing organisms to study too. We rarely perform biopsies unless there is a specific physiological interest, in which case we operate as the physiologist does.

Physiological systems are variable, not uniform. Williams (*11*) amply documents this and points out that implicit in our use of "normal" is reference to some region of a distribution arbitrarily designated as not extreme—for example, the median 50 percent, 95 percent, or 99 percent. We choose such a region for every trait. Among n mathematically independent traits—for example, traits dependent on n different chromosomes—the probability that a randomly selected individual will be normal for all n traits is the value for the size of that region raised to the nth power. Where "normal" is the median 50 percent and $n = 10$, on the average only 1 individual out of 1024 will be normal (for ten traits). When we consider at one time the distributions throughout a population of large numbers of physiological systems, we should expect negative deviates from some distributions to combine with positive deviates from others, both kinds of extreme deviates to combine with centrally located ones, and deviates of similar algebraic sign and magnitude to combine. Each individual's particular balance of physiological endowments will be the developmental result of the genotype he draws in the lotteries of meiosis and the mating ritual. Because of crossing over, most genes assort independently. Hence, we cannot expect high correlations among the systems they generate.

If, underlying every behavior, there were only a single such system—for example, if the male "sexual drive" were mainly dependent on the seminal vesicles (*12*) or if escape behavior were mainly dependent on the adrenals—then the same kind of distribution might be expected for both the behavior and the underlying system. Whatever uniformity might exist at one level would be reflected at the other. The last few decades of research on the biological correlates of behavior have made it increasingly clear that behavior is the integration of most of these systems rather than the expression of any one of them. Therefore, there is little reason to expect that the many possible combinations and integrations of those systems that go to make up the members of a population will yield a homogeneously normal distribution of responses for many behavioral measures. An organism richly endowed with the components of one subset of systems and poorly endowed with those of another is not to be expected to behave in the same manner as an organism with an entirely different balance of endowments. The obviousness of this fact is well illustrated by the differences in behavior among the various breeds of dogs and horses.

Reductionism

Another conviction, strongly held by some, is that *real* explanations must be reductionistic. Those who hold this view in its most extreme form assert that no behavior can be understood until its physical basis has been unraveled. And the search for the physical basis proceeds along physiological, biochemical, biophysical, or genetical lines, depending on the skills and predilections of the investigators.

In laboratory experiments, some rats learn mazes more readily on the basis of visual cues while others do better with predominantly kinesthetic cues (*13*). The kinds of differences in organization that can coexist as alternative forms within a species, as well as some relations between one behavior and the component subsystems that are alternative possibilities, have been further revealed in a series of studies of the effects of domestication. In some domesticated rats, activity in a revolving drum was controlled by the gonads: control rats had daily activity scores as high as 18,000 revolutions, while gonadectomized rats scored only a few hundred revolutions. Cortisone therapy restored a high activity level in the gonadectomized rats. When the same experiment was repeated on wild Norway rats, however, the presence or absence of gonads made no detectable difference in measured activity. Further study of differences between these domesticated and wild rats revealed larger adrenals in the wild rats and larger gonads in the domesticated (*14*). So it appears that activity may be under the control of adrenal output in one case and gonadal output in the other—that behavior is not a univocal index to an organism's balance of endowments. The fallacy of reductionism lies in assuming a one-one relation between different levels of organization. With degeneracy already demonstrated in the genetic code of messenger RNA base triplets for the amino acids of proteins, we should be surprised not to find it at the levels of complexity we are considering (*15*).

Behaviorism and Introspection

According to my naive picture, the pyramid of sciences forges links of knowledge "out" from the periodic table: on the one hand, "down" into atomic structure through advances in physics; on the other, "up" into life through the genetic code and organic structure by advances in biophysics and biochemistry. The place of the behavioral sciences in the outline of that pyramid has been clearly demarcated for some time (*16*). Our models and assumptions must be consistent with the knowledge that is burgeoning at other levels. This means doing our homework and learning (*17*) about developments in fields which may once have seemed remote from behavior, but which clearly are not. Unfortunately, we are still plagued by a legacy of pseudo-problems which, like MacArthur's old soldier, seem to be slowly fading away instead of discreetly dying.

Recently, in *Science* (*18*), immediately following Wilkin's exposition (*19*) of his magnificent work on nucleic acids that led to the Watson-Crick model, Skinner heeded a call to issue "a restatement of radical behaviorism. . . ." It may be recalled that behaviorism bears its title to call attention to the fact that it studies behavior objecively rather than mind subjectively. Under Watson, in 1913, it wished to distinguish itself from unreliable (?) introspectionist psychology, whose findings lacked intersubjective agreement. Under Skinner, 50 years after, it is still worried about "the dimensions of the things studied by psychology and the methods relevant to them."

Starting from the uniformity assumption, the introspectionists were attempting to study the generalized human mind by analyzing the contents of their own consciousness. Of course, the study of mind through analysis by different individuals of the contents of their consciousness inevitably revealed individual differences. Under a given set of stimulating conditions, different people reported different sensations. According to Boring (*20*) "there is always to be remembered that famous session of the Society of Experimental Psychologists in which Titchener, after hot debate with Holt, exclaimed: 'You can see that green is neither yellowish nor bluish!' and Holt replied: 'On the contrary, it is obvious that a green is that yellowish-blue which is just exactly as blue as it is yellow.' That impasse was an ominous portent. . . ."

In over 50 years no one has suggested that Titchener and Holt might *both* have been making reliable observations. The event Boring bemoans would not be looked upon as an "impasse" that represents "an ominous portent" by a behavioral science that understands the structure of the materials it studies. Until recently, some of our best information on the assignment of genes to human chromosomes came from introspective behavioral observation. We know that genes affecting red-green color discrimination are carried on the X chromosome. We know it because some people fail to report differences in sensation easily observed by others, and the determining factors are transmitted to sons by mothers but never by fathers. Furthermore, Graham (*21*) has made excellent use of the introspections of one individual whose two eyes receive different color sensations from the same stimulus. Wouldn't it have been of great interest to learn how colored stimuli appeared to other members of the Titchener and Holt families? How many more potentially fruitful leads have been lost in the behavioral sciences because of rigid adherence to the counterfactual uniformity assumption?

Now, what was really wrong with introspection? Is there any other method by which Penfield could have made the startling discovery that apparently long-forgotten experiences remain stored in specific regions of our brain? He succeeded in restoring "lost" memories to introspectively observed consciousness by electrical stimulation of appropriate regions of exposed human brains (22). If the behaviorists had scrutinized the assumptions, which they shared with introspective psychology, they might not have been so quick to condemn its method. Every method has limitations, which it behooves its users to understand.

Behavior Genetics

There now exists a substantial and rapidly growing literature on the behavior genetics of many organisms, from *Drosophila* to man—what Tryon (23) calls "the basic science of individual differences." It comes from research far less hampered by unsound premises. In Fuller and Thompson's useful summary (24) we can see "its documentation of the fact that two individuals of superficially similar phenotype may be quite different genotypically and respond in completely different fashion when treated alike." This field, like others, is passing through stages.

The goal of the early work was a genetics *of* behavior. It took a while to learn that heritability is a property of populations and never of behaviors: the relation between behavioral variation and relevant genetic variation is never constant. It must be measured in specific populations under specific conditions, because it varies with both. Tolman (24), Tryon (24), and Heron (24) each measured individual differences in rats' ability to learn and then, by selective breeding, produced strains of "maze-bright" and "maze-dull" rats. Hall (*24*) and Broadhurst (*24*) selected for differences in emotional responses. Analogous studies have been made of performance on an animal "intelligence test" (*25*).

Many strains of small mammals (mice, hamsters, rats, guinea pigs, and rabbits) are maintained under varying inbreeding regimens for purposes of medical and other research. When different strains within a species are compared, it actually becomes a challenge not to find differences in one or more behaviors. When strain comparisons are followed by appropriate genetic crosses, genetic correlates of behavioral differences are demonstrated. Such experiments have been performed for a large variety of behaviors: alcohol preference (*26*), hoarding (*24*), mating competition (*24*), susceptibility to audiogenic seizure (*24*), exploratory tendency (*24*), and various learning measures (*24*).

Paralleling the animal research are studies of human pedigrees, studies of family resemblances, twin comparisons, population surveys, and studies of race differences. Again, heritabilities have been demonstrated for many behaviors; for example, nature-nurture ratios were computed for intelligence-test and personality-test performance. Kallmann and his associates have pioneered, and others have joined, in collecting an impressive body of evidence on genetic factors in schizophrenia and other psychopathologies (*27*).

In 1963, with the wisdom of hindsight, we can ask why so many demonstrations were necessary. Should it not have been common knowledge that within each population the variation pattern for most traits will be conditioned by the nature of the gene pool, and that this will differ among populations? The answer lies in one phrase: the heredity-environment controversy.

The "opinion leaders" (*28*) of two generations literally excommunicated heredity from the behavioral sciences. Understandably, they objected to amateurish labeling of behaviors as instincts without proper experimental analyses. Also, they were repelled by the pseudogenetics of Hitler and other purveyors of race prejudice (*29*). On the other hand, impressed with the power of conditioning *procedures,* they proclaimed their faith in analysis of experience as the starting point for behavioral science—as though experience, like the Cheshire Cat's grin, could exist without the organism. "Our conclusion . . . is that we have no real evidence for the inheritance of traits," said Watson (*10*). While acknowledging that there are heritable differences in form and structure, he claimed there is no evidence that those differences are related to function, because "hereditary structure lies ready to be shaped in a thousand different ways" (*30*). Behaviorism still makes the gratuitous uniformity assumption that all genetic combinations are equally plastic and respond in like fashion to environmental influences (*31*).

We are now in a more fruitful period. Experimental analysis is yielding information about genes and chromosomes and how they act. The way is open to understanding molecular—ultimately submolecular (*32*)—mechanisms and to following metabolic pathways between genes and phenotypes. In the honey bee, Rothenbuhler (*24*) found that resistance to foulbrood disease (a bacterial infection of the larvae) depends on homozygosity of the worker bees for recessive alleles of at least two genes: one which enables them to uncap compartments containing infected larvae and another which enables them to remove those larvae from the hive (*33*).

Médioni (*34*), in his studies of phototaxis (light-oriented locomotion) in *Drosophila*, employed genetic, physiological, and stimulus variables in an exquisitely detailed analysis articulating relations between components of behavior, components of the organism, and stimulus properties of the environment. Behaviorally, phototaxis is resolvable into five components: (i) a photopositive phase; (ii) a sensory adaptation factor [Viaud's *capacité photopathique* (*35*)]; (iii) an exploratory phase; (iv) a photokinetic factor; and (v) a photoinhibition phase. The interplay of the behavioral components depend on (i) the intensity and wavelength of light, (ii) the differential effects of stimulation through the ocelli and through the compound eyes, (iii) sex, and (iv) genetic background and geographical region of racial origin. Races in 17 regions of the Northern Hemisphere, from Japan across Eurasia to America, arrange themselves into two distinct North-South clines, an

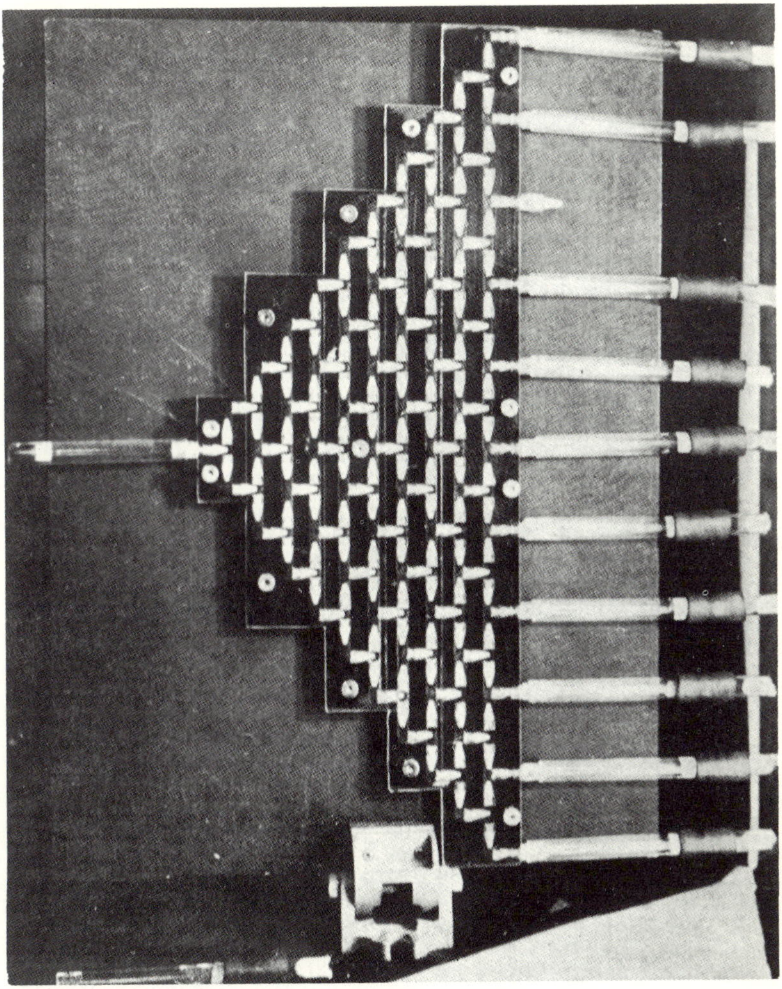

Fig. 1. Vertical ten-unit plastic maze facing a fluorescent tube. Squads of flies introduced in the vial at left are collected from the vials at right. They are attracted through the maze by the odor of food and by light. Small trap-like funnels, having a larger opening continuous with the alley surfaces and a small one debouching in midair, discourage backward movement in the maze. [Hirsch (*51*)]

Eastern and a Western, in which light preference diminishes with latitude of origin.

Our laboratory has made the most detailed analysis, to date, of relations between the genome and a behavioral phenotype in studies of geotaxis (gravity-oriented locomotion) in *Drosophila.* Behavioral distributions for populations are obtained in the apparatus shown in Fig. 1. Selective breeding from a geotactically and genetically heterogeneous foundation population has produced the two strains shown in Fig. 2, which have diametrically opposite response tendencies. Other methods produced three populations differing with respect to both degree and kind of similarity in chromosome constitution among their members. Two parameters of their behavioral distributions were thus controlled. The least dispersion occurred in the population in which all members carried two of the three large chromosomes in identical form. The other two populations, differing from each other with respect to the single chromosome distributed in identical form to all their members, differed in central tendency but not in dispersion, which was twice that of the first population. Figure 3 shows, for this model situation, the kind of prediction and control that an understanding of population structure and its genetic basis may yield.

Erlenmeyer-Kimling's subsequent chromosome analysis shows that genes influencing the response to gravity are distributed throughout the genome. The first two chromosomes in the unselected foundation population contribute to positive geotaxis, and the third to negative. Selection pressure both enhances and reduces their effects, depending on the direction of selection (*36*).

At the molecular level, an exciting development is the measurement, by Hydén and Egyházi (*37*), of changes, with learning, in the RNA base ratios in nuclei of specific mammalian nerve cells and in their glia. This work, if confirmed, represents a major advance in our search for the physical basis of experience. Hydén's speculative, but interesting, suggestion is that the electrical disturbance of the nerve impulse releases, in some as yet unspecified way, a repressed region of chromosomal DNA. This DNA henceforth produces, on demand, its characteristic RNA to code the protein that facilitates forward transmission of the particular temporal pattern of electrical frequencies that first released the DNA. This suggestion is the first to be made that appears capable of reconciling the universal feature of improvement with practice with the idiosyncratic features of individual performance. In this schema the individuality encoded in the chromosomal DNA of each genotype at meiosis and fertilization is propagated directly into the learning and memory mechanism by means of the established sequence of DNA producing RNA producing protein. Such a schema could thus accommodate the distributions of individual differences invariably found in studies of learning and memory.

The study of man is also moving beyond the stage of wondering whether we can find a heritability for this or that behavior. Phenylpyruvic oligophrenia, a form of mental deficiency accompanied by a high concentration of phenylpyruvic acid in the urine, had early been traced to a gene-controlled enzymatic deficiency in phenylalanine metabolism (*27*). Now, Down's syndrome (mongolism) has been associated with the presence of extra chromosomal material (*24*).

Human populations are dimorphic for taste sensitivity to certain bitter compounds. Different races show different distributions with respect to this trait, as well as to almost every other trait that has been genetically analyzed. On the basis of behavioral observations indicating that an individual's ability to taste certain compounds depends upon the presence of his own saliva, Cohen and Ogdon (*38*) suggested that components of the saliva might play a critical role in tasting ability. Lately, Fischer and his coworkers (*39*) have shown in vitro that the bitter-tasting thioureas are oxidized faster by the saliva of nontasters than by that of tasters. Presumably, at low concentrations so much of the compound is oxidized in a nontaster's mouth that the few molecules which might reach their receptor sites remain undetectable. Furthermore, Fischer and his associates have now confirmed Cohen and Ogdon's finding that, in order to taste certain compounds at all, even a taster requires the presence of his own saliva. Superficially at least, it appears that saliva, like many body tissues, cannot be transplanted. A valuable observation here would involve an exchange of saliva between identical twins, who are presumably alike in body chemistry.

The ramifications of the taster phenomenon appear to be legion. There is a significantly higher incidence of nontasting of the bitter compounds among persons with nodular goitre (*40*), among patients with congenital athyreotic cretinism, and among parents of the latter as well (*41*). In another study (*42*) it was found that, among 38 parents of children with Down's syndrome, none was able to taste quinine. Furthermore, all but one of the fathers in that sample were unable to taste a bitter thiourea. Finally, a correlation exists between taste sensitivity and dislike of foods: the more sensitive tasters find more foods objectionable (*43*).

Race Differences

A problem of continuing social importance, for an understanding of which most behavioral scientists have lacked a proper conceptual basis, is the question of race differences. To the liberals this question has been a continuing source of embarrassment (*44*). They have made little progress in answering it since the signing of our Constitution and Bill of Rights, when it was *asserted* that all men are created equal. To the prejudiced the question has presented no difficulties, because they *know* other races are inferior to their own; this seems as obvious to them as the flatness of the earth did to our ancestors.

This question appears in another perspective when it is examined in the light of current knowledge of population structure. Dobzhansky (*45*) has clearly called attention to the difference between equality and identity. Genotypic uniqueness creates biochemical individuality. Without enforcing conformity—irrespective of heredity, training, or ability—a democratic ethicosocial system offers to all equality of opportunity and equal treatment before the law. Genetics explains both individual and population uniqueness. Even though reproductively isolated populations belong to the same species and have the same genes, the relative frequencies of different alleles of

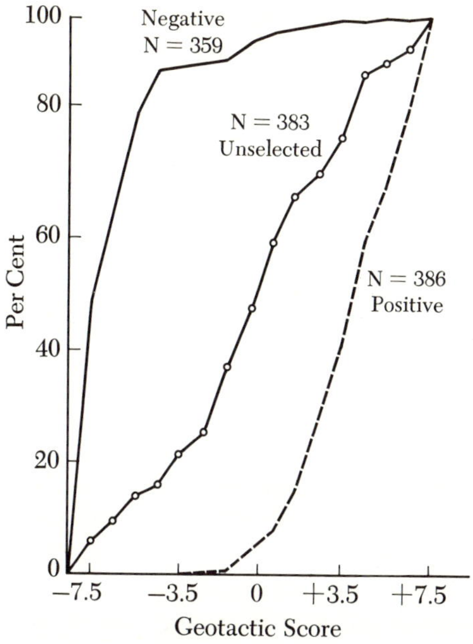

Fig. 2. Cumulated percentages of animals (males and females) that received geotactic scores in a 15-unit maze, from an unselected foundation population (middle curve) and from the two selected strains (outer curves). [Hirsch and Erlenmeyer-Kimling (*52*)]

genes in their gene pools are almost certain to differ. Mutations and recombinations will occur at different places, at different times, and with differing frequencies. Furthermore, selection pressures will also vary (*46*). In analyzing data from such populations we have learned to ask, not whether they are different, but, rather, in what ways they differ.

Races are populations that differ in gene frequencies. Observations on populations are summarized in distributions, so often assumed to be normal (*47*). When we add the assumption of common variance, or make transformations to obtain it, the data fit into the ever popular analysis-of-variance models. *The* difference between two populations must then be a difference between means, because the asumptions of normality and homogeneity of variance for the model leave no other property with respect to which the distributions can differ. The final step in this fantastic chain of reasoning has recently been taken in *Science* by Garrett (*48*). He ignores individual differences and claims that wherever two populations differ on some scale of measurement, no matter how vague, any individual from the population with *the* higher mean is better than any individual in the other population, and that intermarriage will "be not only dysgenic but socially disastrous!"

Distributions have other properties, such as dispersion, skewness, and kurtosis (peakedness), and no single one is exclusively important. Where these other properties have been examined, the inadequacy of a preoccupation with the central tendency and a hasty assumption of normality has been easy to document (*49*). There is no reason to expect two populations with different heredities and different environments to have precisely the same distribution for any trait. We can expect to find varying combinations of similarities and differences in the several properties of distributions when we compare different populations for a given trait, or any set of populations for different traits. Furthermore, the number of traits for which we could make comparisons is effectively unlimited, and many of the traits will be uncorrelated (*11*). Again, a lack of intrinsic correlation would come as no surprise to a behavioral science that understands its materials, because traits are the developmental result of thousands of genes, most of which, because of crossing over, sooner or later undergo independent assortment.

For ease of exposition, I have not considered environment in discussing race. Certainly, it is no less important than genetic endowment. The ontogeny of a responsible and effective citizen requires prolonged socialization, highly dependent upon the socializing agency. A genotype must have an environment in which to develop a phenotype. But the same genotype can produce quite different phenotypes, depending on the environments in which it may develop. Furthermore, a given environment can nurture quite different phenotypes, depending on the genotypes which may develop there. This fact is attested daily by parents and teachers who find that a method of tuition admirably successful with one child may be worth-

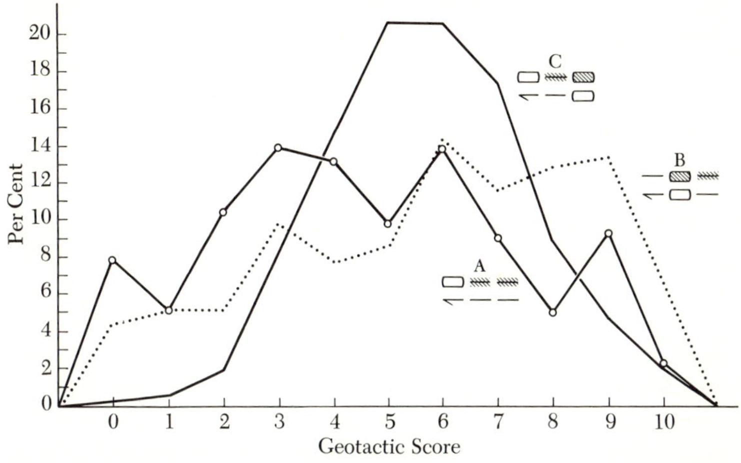

Fig. 3. Distributions of geotactic scores in a ten-unit maze for males of three populations (described in text). (Rectangles) Chromosomes carried in identical form by all members of a population: (dashes) chromosomes varying at random; (hatching) heterozygosity; (half-arrowhead) the Y chromosome of males. (In Fig. 2 the abscissa scale was reversed and the zero point was shifted to the center of the distribution.) [Hirsch (*51*)]

less with another, who nevertheless can learn by a different method. So, while environment makes an undeniably important contribution to the particular values obtained in phenotypic measurements, consideration of particular environments should not change our general picture of population structure. Without an appreciation of the genotypic structure of populations, the behavioral sciences have no basis for distinguishing individual differences that are attributable to differences in previous history from those that are not, and no basis for understanding any differences whatsoever where there is a common history.

Conclusions and Summary

Traditionally, many behavioral scientists have assumed that individuals start life uniformly alike, and that individual differences result only from differentiating experiences. To assume this is as contradictory to the established fact of uniqueness at conception as to assume that entropy is as likely to decrease as it is to increase. Recognition of the contradictory nature of this assumption does not make the role of experience in ontogeny any less important, but we now realize that the effects of experience are conditioned by the genotype. Therefore, a careful reconsideration of our statistical tools, experimental methods, theoretical models, and research goals is in order.

Many problems that have generated violent controversy now appear in totally different perspective. Introspection may provide a legitimate probe into subjective experience, without requiring intersubjective agreement. The concept of a normal individual has no generality. The outlook for understanding the physical basis of behavior has never been more promising. Awareness that a multiplicity of variable systems comprise its substrate, however, emphasizes the integrity and importance of the different levels of biosocial organization at which the several sciences work. In place of reductionism, we may now think of studying correlations between phenomena, reliably observed and analyzed at various levels, and of assessing the correlations over an ever widening range of conditions. The controversial aspects of the heredity-environment question and of the race-differences question arise from failure to understand the genetics of individual and population differences and the rationale of their statistical analysis (*50*).

References and Notes

1. R. R. Bush and F. Mosteller, *Stochastic Models for Learning* (Wiley, New York, 1955).
2. B. S. Rosner, in *Psychology: A Study of a Science,* S. Koch, Ed. (McGraw-Hill, New York, 1962), vol. 4. p. 299.
3. Psychophysics is the study of changes in response associated with changes in physically specified stimuli.
4. K. Breland and M. Breland, *Am. Psychologist 16,* 681 (1961).
5. E. Mayr, *Animal Species and Evolution* (Harvard Univ. Press, Cambridge, 1963), p. 5.

6. T. H. Morgan, A. H. Sturtevant, H. J. Muller, C. B. Bridges, *The Mechanism of Mendelian Heredity* (Holt, New York, 1915).
7. T. Dobzhansky, *Genetics and the Origin of Species* (Columbia Univ. Press, New York, ed. 2, 1941).
8. S. Benzer, *Proc. Natl. Acad. Sci. U.S. 47*, 403 (1961): ——— and S. P. Champe, *ibid.* p. 1025; S. P. Champe and S. Benzer, *ibid. 48*, 532 (1962).
9. J. Hirsch, in *Roots of Behavior*, E. Bliss. Ed. (Hoeber, New York, 1962), p. 6. This calculation has provoked intense resistance. Its implications for well-encrusted modes of thinking in the "establishment" of the behavioral sciences are clearly most unwelcome. In an already legendary correspondence (part of which Mosteller circulated privately without informing me), F. Mosteller and J. Tukey independently attempted to disprove it. Both mathematicians overlooked the simple empirical fact that two sexes are required to produce children in the human species. Of course, my calculation is most conservative: assuming 10,000 human genes and an average of four alleles each gives ten combinations per locus and the astronomical number of $10^{10,000}$ potential human genotypes!
10. J. B. Watson, *Behaviorism* (Univ. of Chicago Press, Chicago, new ed., 1959), pp. 11, 103.
11. R. J. Williams, *Biochemical Individuality* (Wiley, New York, 1956); *Science 126*, 453 (1957); R. J. Williams, R. B. Pelton, F. L. Siegel, *Proc. Natl. Acad. Sci. U.S. 48*, 1461 (1962).
12. F. A. Beach and J. R. Wilson [*Proc. Natl. Acad. Sci. U.S. 49*, 624 (1963)] have demonstrated that this is not the case.
13. I. Krechevsky, *J. Comp. Psychol. 16*, 99 (1933).
14. C. P. Richter, *J. Natl. Cancer Inst. 15*, 727 (1954).
15. M. F. Perutz, *Proteins and Nucleic Acids Structure and Function* (Elsevier, Amsterdam, 1962); R. V. Eck, *Science 140*, 477 (1963).
16. R. J. Williams, *Science 124*, 276 (1956); R. W. Gerard, *Behavioral Sci. 3*, 137 (1958).
17. B. Glass, in *Expanding Goals of Genetics in Psychiatry*, F. J. Kallmann, Ed. (Grune and Stratton, New York, 1962), p. 259.
18. B. F. Skinner, *Science 140*, 951 (1963).
19. M. H. F. Wilkins, *ibid.*, p. 941.

20. E. G. Boring, *Am. J. Psychol. 59*, 173 (1946).
21. C. H. Graham and Y. Hsia, *Proc. Am. Phil. Soc. 102*, 168 (1958).
22. W. Penfield, *Proc. Natl. Acad. Sci. U.S. 44*, 59 (1958).
23. R. C. Tryon, *Am. Psychologist 18*, 134 (1963).
24. J. L. Fuller and W. R. Thompson, *Behavior Genetics* (Wiley, New York, 1960), p. 38.
25. W. R. Thompson and A. Kahn, *Can. J. Psychol. 9*, 173 (1955).
26. G. E. McClearn and D. A. Rodgers, *Quart. J. Studies Alc. 20*, 691 (1959).
27. F. J. Kallmann, Ed., *Expanding Goals of Genetics in Psychiatry* (Grune and Stratton, New York, 1962).
28. G. Lindzey's phrase (private communication).
29. J. H. Steward, *Science 135*, 964 (1962).
30. J. B. Watson, *Behaviorism* (People's Institute Publishing Co., New York, 1925), p. 77.
31. The irresistible attraction that these ideas have had in the behavioral sciences seems all the more appalling today when one reads the excellent systematic exposure of the "fallacies" of behaviorism published in 1930 by Jennings, a well-known and highly respected scientist of that period. [H. S. Jennings, *The Biological Basis of Human Nature* (Norton, New York, 1930); I thank Professor Donald D. Jensen of Indiana University for directing me to Jennings.] What is more, Watson had read Jennings. He cites those parts of the book that suit his purposes (see *10*).
32. A. Szent-Györgyi, *Introduction to a Submolecular Biology* (Academic Press, New York, 1960); M. Kasha and B. Pullman, Eds., *Horizons in Biochemistry Albert Szent-Györgyi Dedicatory Volume* (Academic Press. New York, 1962).
33. To demonstrate the independence of the second gene, Rothenbuhler opens compartments for bees that cannot open them themselves.
34. J. Médioni, thesis, University of Strasbourg (1961), partially summarized in *Ergeb, Biol. 26*, 72 (1963).
35. G. Viaud, *J. Psychol. Normale et Pathologique 42*, 386 (1949).
36. L. Erlenmeyer-Kimling and J. Hirsch, *Science 134*, 1068 (1961); J. Hirsch and L. Erlenmeyer-Kimling, *J. Comp. Physiol. Psychol. 55*, 722 (1962).
37. H. Hydén and E. Egyházi, *Proc. Natl. Acad. Sci. U.S. 48*, 1366 (1962); *49*, 618 (1963).
38. J. Cohen and D. P. Ogdon, *Science 110*, 532 (1949).

39. R. Fischer and F. Griffin, *Behavior Genetics Symposium, 17th International Congress of Psychology Washington, D.C.* (1963).
40. H. Harris, H. Kalmus, W. R. Trotter, *Lancet 1963II,* 1038 (1949): F. D. Kitchen, W. Howel-Evans, C. A. Clarke, R. B. McConnell, *Brit. Med. J. 1959,* 1069 (1959).
41. T. H. Shepard, *J. Clin. Invest. 40,* 1751 (1961).
42. R. Fischer, A. R. Kaplan, F. Griffin, D. W. Sting, *Am. J. Mental Deficiency 67,* 849 (1963).
43. R. Fischer, F. Griffin, S. England, S. M. Garn, *Nature 191,* 1328 (1961).
44. *Science 134,* 1868 (1961).
45. T. Dobzhansky, *ibid. 137,* 112 (1962).
46. R. H. Post, *Eugenics Quart. 9,* 131 (1962).
47. L. S. Minckler, *Science 133,* 202 (1961).
48. H. E. Garrett, *ibid. 135,* 982 (1962); S. Genovés. *ibid.* p. 988.
49. H. G. Yamaguchi, C. L. Hull, J. M. Felsinger, A. I. Gladstone, *Psychol. Rev. 55,* 216 (1948); J. Hirsch, *Am. J. Orthopsychiat. 31,* 478 (1961).
50. I am indebted to E. R. Hilgard for suggesting that I write this article and to the following for commenting on a draft of the manuscript: E. W. Caspari, J. Cohen, D. E. Dulaney, Nikki Erlenmeyer-Kimling, R. W. Frankmann, L. J. Goldsmith, D. A. Hamburg, R. C. Hostetter, C. L. Hulin, L. G. Humphreys, H. L. Jacobs, R. Kesner, G. Ksander, G. E. McClearn, J. E. McGrath, O. H. Mowrer, F. H. Palmer, D. Rosenthal, S. Ross, Leigh M. Triandis, P. Tyler, and M. W. Weir. This work is partially supported by a National Science Foundation grant (G-21238) to the Center for Advanced Study in the Behavioral Sciences, Stanford, Calif.
51. J. Hirsch, *J. Comp. Physiol. Psychol. 52,* 304 (1959).
52. ——— and L. Erlenmeyer-Kimling, *Science 134,* 835 (1961).

The Snark Was a Boojum[1]

FRANK A. BEACH

Those of you who are familiar with the writings of Lewis Carroll will have recognized the title of this address as a quotation from his poem "The Hunting of the Snark." Anyone who has never read that masterpiece of whimsy must now be informed that the hunting party includes a Bellman, a Banker, a Beaver, a Baker and several other equally improbable characters. While they are sailing toward the habitat of their prey the Bellman tells his companions how they can recognize the quarry. The outstanding characters of the genus *Snark* are said to be its taste which is described as "meager but hollow," its habit of getting up late, its very poor sense of humor and its overweening ambition. There are several species of Snarks. Some relatively harmless varieties have feathers and bite, and others have whiskers and scratch. But, the Bellman adds, there are a few Snarks that are Boojums.

When the Baker hears the word, Boojum, he faints dead away, and after his companions have revived him he explains his weakness by recalling for their benefit the parting words of his Uncle.

If your Snark be a Snark, that is right:
Fetch it home by all means—you may serve it
    with greens
And it's handy for striking a light.

****

But oh, beamish nephew, beware of the day,
If your Snark be a Boojum! For then,
You will softly and suddenly vanish away,
And never be met with again!

[1] Presidential address delivered before the Division of Experimental Psychology of the American Psychological Association, September 7, 1949.

From *American Psychologist,* 1950, 5, 115-24. Reprinted by permission of the author and the American Psychological Association.

Much later in the story they finally discover a Snark, and it is the Baker who first sights the beast. But by great misfortune that particular Snark turns out to be a Boojum and so of course the Baker softly and suddenly vanishes away.

Thirty years ago in this country a small group of scientists went Snark hunting. It is convenient to personify them collectively in one imaginary individual who shall be called the Comparative Psychologist. The Comparative Psychologist was hunting a Snark known as Animal Behavior. His techniques were different from those used by the Baker, but he came to the same unhappy end, for his Snark also proved to be a Boojum. Instead of animals in the generic sense he found one animal, the albino rat, and thereupon the Comparative Psychologist suddenly and softly vanished away. I must admit that this description is somewhat overgeneralized. A few American psychologists have done or are doing behavioral research that is broadly comparative. All honor to that tiny band of hardy souls who are herewith excepted from the general indictment that follows.

It is my aim, first, to trace the initial development and subsequent decline of Comparative Psychology in the United States. Secondly, I intend to propose certain explanations for the attitude of American psychologists toward this branch of the discipline. And finally I will outline some of the potential benefits that may be expected to follow a more vigorous and widespread study of animal behavior.

Instead of beginning with the uncritical assumption of a mutual understanding, let me define the basic terms that will be used. Comparative psychology is based upon comparisons of behavior shown by different species of animals including human beings. Comparisons between *Homo sapiens* and other animals are legitimate contributions to comparative psychology, but comparisons between two or more non-human species are equally admissible. Like any other responsible scientist the Comparative Psychologist is concerned with the understanding of his own species and with its welfare; but his primary aim is the exposition of general laws of behavior regardless of their immediate applicability to the problems of human existence. Now this means that he will not be content with discovering the similarities and differences between two or three species. Comparisons between rats and men, for example, do not in and of themselves constitute a comparative psychology although they may well represent an important contribution toward the establishment of such a field. A much broader sort of approach is necessary and it is the failure to recognize this fact that has prevented development of a genuine comparative psychology in this country.

Past and Current Trends

The history of comparative behavior studies in America is reflected in the contents of our journals that are expressly devoted to articles in this field. They have been the *Journal of Animal Behavior* and its successor, the *Journal of Comparative and Physiological Psychology.* Animal studies have, of course, been reported in other publications but the ones mentioned here adequately and accurately represent the general interests and attitudes of Americans toward the behavior of nonhuman animals. I have analyzed a large sample of the volumes of these journals, starting with Volume I and including all odd-numbered volumes through 1948. I have classified the contents of these volumes in two ways—first in terms of the species of animal used, and second in terms of the type of behavior studied. Only research reports have been classified; summaries of the literature and theoretical articles have been excluded from this analysis.

Types of animals studied. Figure 1 shows the number of articles published and the total number of species dealt with in these articles. The number of articles has tended to increase, particularly in the last decade; but the variety of animals studied began to decrease about 30 years ago and has remained low ever since. In other words, contributors to these journals have been inclined to do more and more experiments on fewer and fewer species.

Data represented in Figure 2 further emphasize the progressive reduction in the number of species studied. Here we see that the *Journal of Animal Behavior* contained nearly as

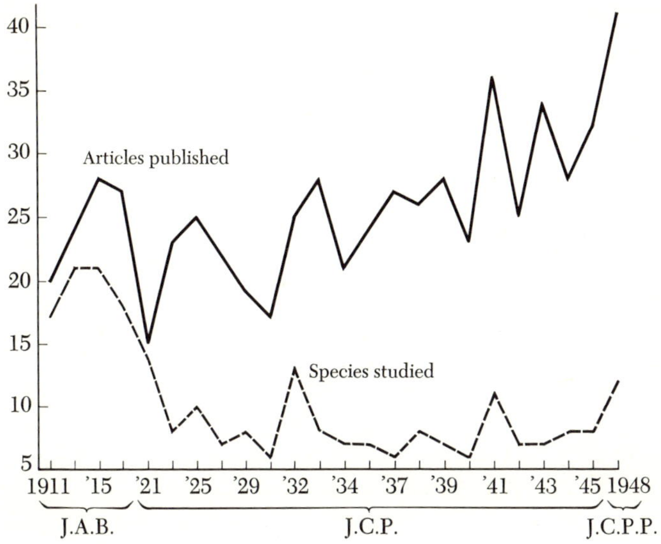

Fig. 1. Number of articles published and variety of species used as subjects.

many articles dealing with invertebrates as with vertebrates; but interest in invertebrate behavior fell off sharply after World War I and, as far as this type of analysis is capable of indicating, it never rose appreciably thereafter. The attention paid to behavior of invertebrates during the second decade of this century is also reflected in the policy of publishing annual surveys of recent research. Each volume of the *Journal of Animal Behavior* contains one systematic review devoted to lower invertebrates, another dealing with spiders and insects with the exception of ants, a third summarizing work on ants and a single section covering all studies of vertebrates.

Figure 2 shows that in the early years of animal experimentation sub-mammalian vertebrates, which include all fishes, amphibians, reptiles, and birds, were used as experimental subjects more often than mammals. But a few mammalian species rapidly gained popularity and by approximately 1920, more work was being done on mammals than on all other classes combined. Now there are approximately 3,500

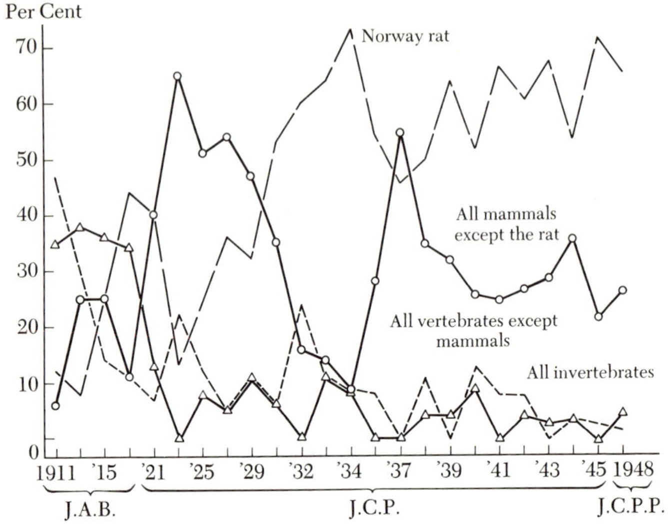

Fig. 2. Per cent of all articles devoted to various phyla, classes or species.

extant species of mammals, but taken together they make up less than one-half of one per cent of all animal species now living. A psychology based primarily upon studies of mammals can, therefore, be regarded as comparative only in a very restricted sense. Moreover the focus of interest has actually been even more narrow than this description implies because only a few kinds of mammals have been used in psychological investigations. The Norway rat has been the prime favorite of psychologists working with animals, and from 1930 until the present more than half of the articles in nearly every volume of the journal are devoted to this one species.

During the entire period covered by this survey the odd-numbered volumes of the journals examined includes 613 experimental articles. Nine per cent of the total deal with invertebrates; 10 per cent with vertebrates other than mammals; 31 per cent with mammals other than the rat; and 50 per cent are based exclusively upon the Norway rat. There is no reason why psychologists should not use rats as subjects in some of their experiments, but this excessive concentration upon a single species has precluded the development of a comparative psychology worthy of the name. Of the known species of animals more than 96 per cent are invertebrates. Vertebrates below the mammals make up 3.2 per cent of the total; and the Norway rat represents .001 per cent of the types of living creatures that might be studied. I do not propose that the number of species found in a particular phyletic class determines the importance of the group as far as psychology is concerned; but it is definitely disturbing to discover that 50 per cent of the experiments analyzed here have been conducted on one one-thousandth of one per cent of the known species.

Some studies of animal behavior are reported in journals other than the ones I have examined but the number of different animals used in experiments published elsewhere is even fewer. The six issues of the *Journal of Experimental Psychology* published in 1948 contain 67 reports of original research. Fifty of these articles deal with human subjects and this is in accord with the stated editorial policy of favoring studies of human behavior above investigations of other species. However, 15 of the 17 reports describing work on non-human organisms are devoted to the Norway rat.

During the current meetings of the APA, 47 experimental reports are being given under the auspices of the Division of Experimental Psychology. The published abstracts show that in half of these studies human subjects were employed while nearly one-third of the investigations were based on the rat.

Is the Experimental Psychologist going to softly and suddenly vanish away in the same fashion as his one-time brother, the Comparative Psychologist? If you permit me to change the literary allusion from the poetry of Lewis Carroll to that of Robert Browning, I will venture a prediction. You will recall that the Pied Piper rid Hamelin Town of a plague of rats by luring the pests into the river with the music of his magic flute. Now the tables are turned. The rat plays the tune and a large group of human beings follow. My prediction is indicated in Figure 3. Unless they escape the spell that *Rattus norvegicus* is casting over them, Experimentalists are in danger of extinction.

Fig. 3. Current position of many experimental psychologists.

Types of Behavior Studied. I trust that you will forgive me for having demonstrated what to many of you must have been obvious from the beginning—namely, that we have been extremely narrow in our selection of types of animals to be studied. Now let us turn our attention to the types of behavior with which psychologists have concerned themselves.

Articles appearing in our sample of volumes of the journals can be classified under seven general headings: (1) conditioning and learning; (2) sensory capacities, including psychophysical measurements, effects of drugs on thresholds, etc.; (3) general habits and life histories; (4) reproductive behavior, including courtship, mating, migration, and parental responses; (5) feeding behavior, including diet selection and reactions to living prey; (6) emotional behavior, as reflected in savageness and wildness, timidity and aggressive reactions; and (7) social behavior, which involves studies of dominance and submission, social hierarchies, and interspecies symbiotic relations.

In classifying articles according to type of behavior studied I have disregarded the techniques employed by the investigator. It is often necessary for an animal to learn to respond differentially to two stimuli before its sensory capacities can be measured; but in such a case the article was listed as dealing with sensory capacity rather than learning. The aim has been to indicate as accurately as possible the kind of behavior in which the experimenter was interested rather than his methods of studying it.

It proved possible to categorize 587 of the 613 articles. Of this total, 8.6 per cent dealt with reproductive behavior, 3.7 per cent with emotional reactions, 3.2 per cent with social behavior, 3.0 per cent with feeding, and 2.8 per cent with general habits. The three most commonly-treated types of behavior were (1) reflexes and simple reaction patterns, (2) sensory capacities, and (3) learning and conditioning. Figure 4 shows the proportion of all articles devoted to each of these three major categories.

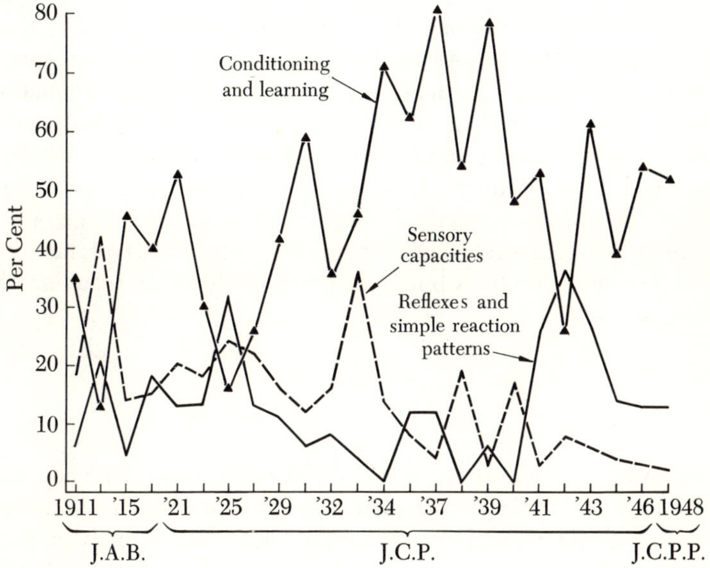

Fig. 4. Per cent of all articles concerned with various psychological functions.

The figure makes it clear that conditioning and learning have always been of considerable interest to authors whose work appears in the journals I have examined. As a matter of fact slightly more than 50 per cent of all articles categorized in this analysis deal with this type of behavior. The popularity of the subject has increased appreciably during the last 15 years, and only once since 1927 has any other kind of behavior been accorded as many articles per volume. This occurred in 1942 when the number of studies dealing with reflexes and simple reaction patterns was unusually large. The temporary shift in relative emphasis was due almost entirely to a burst of interest in so-called "neurotic behavior" or "audiogenic seizures."

Combining the findings incorporated in Figures 2 and 4, one cannot escape the conclusion that psychologists publishing in these journals have tended to concentrate upon one animal species and one type of behavior in that species. Perhaps it would be appropriate to change the title of our journal to read "The Journal of Rat Learning," but there are many who would object to this procedure because they appear to believe that in studying the rat they are studying all or nearly all that is important in behavior. At least I suspect that this is the case. How else can one explain the fact that Professor Tolman's book, "Purposive Behavior in Animals and Men," deals primarily with learning and is dedicated to the white rat, "where, perhaps, most of all, the final credit or discredit belongs." And how else are we to interpret Professor Skinner's 457-page opus which is based exclusively upon the performance of rats in bar-pressing situations but is entitled simply "The Behavior of Organisms"?

Interpretation of Trends

In seeking an interpretation of the demonstrated tendency on the part of so many experimentalists to restrict their attention to a small number of species and a small number of behavior patterns, one comes to the conclusion that the current state of affairs is in large measure a product of tradition. From its inception, American psychology has been strongly anthropocentric. Human behavior has been accepted as the primary object of study and the reactions of other animals have been of interest only insofar as they seemed to throw light upon the psychology of our own species. There has been no concerted effort to establish a genuine comparative psychology in this country for the simple reason that with few exceptions American psychologists have no interest in animal behavior *per se*.

Someone, I believe it was W. S. Small at Clark University in 1899, happened to use white rats in a semi-experimental study. The species "caught on," so to speak, as a laboratory subject, and gradually displaced other organisms that were then being examined. Psychologists soon discovered that rats are hardy, cheap, easy to rear, and well adapted to a laboratory existence. Because of certain resemblances between the associative learning of rats and human beings, *Rattus norvegicus* soon came to be accepted as a substitute for *Homo sapiens* in many psychological investigations. Lack of acquaintance with the behavioral potentialities of other animal species and rapid increase in the body of data derived from rat studies combined to progressively reduce the amount of attention paid to other mammals, to sub-mammalian vertebrates and to invertebrate organisms. Today the trend has reached a point where the average graduate student who intends to do a thesis problem with animals turns automatically to the white rat as his experimental subject; and all too often his professor is unable to suggest any alternative.

To sum up, I suggest that the current popularity of rats as experimental subjects is in large measure the consequence of historical accident. Certainly it is not the result of systematic examination of the available species with subsequent selection of this particular animal as the one best suited to the problems under study.

Concentration of experimental work upon learning seems to stem almost exclusively from the anthropocentric orientation of American psychology. Learning was very early accepted as embodying the most important problems of human behavior; and accordingly the majority of animal investigations have been concerned with this type of activity.

Advantages and Disadvantages of Concentration

I have no wish to discount the desirable aspects of the course which experimental psychology has been pursuing. There are many important advantages to be gained when many independent research workers attack similar problems using the same kinds of organisms. We see this to be true in connection with various biological sciences. Hundreds of geneticists have worked with the fruitfly, *Drosophila.* And by comparing, combining, and correlating the results of their investigations, it has been possible to check the accuracy of the findings, to accelerate the acquisition of new data, and to formulate more valid and general conclusions than could have been derived if each worker dealt with a different species. Something of the same kind is happening in psychology as a result of the fact that many investigators are studying learning in the rat, and I repeat that this is a highly desirable objective.

Another valuable result achieved by the methods currently employed in experimental psychology is the massing of information and techniques pertaining to rat behavior to a point which permits use of this animal as a pedagogical tool. A recent article in the *American Psychologist* reveals that each student in the first course in psychology at Columbia University is given one or two white rats which he will study throughout the semester. This, it seems to me, is an excellent procedure. The beginning student in physiology carries out his first laboratory exercises with the common frog. The first course in anatomy often uses the dogfish or the cat as a sample organism. And college undergraduates learn about genetics by breeding fruitflies. But the usefulness of the rat as a standardized animal for undergraduate instruction, and the preoccupation of mature research workers with the same, single species are two quite different things.

Advanced research in physiology is not restricted to studies of the frog and although many geneticists may confine their personal investigations to *Drosophila,* an even larger number deals with other animal species or with plans. As a matter of fact, the benefits that students can derive from studying one kind of animal as a sample species must always stand in direct proportion to the amount of information research workers have gathered in connection with other species. The rat's value as a teaching aid in psychology depends in part upon the certainty with which the student can generalize from the behavior he observes in this one animal; and this in turn is a function of available knowledge concerning other species.

There is another obvious argument in favor of concentrating our efforts on the study of a single animal species. It is well expressed in Professor Skinner's book, "The Behavior of Organisms."

> In the broadest sense a science of behavior should be concerned with all kinds of organisms, but it is reasonable to limit oneself, at least in the beginning, to a single representative species.

I cannot imagine that anyone would quarrel with Skinner on this point and I am convinced that many of the psychologists currently using rats in their investigational programs would agree with him in his implicit assumption that the Norway rat *is* a "representative species." But in what way is it "representative," and how has this "representativeness" been demonstrated? These questions lead at once to a consideration of the disadvantages of overspecialization in terms of animals used and types of behavior studied.

To put the question bluntly: Are we building a general science of behavior or merely a science of rat learning? The answer is not obvious to me. Admittedly there are many similarities between the associative learning of lower animals and what is often referred to as rôte learning in man. But the variety of organisms which have been studied, and the number of techniques which have been employed are so limited, it is difficult to believe that we can be approaching a comprehensive understanding of the basic phenomena of learning. It may be that much remains to be discovered by watching rats in mazes and problem boxes, but it is time to ask an important question. How close are we getting to that well-known point of diminishing returns? Would we not be wise to turn our attention to other organisms and to devise new methods of testing behavior before we proceed to formulate elaborate theories of learning which may or may not apply to other species and other situations.

Another very important disadvantage of the present method in animal studies is that because of their preoccupation with a few species and a few types of behavior, psychologists are led to neglect many complex patterns of response that stand in urgent need of systematic analysis. The best example of this tendency is seen in the current attitude toward so-called "instinctive" behavior.

The growing emphasis upon learning has produced a complementary reduction in the amount of study devoted to what is generally referred to as "unlearned behavior." Any pattern of response that does not fit into the category of learned behavior as currently defined is usually classified as "unlearned" even though it has not been analyzed directly. Please note that the classification is made in strictly negative terms *in spite of the fact that the positive side of the implied dichotomy is very poorly defined.* Specialists in learning are not in accord as to the nature of the processes involved, nor can they agree concerning the number and kinds of learning that may occur. But in spite of this uncertainty most "learning psychologists" confidently identify a number of complex behavior patterns as "unlearned." Now the obvious question arises: Unless we know what learning is—unless we can recognize it in all of its manifestations—how in the name of common sense can we identify any reaction as "unlearned"?

The fact of the matter is that none of the responses generally classified as "instinctive" have been studied as extensively as maze learning or problem-solving behavior. Data relevant to all but a few "unlearned" reactions are too scanty to permit any definite conclusion concerning the role of experience in the shaping of the response. And those few cases in which an exhaustive analysis has been attempted show that the development of the behavior under scrutiny is usually more complicated than a superficial examination could possibly indicate.

For example, there is a moth which always lays its eggs on hackberry leaves. Females of each new generation select hackberry as an oviposition site and ignore other potential host plants. However, the eggs can be transferred to apple leaves, and when this is done the larvae develop normally. Then when adult females that have spent their larval stages on apple leaves are given a choice of materials upon which to deposit their eggs, a high proportion of them select apple leaves in preference to hackberry. This control of adult behavior by the larval environment does not fit into the conventional pigeonhole labeled "instinct," and neither can it be placed in the category of "learning." Perhaps we need more categories. Certainly we need more data on more species and more kinds of behavior.

Primiparous female rats that have been reared in isolation usually display biologically effective maternal behavior when their first litter is born. The young ones are cleaned of fetal membranes, retrieved to the nest, and suckled regularly. However, females that have been reared under conditions in which it was impossible for them to groom their own bodies often fail to clean and care for their newborn offspring. Observations of this nature cannot be disposed of by saying that the maternal reactions are "learned" rather than "instinctive." The situation is not so simple as that. In some way the early experience of the animal prepares her for effective maternal performance even though none of the specifically maternal responses are practiced before parturition.

It seems highly probable that when sufficient attention is paid to the so-called "instinctive" patterns, we will find that their development involves processes of which current theories take no account. What these processes may be we shall not discover by continuing to concentrate on learning as we are studying it. And yet it is difficult to see how a valid theory of learning can be formulated without a better understanding of the behavior that learning theorists are presently categorizing as "unlearned."

Potential Returns from the Comparative Approach

If more experimental psychologists would adopt a broadly comparative approach, several important goals might be achieved. Some of the returns are fairly specific and can be described in concrete terms. Others are more general though no less important.

Specific Advantages. I have time to list only a few of the specific advantages which can legitimately be expected to result from the application of comparative methods in experimental psychology. In general, it can safely be predicted that some of the most pressing questions that we are now attempting to answer by studying a few species and by employing only a few experimental methods would be answered more rapidly and adequately if the approach were broadened.

Let us consider learning as one example. Comparative psychology offers many opportunities for examination of the question as to whether there are one or many kinds of learning and for understanding the rôle of learning in the natural lives of different species. Tinbergen (1942) has reported evidence indicating the occurrence of one-trial learning in the behavior of hunting wasps. He surrounded the opening of the insect's burrow with small objects arranged in a particular pattern. When she emerged, the wasp circled above the nest opening for a few seconds in the usual fashion and then departed on a hunting foray. Returning after more than an hour, the insect oriented directly to the pattern stimulus to which she had been exposed only once. If the pattern was moved during the female's absence she was able to recognize it immediately in its new location.

Lorenz's concept of "imprinting" offers the learning psychologist material for new and rewarding study. Lorenz (1935) has observed that young birds of species that are both precocial and social quickly become attached to adults of their own kind and tend to follow them constantly. Newly-hatched birds that are reared by parents of a foreign species often form associations with others of the foster species and never seek the company of their own kind. A series of experiments with incubator-reared birds convinced Lorenz that the processes underlying this sort of behavior must occur very early in life, perhaps during the first day or two after hatching, and that they are irreversible, or, to phrase it in other terms, that they are not extinguished by removal of reinforcement.

J. P. Scott's studies (1945) of domestic sheep reveal the importance of early learning in the formation of gregarious habits. Conventional learning theories appear adequate to account for the phenomena, but it is instructive to observe the manner in which the typical species pattern of social behavior is built up as a result of reinforcement afforded by maternal attentions during the nursing period.

The general importance of drives in any sort of learning is widely emphasized. Therefore it would seem worth while to study the kinds of drives that appear to motivate different kinds of animals. In unpublished observations upon the ferret, Walter Miles found that hunger was not sufficient to produce maze learning. Despite prolonged periods of food deprivation, animals of this species continue to explore every blind alley on the way to the goal box.

Additional evidence in the same direction is found in the studies of Gordon (1943) who reports that non-hungry chipmunks will solve mazes and problem boxes when rewarded with peanuts which the animals store in their burrows but do not eat immediately. Does this represent a "primary" drive to hoard food or an "acquired" one based upon learning?

Many experimentalists are concerned with problems of sensation and perception; and here too there is much to be gained from the comparative approach. Fring's studies (1948) of chemical sensitivity in caterpillars, rabbits and men promise to increase our understanding of the physiological basis for gustatory sensations. In all three species there appears to be a constant relationship between the ionic characteristics of the stimulus material and its effectiveness in evoking a sensory discharge. The investigations of Miles and Beck (1949) on reception of chemical stimuli by honey bees and cockroaches provides a test for the theory of these workers concerning the human sense of smell.

The physical basis for vision and the role of experience in visual perception have been studied in a few species but eventually it must be investigated on a broader comparative basis if we are to arrive at any general understanding of the basic principles involved. Lashley and Russell (1934) found that rats reared in darkness give evidence of distance perception without practice; and Hebb (1937) added the fact that figure-ground relationships are perceived by visually-naive animals of this species. Riesen's (1947) report of functional blindness in apes reared in darkness with gradual acquisition of visually-directed habits argues for a marked difference between rodents and anthropoids; and Senden's (1932) descriptions of the limited visual capacities of human patients after removal of congenital cataract appear to support the findings on apes. But the difference, if it proves to be a real one, is not purely a function of evolutionary status of the species involved. Breder and Rasquin (1947) noted that fish with normal eyes but without any visual experience are unable to respond to food particles on the basis of vision.

I have already mentioned the necessity for more extensive examination of those patterns of behavior that are currently classified as "instinctive." There is only one way to approach this particular problem and that is through comparative psychology. The work that has been done thus far on sexual and parental behavior testifies, I believe, to the potential returns that can be expected if a more vigorous attack is launched on a broader front.

We are just beginning to appreciate the usefulness of a comparative study of social behavior. The findings of Scott which I mentioned earlier point to the potential advantages of using a variety of animal species in our investigation of interaction between members of a social group. Carpenter's (1942) admirable descriptions of group behavior in free-living monkeys point the way to a better understanding of dominance, submission, and leadership.

One more fairly specific advantage of exploring the comparative method in psychology lies in the possibility that by this means the experimentalist can often discover a particular animal species that is specially suited to the problem with which he is concerned. For example, in recent years a considerable amount of work has been done on hoarding behavior in the laboratory rat. The results are interesting, but they indicate that some rats must learn to hoard and some never do so. Now this is not surprising since Norway rats rarely hoard food under natural conditions. Would it not seem reasonable to begin the work with an animal that is a natural hoarder? Chipmunks, squirrels, mice of the genus *Peromyscus,* or any one of several other experimental subjects would seem to be much more appropriate.

And now, as a final word, I want to mention briefly a few of the more general facts that indicate the importance of developing comparative psychology.

General Advantages. For some time it has been obvious that psychology in this country is a rapidly expanding discipline. Examination of the membership roles of the several Divisions of this Association shows two things. First, that the number of psychologists is increasing at a prodigious rate; and second that the growth is asymmetrical in the sense that the vase majority of new workers are turning to various applied areas such as industrial and clinical psychology.

It is generally recognized that the applied workers in any science are bound to rely heavily upon "pure" or "fundamental" research for basic theories, for general methodology and for new points of view. I do not suggest that we, as experimentalists, should concern ourselves with a comparative approach to practical problems of applied psychology. But I do mean to imply that if we intend to maintain our status as indispensable contributors to the science of behavior, we will have to broaden our attack upon the basic problems of the discipline. This will sometimes mean sacrificing some of the niceties of laboratory research in order to deal with human beings under less artificial conditions. It may also mean expanding the number of non-human species studied and the variety of behavior patterns investigated.

Only by encouraging and supporting a larger number of comparative investigations can psychology justify its claim to being a true science of behavior. European students in this field have justly condemned Americans for the failure to study behavior in a sufficiently large number of representative species. And non-psychologists in this country are so well aware of our failure to develop the field that they think of animal behavior as a province of general zoology rather than psychology. Top-rank professional positions that might have been filled by psychologically trained investigators are today occupied by biologists. Several large research foundations are presently supporting extensive programs of investigation into the behavior of sub-human animals, and only in one instance is the program directed by a psychologist.

Conclusion

If we as experimental psychologists are missing an opportunity to make significant contributions to natural science—if we are failing to assume leadership in an area of behavior investigation where we might be useful and effective—if these things are true, and I believe that they are, then we have no one but ourselves to blame. We insist that our students become well versed in experimental design. We drill them in objective and quantitative methods. We do everything we can to make them into first rate experimentalists. And then we give them so narrow a view of the field of behavior that they are satisfied to work on the same kinds of problems and to employ the same methods that have been used for the past quarter of a century. It would be much better if some of our well-trained experimentalists were encouraged to do a little pioneering. We have a great deal to offer in the way of professional preparation that the average biologist lacks. And the field of animal behavior offers rich returns to the psychologist who will devote himself to its exploration.

I do not anticipate that the advanced research worker whose main experimental program is already mapped out will be tempted by any argument to shift to an entirely new field. But those of us who have regular contact with graduate students can do them a service by pointing out the possibilities of making a real contribution to the science of psychology through the medium of comparative studies. And even in the absence of professorial guidance the alert beginner who is looking for unexplored areas in which he can find new problems and develop new methods of attacking unsettled issues would be wise to give serious consideration to comparative psychology as a field of professional specialization.

References

1. Breder, C. M. and Rasquin, P. Comparative studies in the light sensitivity of blind characins from a series of Mexican caves. *Bulletin Amer. Mus. Natl. Hist.*, 1947, *89*, Article 5, 325-351.
2. Carpenter, C. R. Characteristics of social behavior in non-human primates. *Trans. N. Y. Acad. Sci.*, 1942, Ser. 2, *4*, No. 8, 248.
3. Frings, H. A contribution to the comparative physiology of contact chemoreception. *J. comp. physiol. Psychol.*, 1948, *41*, No. 1, 25-35.
4. Gordon, K. The natural history and behavior of the western chipmunk and the mantled ground squirrel. *Oregon St. Monogr. Studies in Zool.*, 1943, No. 5, 7-104.
5. Hebb, D. O. The innate organization of visual activity. I. Perception of figures by rats reared in total darkness. *J. gen. Psychol.*, 1937, *51*, 101-126.
6. Lashley, K. S. and Russell, J. T. The mechanism of vision. XI. A preliminary test of innate organization. *J. genet. Psychol.*, 1934, *45*, No. 1, 136-144.
7. Lorenz, K. Der Kumpan in der Umwelt des Vogels. *J. f. Ornith.*, 1935, *83*, 137-213.
8. Miles, W. R. and Beck, L. H. Infrared absorption in field studies of olfaction in honeybees. *Proceed. Natl. Acad. Sci.*, 1949, *35*, No. 6, 292-310.
9. Riesen, A. H. The development of visual perception in man and chimpanzee. *Science*, 1947, *106*, 107-108.
10. Scott, J. P. Social behavior, organization and leadership in a small flock of domestic sheep. *Comp. Psychol. Monogr.*, 1945, *18*, No. 4, 1-29.
11. Senden, M. v. *Raum- und Gestaltauffassung bei operierten Blindgeborenen vor und nach der Operation.* Leipzig: Barth, 1932.
12. Tinbergen, N. An objectivistic study of the innate behaviour of animals. *Biblio. Biotheoret.*, 1942, *1*, Pt. 2, 39-98.

Critical Periods in Behavioral Development

J. P. SCOTT

A number of years ago I was given a female lamb taken from its mother at birth. My wife and I raised it on the bottle for the first 10 days of life and then placed it out in the pasture with a small flock of domestic sheep. As might have been expected from folklore, the lamb became attached to people and followed the persons who fed it. More surprisingly, the lamb remained independent of the rest of the flock when we restored it to the pasture. Three years later it was still following an independent grazing pattern. In addition, when it was mated and had lambs of its own it became a very indifferent mother, allowing its offspring to nurse but showing no concern when the lamb moved away with the other members of the flock (*1*).

Since following the flock is such a universal characteristic of normal sheep, I was impressed by the extensive and permanent modification of this behavior that resulted from a brief early experience. The results suggested that Freud was right concerning the importance of early experience, and pointed toward the existence of critical periods in behavioral development. As I soon discovered, there is considerable evidence that a critical period for determining early social relationships is a widespread phenomenon in vertebrates; such a critical period had long been known in ants (*2*).

The theory of critical periods is not a new one in either biology or psychology. It was strongly stated by Stockard in 1921, in connection with his experiments on the induction of monstrosities in fish embryos, although he gave credit to Dareste for originating the basic idea 30 years earlier (*3*). In experimenting with the effects of various inorganic chemicals upon the development of *Fundulus* eggs, Stockard at first

From *Science*, 1962, *138*, 949-58. Reprinted by permission of the author and publisher.

thought one-eyed monsters were specifically caused by the magnesium ion. Further experiments showed him that almost any chemical would produce the same effect, provided it was applied at the proper time during development. These experiments and those of Child (*4*) and his students established the fact that the most rapidly growing tissues in an embryo are the most sensitive to any change in conditions, thus accounting for the specificity of effects at particular times.

Meanwhile Freud had attempted to explain the origin of neuroses in human patients as the result of early experience and had implied that certain periods in the life of an infant are times of particular sensitivity. In 1935, Lorenz (*5*) emphasized the importance of critical periods for the formation of primary social bonds (imprinting) in birds, remarking on their similarity to critical periods in the development of the embryo, and McGraw soon afterward (*6*) pointed out the existence of critical periods for optimal learning of motor skills in the human infant.

Since then, the phenomenon of critical periods has excited the imagination of a large group of experimenters interested in human and animal development. In describing this fast-moving scientific field, I shall point out some of the most significant current developments. More detailed information is available in some excellent recent reviews (*7, 8*).

To begin with, three major kinds of critical-period phenomena have been discovered. These involve optimal periods for learning, for infantile stimulation, and for the formation of basic social relationships. The last of these has been established as a widespread phenomenon in the animal kingdom and consequently receives major attention in this article.

Periods Are Based on Processes

In the dog, the development of behavior may be divided into several natural periods marked off by important changes in social relationships (Table 1). Only a few other species have been studied in sufficient detail for making adequate comparisons, but enough data have been accumulated to show that similar periods can be identified in other mammals and in birds (*9, 10*). I originally expected to find that the course of postnatal development, like that of embryonic development, would be essentially similar in all vertebrates, and that while the periods might be extended or shortened, the same pattern of development would be evident in all (*11*). However, comparison of only two species, man and the dog, shows that the periods can actually occur in reverse order, and that there is an astonishing degree of flexibility in behavioral development (*12*).

This leads to the conclusion that the important aspect of each developmental period is not time sequence but the fact that each represents a major developmental process. Thus, the neonatal period is chiefly characterized by the process of neonatal nutrition—nursing in mammals and parental feeding in many birds. The transition period is characterized by the process of transition to adult methods of nutrition and locomotion and the appearance of adult patterns of social behavior, at least in immature form. The period of socialization is the period in which primary social bonds are formed. If we consider processes alone, it is apparent that they are not completely dependent on each other and that they can therefore be arranged in different orders. It is also apparent that certain of these processes persist beyond the periods characterized by them. For example, a mammal usually retains throughout life the ability to suck which characterizes the neonatal period, although in most cases this ability is little used.

Process of Primary Socialization

Since one of the first acts of a young mammal is to nurse, and since food rewards are known to modify the behavior of adult animals, it once seemed logical to suppose that the process of forming a social attachment begins with food rewards and develops as an acquired drive. However, the experimental evidence does not support this extreme viewpoint. Brodbeck reared a group of puppies during the critical period of socialization, feeding half of them by hand and the other half by machine, but giving all of them the same degree of human contact (*13*). He found that the two sets of puppies became equally attached to people. This result was later confirmed by Stanley and his co-workers (*14*), who found that the only difference in response between the machine-fed and the hand-fed puppies was that the latter yelped more when they saw the experimenter. Elliot and King (*15*) fed all their puppies by hand but overfed one group and underfed another. The hungry puppies became more rapidly attached to the handlers. We can conclude that, in the dog, food rewards per se are not necessary for the process of socialization, but that hunger will speed it up.

Fisher (*16*) reared fox terrier puppies in isolation boxes through the entire socialization period. The puppies were fed mechanically (thus, food was entirely eliminated as a factor in the experiment), but they were removed from the boxes for regular contacts with the experimenter. One group of puppies was always rewarded by kind social treatment. A second group was sometimes rewarded and sometimes punished, but in a purely random way. Still a third group was always punished for any positive approach to the experimenter. The puppies that were both rewarded and punished showed most attraction and dependency behavior with respect to the experimenter, and the puppies that were always punished showed the least. After the treatment was discontinued, all the puppies began coming toward the experimenter, and the differences rapidly disappeared. This leads to the surprising conclusion that the process of socialization is not inhibited by punishment and may even be speeded up by it.

At approximately 3 weeks of age—that is, at the beginning of the period of socialization—young puppies begin to bark or whine when isolated or placed in strange places. Elliot and Scott (*17*) showed that the reaction to isolation in a strange place reaches a peak at 6 to 7 weeks of age, approximately the midpoint of the critical period, and begins to decline thereafter. Scott, Deshaies, and Morris (*18*) found that separating young puppies overnight from their mother and litter mates in a strange pen for 20 hours per day produced a strong emotional reaction and speeded up the process of socialization to human handlers. All this evidence indicates that any sort of strong emotion, whether hunger, fear, pain, or loneliness, will speed up the process of socialization. No experiments have been carried out to determine the effects of pleasant types of emotion, such as might be aroused by play and handling, but these were probably a factor in Brodbeck's experiment with machine-fed puppies.

The results of these experiments on dogs agree with evidence from other species. While they were going on, Harlow (*19*) was performing his famous experiments with rhesus monkeys isolated at birth and supplied with dummy "mothers." When given the choice between a comfortable cloth-covered mother without a nipple and an uncomfortable mother made of wire screening but equipped with a functional nursing bottle, the young rhesus monkeys definitely preferred the cloth-covered models from which they had received no food rewards. Harlow concluded that the acquired-drive theory of the origin of social attachment could be discarded.

Later, Igel and Calvin (*20*) performed a similar but more elaborate experiment with puppies. These animals had more opportunity to choose, being provided with four kinds of mother models: comfortable and uncomfortable, each type with and without nipples. Like rhesus monkeys, the puppies preferred the comfortable "mother" but usually chose one with a nipple. Thus, it appears that food rewards do contribute something to the social relationship, although they do not form its prime basis.

Since then Harlow (*21*) has raised to maturity the monkeys raised on dummy mothers, has mated them, and has observed their behavior toward their own young. They became uniformly poor mothers, neglecting their offspring and often punishing them when they cry. In spite of such rejection, the young rhesus infants desperately crawl toward their mothers and give every evidence of becoming attached to them, although perhaps not as strongly as in the normal relationship. Here again punishment does not inhibit the formation of a social bond.

The hypothesis that the primary social bond originates through food rewards had already been shown to be invalid in the precocial birds, many of which form attachments prior to the time when they begin to feed. Lorenz (*5*) was the first to point out the significance of this phenomenon, which he called "imprinting." He also stated that it differed from conditioning, primarily in that it was very rapid and apparently irreversible. However, rapid formation and great persistence are also characteristic of many conditioned responses and other learned behavior. Fabricius (*22*) pointed out that no sharp line can be drawn between imprinting and conditioning, and Collias (*23*) concluded that imprinting is a form of learned behavior that is self-reinforcing.

The process of imprinting in young ducklings and chicks has since been experimentally analyzed in much detail, with results that invariably confirm the conclusion that it takes place without any obvious external rewards or reinforcement. Hess (*24*) found that if he caused young ducklings to follow a model over varying distances or over hurdles, the ducklings which had to make the greater effort became more strongly imprinted. He also found that the drug meprobamate and its congener carisoprodol, which are muscle relaxants as well as tranquilizers, greatly reduce imprinting if given during the critical period. James (*25*) found that chicks would become attached to an object illuminated by a flickering light, even

Fig. 1. Hess's apparatus for measuring the following response in ducklings and chicks. A decoy revolves on a circular path, the young duckling staying nearby. Other revolving objects may be substituted for the decoy, which is wired for sound. The following response is a major positive timing mechanism that initiates the critical period for imprinting; it is also an indicator that an attachment has been formed. [From a photo by E. H. Hess]

though they were not allowed to follow, and Gray (*26*) later showed that they will become attached to a motionless object illuminated by a steady light and viewed from an isolation box. It is therefore apparent that chicks can become imprinted without following, although muscular tension may still be important.

Guiton (*27*) found that chicks allowed to follow a model in a group become less strongly imprinted than chicks exposed singly, and he attributed the results to the greater fear shown by the isolated chicks. Recently, Pitz and Ross (*28*) subjected young chicks following a model to a loud sound and found that this increased the speed with which they formed a social bond.

TABLE 1

Periods of development in the puppy and song sparrow. The six periods of development described by Nice (*10*) for the song sparrow correspond to the first four periods in the puppy, as indicated in the table. The young of the two species are born or hatched in an immature state, require intensive parental care and feeding, and go through much the same stages before becoming independent. Development is much more rapid in the bird than in the puppy, although small mammals such as mice mature at about the same rate as birds.

Puppy			Song sparrow		
Name of period	Length of period (weeks)	Initial event	Name of period	Length of period (days)	Initial event
I. Neonatal	0-2	Birth, nursing	Stage 1 (nestling)	0-4	Hatching, gaping
II. Transition	2-3	Eyes open	Stage 2	5-6	Eyes open
III. Socialization	3-10	Startle to sound	Stage 3	7-9	Cowering—first fear reactions
			Stage 4 (fledgling)	10-16	Leaving nest—first flight
			Stage 5	17-28	Full flight
IV. Juvenile	10-	Final weaning	Stage 6 (juvenile)	29-	Independent feeding

Hess (*29*) (with the apparatus shown in Fig. 1) has given a mild electric shock to chicks following a model and finds that this also increases the strength of imprinting. Instead of avoiding the model, the distressed chick runs after it more closely.

We may conclude that these young birds become attached to any object to which they are long exposed during the critical period, even when their contact is only visual. We may also conclude that the speed of formation of a social bond is dependent upon the degree of emotional arousal, irrespective of the nature of that arousal. Whether attachment is the result of the emotion itself or of the reduction of emotion as the chick or duckling approaches the model is still a matter of conjecture (*30*).

Timing Mechanisms

The basic timing mechanisms for developmental periods are obviously the biological processes of growth and differentiation, usually called maturation. For various reasons, these are not precisely correlated with age from birth or hatching. For example, birds often retain newly formed eggs in their bodies overnight, thus incubating them for several hours before laying. By chilling duck eggs just before placing them in an incubator (thus killing all embryos except those in the earliest stages of development) Gottlieb (*31*) was able to time the age of ducklings from the onset of incubation rather than from hatching and found that variation in the timing for the critical period was much reduced. No such exact timing studies have been made in mammals, but I have estimated that there is at least a week's variation in development among puppies at 3 weeks of age, and the variation among human infants must be considerably greater (*32*).

Another approach to the problem is to try to identify the actual mechanisms which open and close a period. Since an important part of forming a primary social relationship appears to be emotional arousal while the young animal is in contact with another, it is obvious that the critical period for socialization could be timed by the appearance of behavioral

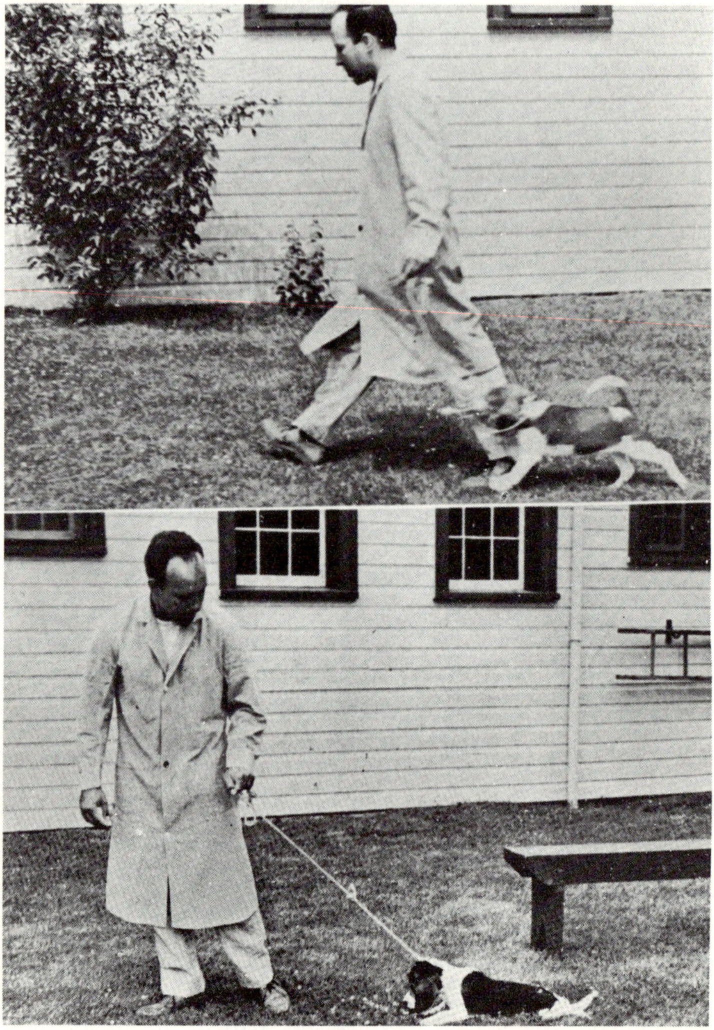

Fig. 2. Puppies reacting to leash training, according to their experience during the critical period (see *35*). (Top) Reaction of a puppy that had had contact with people for 1 week during the peak of the critical period. (Bottom) Reaction of a puppy that had had contact with other dogs prior to the end of the critical period. This puppy shows an extreme fear reaction and refuses to follow.

mechanisms which maintain or prevent contact, and this indeed is the case. There are demonstrable positive mechanisms, varying from species to species, which bring young animals close to other members of their kind: the clinging response of young rhesus monkeys; the following response of chicks, ducklings, and lambs and other herd animals; the social investigation, tail wagging, and playful fighting of puppies; and the visual investigation and smiling of the human infant (*33*). These are, of course, accompanied by interacting responses from adult and immature members of the species: holding and clasping by primate mothers, brooding of mother hens and other birds, calling by mother sheep, investigation and play on the part of other young puppies, and the various supporting and nurturing activities of human mothers.

If contact and emotional arousal result in social attachment, there must be negative mechanisms which prevent such attachment once the critical period is past. Perhaps the most widespread of these is the development of a fear response which causes the young animal to immediately leave the vicinity of a stranger and hence avoid contact. This developing fear response is found in young chicks (*7*), ducklings (*22, 34*), dogs (*35*; Fig. 2), rhesus monkeys (*36*), and in many other birds and mammals. Even in children there is a period between the ages of 5 and 12 months in which there is a mounting fear of strangers (*37*), sometimes called "8-months anxiety" (*38*). As already pointed out, there is a time in development when certain fear responses actually facilitate imprinting, but, as they grow stronger, the escape reaction follows so quickly that it prevents contact altogether.

Another sort of negative mechanism is the rejection of strange young by adult sheep, goats, and many other herd animals (*39*). In these species the mothers become strongly attached to the young within a few hours after birth and refuse to accept strangers thereafter (Fig. 3). This indicates that the rapid formation of emotional bonds is not limited to young animals.

These timing mechanisms all depend primarily on the development of social behavior patterns, but both sensory and motor development can also influence timing. For example, a

Fig. 3. Rejection by a mother goat of a kid not her own. The behavior of adults terminates the critical period for primary socialization in sheep and goats. Immediately after giving birth the mother will accept any young kid, but within a few hours she will reject any strange kid that approaches. [From a photo by A. U. Moore]

very immature animal cannot maintain contact by following, and in slowly developing altricial birds such as jackdaws and doves (*5, 40*), the period of imprinting comes much later than it does in the precocial species. In the human infant the process of socialization begins before the adult motor patterns develop, but contact is maintained by visual exploration and by the smiling response to human faces (*33*). Thus, understanding the process of socialization and its timing mechanisms in any particular species requires a systematic study of the development of the various capacities which affect the time of onset and the duration of the critical period. These include sensory, motor, and learning capacities as well as the ability to perform essential patterns of social behavior.

The fact that emotional arousal is so strongly connected with the process of primary socialization suggests that the capacity to produce emotional reactions may also govern the time of onset of a critical period. Figure 4 summarizes the results of a study of emotional development in the dog during the critical period. If puppies are kept in large fields, totally isolated from people, fear and escape responses toward human beings very nearly reach a maximum by the time the puppies are 14 weeks old—a finding that fixes the upper limit of the period of socialization (*35*). On the other hand, the peak of the emotional response to isolation in a strange place occurs when puppies are approximately 6 to 7 weeks old, as does the peak of the heart-rate response to handling. At this age, such emotional arousal actually contributes to the strength of the social bond. Fuller (*41*) was unable to condition the heart-rate response consistently until puppies were 5 weeks old. This indicates that one of the factors that brings the critical period to a close may be the developing ability of the young puppy to associate fear responses with particular stimuli.

All this suggests that if the development of the escape response to strangers could be held in check, the critical period

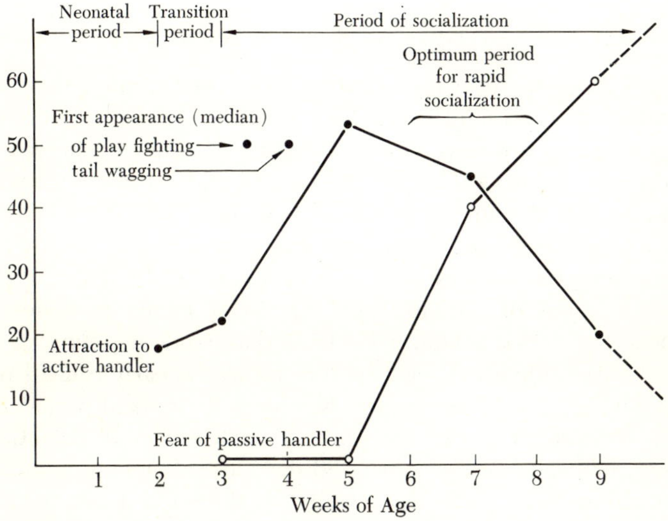

Fig. 4. Timing mechanisms for the critical period in puppies (see *35*). The period is initiated by positive behavior mechanisms, such as playful fighting, which result in attraction to a strange individual, and it is brought to a close by the development of a fear response which causes the attraction to decline. The optimum period for rapid and permanent socialization comes shortly after the appearance of prolonged avoidance reactions.

might be extended indefinitely. Raising puppies in small isolation boxes during the critical period inhibits the development of the escape response but they still show obvious signs of fear when they are first removed from their cages. Fuller (*42*) reports some success in socializing these older pups by overcoming their fear responses, either by careful handling or through the use of tranquilizing drugs.

Fear responses thus have the dual effect of facilitating the formation of the social bond during the critical period (along with other emotions) and of bringing the period to a close. This is understandable because the type of fear which terminates the critical period is a developing fear of strange animals. In the early part of the critical period the escape reaction is either lacking or is momentary and weak. At the close of the period it is strong enough to prevent contact altogether.

Formation of Affectional Bonds in Adult Life

Until recently, most investigators have concentrated their attention on the critical period for primary socialization or imprinting and few have gone on to study similar phenomena in later development. This field of investigation is just beginning to open up, though many related facts have long been known. For example, many birds form strong pair bonds which are maintained as long as both members survive. In studying the development of various types of social bonds in different species of ducks, Schultz (*43*) finds that, while attachments to particular individuals may be formed in the early critical period from 12 to 17 hours after hatching, the critical period for the attachment to the species may not come until sometime later, in some cases as late as 30 days after hatching, and the attachment to a particular member of the opposite sex, or the pair bond, does not come until the age of 5 months or so. Schultz also finds that female mallards cannot be sexually imprinted with respect to other species but always mate with other mallards no matter what their earliest experience has been. A similar phenomenon is reported by Warriner (*44*), who finds that male pigeons prefer to mate with birds whose color is similar to that of the parents who reared them, whether of the same or another color from themselves, but females show no preference.

Certain species of mammals, such as foxes (*45*), form long-lasting mating bonds. It is possible that the violence of the sexual emotions contributes to the formation of the adult bond, just as other sorts of emotional arousal are important to the primary socialization of the infant. Klopfer (*46*) has suggested that the rapid formation of the social bond in a mother goat toward her kid is the result of the high degree of emotional arousal which accompanies the birth of the offspring.

In short, it seems likely that the formation of a social attachment through contact and emotional arousal is a process that may take place throughout life, and that although it may take place more slowly outside of certain critical periods, the capacity for such an attachment is never completely lost.

At this point it may be remarked that, in attempting to analyze the development of affection and social bonds objectively, scientists have often tried to simplify the problem by postulating various unitary, unromantic, and sometimes unesthetic explanations. One of these was the "acquired drive" hypothesis—that children love you because you feed them. Taking a more moderate view Harlow (*19*) has emphasized "contact comfort" as a major variable—that the young monkey begins to love its mother because she feels warm and comfortable—but that a number of other factors are involved. As this article indicates, evidence is accumulating that there is a much less specific, although equally unromantic, general mechanism involved—that given any kind of emotional arousal a young animal will become attached to any individual or object with which it is in contact for a sufficiently long time. The necessary arousal would, of course, include various specific kinds of emotions associated with food rewards and contact comfort.

It should not be surprising that many kinds of emotional reactions contribute to a social relationship. The surprising thing is that emotions which we normally consider aversive should produce the same effect as those which appear to be rewarding. This apparent paradox is partially resolved by evidence that the positive effect of unpleasant emotions is normally limited to early infancy by the development of escape reactions.

Nevertheless, this concept leads to the somewhat alarming conclusion that an animal (and perhaps a person) of any age, exposed to certain individuals or physical surroundings for any length of time, will inevitably become attached to them, the rapidity of the process being governed by the degree of emotional arousal associated with them. I need not dwell on the consequences for human behavior, if this conclusion should apply to our species as well as to other animals, except to point out that it provides an explanation of certain well-known clinical observations such as the development by neglected children of strong affection for cruel and abusive parents, and the various peculiar affectional relationships that develop between prisoners and jailors, slaves and masters, and so on. Perhaps the general adaptive nature of this mechanism is that since the survival of any member of a highly social species depends upon the rapid development of social relationships, a mechanism has evolved which makes it almost impossible to inhibit the formation of social bonds.

Critical Periods of Learning

Unlike the process of socialization, the phenomenon of critical periods of learning was first noticed in children rather than in lower animals. McGraw's (*47*) famous experiment with the twins Johnny and Jimmy was a deliberate attempt to modify behavioral development by giving one of a pair of identical twins special early training. The result varied according to the activity involved. The onset of walking, for example, was not affected by previous practice or help. Other activities, however, could be greatly speeded up—notably roller skating, in which the favored twin became adept almost as soon as he could walk. In other activities performance was actually made worse by early practice, simply because of the formation of unskilled habits. McGraw (*6*) concluded that there are critical periods for learning which vary from activity to activity; for each kind of coordinated muscular activity there is an optimum period for rapid and skillful learning.

In an experiment with rats, Hebb (*48*) used the technique of providing young animals with many opportunities for spontaneous learning rather than formal training. Pet rats raised in the rich environment of a home performed much better on learning tasks than rats reared in barren laboratory cages. Since then, other experimenters (*49*) have standardized the "rich" environment as a large cage including many objects and playthings and have gotten similar effects.

Forgays (see *50*) finds that the age at which the maximum effect is produced is limited to the period from approximately 20 to 30 days of age, immediately after weaning. A similar experience in adult life produces no effect. In rats, at any rate, the critical period of learning seems to coincide with the critical period of primary socialization, and it may be that the two are in some way related. Candland and Campbell (*51*) find that fearful behavior in response to a strange situation begins to increase in rats between 20 and 30 days after birth, and Bernstein (*52*) showed earlier that discrimination learning could be improved by gentle handling beginning at 20 days. It may well be that the development of fear limits the capacity for future learning as well as the formation of social relationships.

In addition to these studies on motor learning and problem solving, there are many experiments demonstrating the existence of critical periods for the learning of social behavior patterns. It has long been known that many kinds of birds do not develop the characteristic songs of their species if they are reared apart from their own kind (*53*). More recently, Thorpe (*54*) discovered a critical period for this effect in the chaffinch. If isolated at 3 or 4 days of age, a young male chaffinch produces an incomplete song, but if he hears adults singing, as a fledgling 2 or 3 weeks old or in early juvenile life before he sings himself, he will the next year produce the song characteristic of the species, even if he has been kept in isolation. In nature, the fine details of the song are added at the time of competition over territory, within a period of 2 or 3 weeks, when the bird is about a year old. At this time it learns the songs of two or three of its neighbors, and never learns any others in subsequent years. The critical period for song learning is thus a relatively long one, but it is definitely over by the time the bird is a year old. There is no obvious explanation for its ending at this particular time, but it is possible that learning a complete song pattern in some way interferes with further learning.

King and Gurney (*55*) found that adult mice reared in groups during youth fought more readily than animals isolated at 20 days of age. Latter experiments showed that most of the effect was produced in a 10-day period just after weaning, and that similar experience as adults produced little or no effect (*56*). Thus, there appears to be a critical period for learning to fight through social experience, and this experience need be no more than contact through a wire. In this case the effect is probably produced by association with other mice before the fear response has been completely developed. Similarly, Fisher (*16*) and Fuller (*57*) inhibited the development of attacking behavior in fox terriers by raising them in isolation through the critical period for socialization. The animals would fight back somewhat ineffectually if attacked, but did not initiate conflicts. Tinbergen (*58*) found a critical period in dogs for learning territorial boundaries, coinciding with sexual maturity.

The results of corresponding experiments on sexual behavior vary from species to species. In mice, rearing in isolation produced no effects (*59*). Beach (*60*) found that male rats reared with either females or males were actually slower to respond to sexual behavior than isolated males, and he suggested that habits of playful fighting established by the group-reared animals interfered with sexual behavior later on. In guinea pigs, contact with other young animals improves sexual performance (*61*).

On the other hand, young chimpanzees (*62*) reared apart from their kind can only be mated with experienced animals. Harlow (*21*) discovered that his rhesus infants reared on dummy mothers did not develop normal patterns of sexual behavior, and he was able to obtain matings only by exposing females to experienced males (Fig. 5). Normal behavior can be developed by allowing 20-minute daily play periods with other young monkeys, but if rhesus infants are reared apart

Fig. 5. Through play behavior during a critical period young rhesus monkeys develop the capacity for adult sexual behavior. The two in bottom photo exhibit an approximation of the adult posture. It is almost impossible to mate monkeys raised with cloth "mothers" without an opportunity for play with other infants. [H. F. Harlow]

from all other monkeys beyond the period when they spontaneously play with their fellows, patterns of both sexual and maternal behavior fail to develop normally. These results suggest that play has an important role in developing adult patterns of social behavior in these primates, and that the decline of play behavior sets the upper limit of the critical period during which normal adult behavior may be developed.

Such great changes in the social environment rarely occur in humans even by accident, but Money, Hampson, and Hampson (*63*) have studied the development of hermaphroditic children who have been reared as one sex and then changed to the other. They find that if this occurs before 2½ years of age, very little emotional disturbance results. Thus, there is a critical period for learning the sex role, this capacity persisting unchanged up to a point in development which roughly corresponds to the age when children begin to use and understand language. Perhaps more important, this is the age when children first begin to take an interest in, and play with, members of their own age group.

It is difficult to find a common factor in these critical periods for learning. In some species, such as rats, mice, dogs, and sheep, certain critical periods for learning coincide with the period for primary socialization and seem to be similarly brought to a close by the development of fear reactions. Other critical periods, in chaffinches and dogs, coincide with the formation of adult mating bonds. However, the critical period for sexual learning in the rhesus monkey comes later than that for primary socialization (*64*), as do critical periods for various kinds of learning in human beings.

Part of this apparent inconsistency arises from our ignorance regarding timing mechanisms. One such mechanism must be the development of learning capacities, and we have evidence in dogs (*65*), rhesus monkeys (*6*), and human infants (*12*) that learning capacities change during development, sometimes in a stepwise fashion. One element in these capacities is the ability to learn things which facilitate subsequent learning.

It is equally possible, however, to "learn not to learn," and such a negative learning set may act to bring the critical period to a close. At this point, we can only state a provisional general hypothesis: that the critical period for any specific sort of learning is that time when maximum capacities—sensory, motor, and motivational, as well as psychological ones—are first present.

Critical Periods for Early Stimulation

Experiments to determine the effects of early stimulation have been mainly performed on infant mice and rats, which are usually weaned at about 21 days at the earliest, and have been concerned with the effect of stimulation during this pre-weaning period. All investigators beginning with Levine (*67*) and Schaefer (*68*), agree that rats handled during the first week or 10 days of life have a lessened tendency to urinate and defecate in a strange "open field" situation, learn avoidance behavior more readily, and survive longer when deprived of food and water. In short, early stimulation produces an animal that is less timorous, learns more quickly, and is more vigorous. Levine found that the effect could be obtained by a variety of stimuli, including electric shock and mechanical shaking as well as handling. This ruled out learned behavior as an explanation of the effect, and Levine, Alpert, and Lewis (*69*) discovered that animals handled in the early period showed a much earlier maturation of the adrenocortical response to stress. Levine interpreted these results as indicating that the laboratory environment did not provide sufficient stimulation for the proper development of the hormonal systems of the animals. This interpretation is in agreement with Richter's finding (*70*) that laboratory rats are quite deficient in adrenocortical response as compared with the wild variety. Schaefer, Weingarten, and Towne (*71*) have duplicated Levine's results by the use of cold alone, and have suggested temperature as a possible unitary mechanism. However, their findings are not necessarily in disagreement with those of Levine, as the hormonal stress response can be elicited by a variety of stimuli, and temperature may simply be another of the many kinds of stimuli which produce the effect.

According to Thompson and Schaefer (*72*) the earlier the stimulation the greater the effect. If the hormonal mechanism is the chief phenomenon involved, we can say that there is a critical period during the first week or 10 days of life, since the adrenal response in any case matures and becomes fixed by 16 days of age.

Denenberg (*73*) takes a somewhat different approach, pointing out that there should be optimal levels of stimulation, so that either very weak or very strong stimulation would produce poor results. He suggests that there are different critical periods for the effect of early stimulation, depending on the intensity of stimulation and the kind of later behavior measured. Working within the critical first 10 days, Denenberg found that the best avoidance learning was produced by stimulation in the second half of the period, whereas the best survival rates were produced by stimulation in the first half. Weight was approximately equally affected, except that there was little effect in the first 3 days (*74*).

Analyzing the effect on avoidance learning, Denenberg (*75*) and his associates found that both unhandled controls and rats handled for the first 20 days performed poorly, the former because they were too emotional and the latter because they were too calm to react quickly. An intermediate amount of emotional response produces the best learning, and this can be produced by handling only in the first 10 days of life; handling during the second 10 days has a lesser effect. No handling produces too much emotionality, and handling for 20 days results in too little. Irrespective of the effect on learning, the data lead to the important conclusion that emotional stimulation during a critical period early in life can lead to the reduction of emotional responses in later life.

More precisely, there appear to be two critical periods revealed by research on early stimulation of rats, one based on a physiological process (the development of the adrenal cortical stress mechanism) and extending to 16 days of age at the latest, the other based on a psychological process (the reduction of fear through familiarity) (*51*), beginning about 17 days when the eyes first open and extending to 30 days. The effects of handling during these two periods are additive, and many experiments based on arbitrary time rather than developmental periods undoubtedly include both.

The deleterious effects of excessive stimulation in the life of the infant may also be interpreted as a traumatic emotional experience. Bowlby (*76*), in studying a group of juvenile thieves, found that a large proportion of them had been separated from their mothers in early infancy, and he postulated that this traumatic emotional experience had affected their later behavior. Since this conclusion was based on retrospective information, he and his co-workers have since studied the primary symptoms of separation and have described in detail the emotional reactions of infants sent to hospitals, and thus separated from their mothers (*77*). Schaffer (*78*) found a difference in reaction to separation before 7 months and separation afterward. Both sets of infants were disturbed, but they were disturbed in different ways. Infants show increasingly severe emotional reactions to adoption from 3 through 12 months of age (*33*). It seems logical to place the beginning of the critical period for maximum emotional disturbance at approximately 7 months—at the end of the critical period for primary socialization, which Gray (*79*) places at approximately 6 weeks to 6 months. Infants whose social relationships have been thoroughly established and whose fear responses toward strangers have been fully developed are much more likely to be upset by changes than infants in which these relationships and responses have not yet been developed.

However, not all apparently "traumatic" early experiences have such a lasting effect. Experimental work shows that young animals have a considerable capacity to recover from unpleasant emotions experienced in a limited period in early life (*80*), and that what is traumatic in one species may not be in another. While young rats become calmer after infantile stimulation, young mice subjected to excessive auditory stimulation later become more emotional (*81*). At this point it is appropriate to point out that critical periods are not necessarily involved in every kind of early experience. Raising young chimpanzees in the dark produces degeneration of the retina, but this is a long and gradual process (*82*).

Another approach to the problem is to stimulate emotional

responses in mothers and observe the effect on the offspring. Thompson (*83*) and other authors (*84*) have shown that the offspring of rats made fearful while pregnant are more likely to be overemotional in the open-field situation than the offspring of animals not so stimulated. Since any direct influence of maternal behavior was ruled out by cross-fostering experiments, it seems likely that the result is produced by modification of the adrenocortical stress mechanism—in this case, by secretion of maternal hormones acting on the embryo rather than by stimulation after birth of the young animal itself. No precise critical period for the effect has been established, but it is probably confined to the latter part of pregnancy. Similar effects have been obtained in mice (*85*), and if such effects can be demonstrated in other mammals, the implications for prenatal care in human beings are obvious.

It is interesting to note that, whereas shocking the mother both before and after parturition has the effect of increasing emotional responses in the young, the emotional responses of young rats are *decreased* when the treatment is applied directly to them. The explanation of this contradiction must await direct experiments on the endocrine system.

General Theory of Critical Periods

There are at least two ways in which experience during critical periods may act on behavioral development. The critical period for primary socialization constitutes a turning point. Experience during a short period early in life determines which shall be the close relatives of the young animal, and this, in turn, leads the animal to develop in one of two directions—the normal one, in which it becomes attached to and mates with a member of its own species, or an abnormal one, in which it becomes attached to a different species with consequent disrupting effects upon sexual and other social relationships with members of its own kind.

The concept of a turning point applies equally well to most examples of critical periods for learning. Up to a certain point in development a chaffinch can learn several varieties of song, but once it has learned one of them it no longer has a choice. Similarly, the human infant can learn either sex role up to a certain age, but once it has learned one or the other, changing over becomes increasingly difficult. What is learned at particular points limits and interferes with subsequent learning, and Schneirla and Rosenblatt (*86*) have suggested that there are critical stages of learning—that what has been learned at a particular time in development may be critical for whatever follows.

A second sort of action during a critical period consists of a nonspecific stimulus producing an irrevocable result, not modifiable in subsequent development. Thus, almost any sort of stimulus has the effect of modifying the development of the endocrine stress mechanism of young rats in early infancy.

Is there any underlying common principle? Each of these effects has its counterpart in embryonic development. Up to a certain point a cell taken from an amphibian embryo and transplanted to a new location will develop in accordance with its new environment. Beyond this turning point it develops in accordance with its previous location. Some cells retain a degree of lability, but none retain the breadth of choice they had before. Similarly, specific injuries produced by nonspecific causes are also found in embryonic development: damage to an embryonic optic vesicle results in a defective eye, no matter what sort of chemical produces the injury. It is obvious that the similarity between this case and the critical period for early stimulation can be accounted for by the single common process of growth, occurring relatively late in development in the case of the endocrine stress mechanism and much earlier in the development of the eye. The effects are nonspecific because of the fact that growth can be modified in only very limited ways, by being either slowed down or speeded up.

Both growth and behavioral differentiation are based on organizing processes. This suggests a general principle of organization: that once a system becomes organized, whether it is the cells of the embryo that are multiplying and differentiating or the behavior patterns of a young animal that are becoming organized through learning, it becomes progessively more difficult to reorganize the system. That is, organization inhibits reorganization. Further, organization can be strongly modified only when active processes of organization are going on, and this accounts for critical periods of development.

Conclusion

The concept of critical periods is a highly important one for human and animal welfare. Once the dangers and potential benefits for each period of life are known, it should be possible to avoid the former and take advantage of the latter.

The discovery of critical periods immediately focuses attention on the developmental processes which cause them. As these processes become understood, it is increasingly possible to deliberately modify critical periods and their results. For example, since the development of fear responses limits the period of primary socialization, we can deliberately extend the period by reducing fear reactions, either by psychological methods or by the use of tranquilizing drugs. Or, if it seems desirable, we can increase the degree of dependency of a child or pet animal by purposely increasing his emotional reactions during the critical period. Again, if infantile stimulation is desirable, parents can be taught to provide it in appropriate amounts at the proper time.

Some data suggest that for each behavioral and physiological phenomenon there is a different critical period in development. If this were literally true, the process of development, complicated by individual variability, would be so complex that the concept of critical periods would serve little useful purpose. Some sort of order can be obtained by dealing with different classes of behavioral phenomena. For example, it can be stated that the period in life in which each new social relationship is initiated is a critical one for the determination of that relationship. Furthermore, there is evidence that critical-period effects are more common early in life than they are later on, and that the critical period for primary socialization is also critical for other effects, such as the attachment to particular places (87), and may overlap with a critical period for the formation of basic food habits (88).

We may expect to find that the periods in which actual physiological damage through environmental stimulation is possible will turn out to be similarly specific and concentrated in early life.

A great deal of needed information regarding the optimum periods for acquiring motor and intellectual skills is still lacking. These skills are based not merely on age but on the relative rate of maturation of various organs. Any attempt to teach a child or animal at too early a period of development may result in his learning bad habits, or simply in his learning "not to learn," either of which results may greatly handicap him in later life. In the long run, this line of experimental work should lead to greater realization of the capacities possessed by human beings, both through avoidance of damaging experiences and through correction of damage from unavoidable accidents (89).

References and Notes

1. J. P. Scott, *Comp. Psychol. Monogr. 18*, 1 (1945).
2. A. M. Fielde, *Biol. Bull.* 7, 227 (1904).
3. C. R. Stockard, *Am. J. Anat. 28*, 115 (1921).
4. C. M. Child, *Patterns and Problems of Development* (Univ. of Chicago Press, Chicago, 1941).
5. K. Lorenz, *J. Ornithol. 83*, 137, 289 (1935).
6. M. B. McGraw, in *Manual of Child Psychology*, L. C. Carmichael, Ed. (Wiley, New York, 1946), pp. 332-369.
7. E. H. Hess, in *Nebraska Symposium on Motivation* (Univ. of Nebraska Press, Lincoln, 1959), pp. 44-77.
8. H. Moltz, *Psychol. Bull.* 57, 291 (1960); J. L. Gewirtz, in *Determinants of Infant Behaviour*, B. M. Foss, Ed. (Methuen, London, 1961), pp. 213-299.
9. J. P. Scott, in *Social Behavior and Organization in Vertebrates*, W. Etkin, Ed. (Univ. of Chicago Press, Chicago, in press).
10. M. M. Nice, *Trans. Linnaean Soc. N.Y. 6*, 1 (1943).
11. J. P. Scott and M. V. Marston, *J. Genet. Psychol.* 77, 25 (1950).
12. J. P. Scott, *Child Develop. Monogr.*, in press.
13. A. J. Brodbeck, *Bull. Ecol. Soc. Am. 35*, 73 (1954).
14. W. C. Stanley, private communication (1962).
15. O. Elliot and J. A. King, *Psychol. Repts. 6*, 391 (1960).
16. A. E. Fisher, thesis, Pennsylvania State Univ. (1955).

17. O. Elliot and J. P. Scott, *J. Genet, Psychol. 99,* 3 (1961).
18. J. P. Scott, D. Deshaies, D. D. Morris, "Effect of emotional arousal on primary socialization in the dog," address to the New York State Branch of the American Psychiatric Association, 11 Nov. 1961.
19. H. Harlow, *Am. Psychologist 13,* 673 (1958).
20. G. J. Igel and A. D. Calvin, *J. Comp. Physiol. Psychol. 53,* 302 (1960).
21. H. F. Harlow and M. K. Harlow, personal communication (1962).
22. E. Fabricius, *Acta Zool. Fennica 68,* 1 (1951).
23. N. Collias, in *Roots of Behavior,* E. L. Bliss, Ed. (Harper, New York, 1962), pp. 264-273.
24. E. H. Hess, *Ann. N.Y. Acad. Sci. 67,* 724 (1957); in *Drugs and Behavior,* L. Uhr and J. G. Miller, *Eds.* (Wiley, New York, 1960), pp. 268-271.
25. H. James, *Can. J. Psychol. 13,* 59 (1959).
26. P. H. Gray, *Science 132,* 1834 (1960).
27. P. Guiton, *Animal Behavior 9,* 167 (1961).
28. G. F. Pitz and R. B. Ross, *J. Comp. Physiol. Psychol. 54,* 602 (1961).
29. E. H. Hess, "Influence of early experience on behavior," paper presented before the American Psychiatric Association, New York State Divisional Meeting, 1961.
30. H. Moltz, L. Rosenblum, N. Halikas, *J. Comp. Physiol. Psychol. 52,* 240 (1959).
31. G. Gottlieb, *ibid. 54,* 422 (1961).
32. J. P. Scott, *Psychosomat. Med. 20,* 42 (1958).
33. B. M. Caldwell. *Am. Psychol. 16,* 377 (1961).
34. R. A. Hinde, W. H. Thorpe, M. A. Vince, *Behaviour 9,* 214 (1956).
35. D. G. Freedman, J. A. King, O. Elliot, *Science 133,* 1016 (1961).
36. H. F. Harlow and R. R. Zimmermann, *ibid. 130,* 421 (1959).
37. D. G. Freedman, *J. Child Psychol. Psychiat. 1961,* 242 (1961).
38. R. A. Spitz, *Intern. J. Psychoanalysis 31,* 138 (1950).
39. N. E. Collias, *Ecology 37,* 228 (1956).
40. W. Craig, *J. Animal Behavior 4,* 121 (1914).
41. J. L. Fuller and A. Christake, *Federation Proc. 18,* 49 (1959).
42. J. L. Fuller, private communication.
43. F. Schutz, private communication.
44. C. C. Warriner, thesis, Univ. of Oklahoma (1960).

45. R. K. Enders, *Sociometry 8*, 53-55 (1945).
46. P. H. Klopfer, *Behavioral Aspects of Ecology* (Prentice-Hall, New York, in press).
47. M. B. McGraw, *Growth: a Study of Johnny and Jimmy* (Appleton-Century, New York, 1935).
48. D. O. Hebb, *Am. Psychologist 2*, 306 (1947).
49. D. G. Forgays and J. W. Forgays, *J. Comp. Physiol. Psychol. 45*, 322 (1952).
50. D. G. Forgays, "The importance of experience at specific times in the development of an organism," address before the Eastern Psychological Association (1962).
51. D. K. Candland and B. A. Campbell, private communication (1962).
52. L. Bernstein, *J. Comp. Physiol. Psychol. 50*, 162 (1957).
53. W. E. D. Scott, *Science 14*, 522 (1901).
54. W. H. Thorpe, in *Current Problems in Animal Behaviour*, W. H. Thorpe and O. L. Zangwill, Eds. (Cambridge Univ. Press, Cambridge, 1961).
55. J. A. King and N. L. Gurney, *J. Comp. Physiol. Psychol. 47*, 326 (1954).
56. J. A. King, *J. Genet. Psychol. 90*, 151 (1957).
57. J. L. Fuller, "Proceedings, International Psychiatric Congress, Montreal," in press.
58. N. Tinbergen, *The Study of Instinct* (Oxford Univ. Press, Oxford, 1951).
59. J. A. King, *J. Genet. Psychol. 88*, 223 (1956).
60. F. A. Beach, *ibid. 60*, 121 (1942).
61. E. S. Valenstein, W. Riss, W. C. Young, *J. Comp. Physiol. Psychol. 47*, 162 (1954).
62. H. Nissen, *Symposium on Sexual Behavior in Mammals, Amherst, Mass.* (1954), pp. 204-227.
63. J. Money, J. G. Hampson, J. L. Hampson, *Arch. Neurol. Psychiat. 77*, 333 (1957).
64. H. Harlow, in *Determinants of Infant Behaviour*, B. M. Foss, Ed. (Wiley, New York, 1961), pp. 75-97.
65. J. L. Fuller, C. A. Easler, E. M. Banks, *Am. J. Physiol. 160*, 462 (1950); A. C. Cornwell and J. L. Fuller, *J. Comp. Physiol. Psychol. 54*, 13 (1961).
66. H. F. Harlow, M. K. Harlow, R. R. Rueping, W. A. Mason, *J. Comp. Physiol. Psychol. 53*, 113 (1960).
67. S. Levine, J. A. Chevalier, S. J. Korchin, *J. Personality 24*, 475 (1956).

68. T. Schaefer, thesis, Univ. of Chicago (1957).
69. S. Levine, M. Alpert, G. W. Lewis, *Science 126*, 1347 (1957).
70. C. P. Richter, *Am. J. Human Genet. 4*, 273. (1952).
71. T. Schaefer, Jr., F. S. Weingarten, J. C. Towne, *Science 135*, 41 (1962).
72. W. R. Thompson and T. Schaefer, in *Functions of Varied Experience*, D. W. Fiske and S. R. Maddi, Eds. (Dorsey, Homewood, Ill., 1961), pp. 81-105.
73. V. H. Denenberg, in *The Behaviour of Domestic Animals*, E. S. E. Hafez, Ed. (Bailliere, Tindall and Cox, London), 109-138.
74. ———, *J. Comp. Physiol. Psychol.*, in press.
75. ——— and G. G. Karas, *Psychol. Repts. 7*, 313 (1960).
76. J. Bowlby, *Intern. J. Psychoanalysis 25*, 19, 107 (1944).
77. C. M. Heinicke, *Human Relations 9*, 105 (1956).
78. H. R. Schaffer, *Brit. J. Med. Psychol. 31*, 174 (1950).
79. P. H. Gray, *J. Psychol. 46*, 155 (1958).
80. M. W. Kahn, *J. Genet. Psychol. 79*, 117 (1951). A. Baron, K. H. Brookshire, R. A. Littman, *J. Comp. Physiol. Psychol. 50*, 530 (1957).
81. G. Lindzey, D. T. Lykken, H. D. Winston, *J. Abnormal Soc. Psychol. 61*, 7 (1960).
82. A. H. Riesen, in *Functions of Varied Experience*, D. W. Fiske and S. R. Maddi, Eds. (Dorsey, Homewood, Ill., 1961), pp. 57-80.
83. W. R. Thompson, *Science 125*, 698 (1957).
84. C. H. Hockman, *J. Comp. Physiol. Psychol. 54*, 679 (1961); R. Ader and M. L. Belfet, *Psychol. Repts. 10*, 711 (1962).
85. K. Keeley, *Science 135*, 44 (1962).
86. T. C. Schneirla and J. S. Rosenblatt, *Am. J. Orthopsychiat. 31*, 223 (1960).
87. W. H. Thorpe, *Learning and Instinct in Animals* (Methuen, London, 1956).
88. E. H. Hess, in *Roots of Behavior*, E. L. Bliss, Ed. (Harper, New York, 1962), pp. 254-263.
89. Part of the research described in this article was supported by a Public Health Service research grant (No. M-4481) from the National Institute of Mental Health.

Of Flies and Men[1]

THEODOSIUS DOBZHANSKY

One of the assertions which have gained acceptance by dint of frequent repetition is that science is competent to deal only with what recurs, returns, repeats itself. To study something scientifically, this something must be made representative of a class, group, or assemblage. A single *Drosophila* fly is of no interest whatsoever. A fly may merit some attention only if it is taken as a representative of its species. An individual person may, to be sure, merit attention. However, it is allegedly not in the province of science, but of insight, empathy, art, and literature to study and understand a person in his uniqueness.

I wish to challenge this view. Individuality, uniqueness, is not outside the competence of science. It may, in fact it must, be understood scientifically. In particular, the science of genetics investigates individuality and its causes. The singularity of the human self becomes comprehensible in the light of genetics. You may, of course, object that what science comprehends is not really a singularity but a plurality of singularities. However, an artist, no less than a biologist, becomes aware of the plurality because he has observed some singularities.

In the main, genetics is a study of differences among living beings. Genetics would be superfluous if all living beings were exactly alike. If all members of a species were exactly alike genetics could do very little. Since Mendel, the most powerful method of genetics is to observe differences among individuals in the progenies of parents which differed in some ways. Heredity and variation are the two sides of the same coin. Ge-

[1] Invited address presented at American Psychological Association, New York, September 1966. Some of the experimental work referred to in the text supported under Contract No. AT-(30-1)-3096, United States Atomic Energy Commission.

From *American Psychologist*, 1967, 22, 41-48. Reprinted by permission of the author and the American Psychological Association.

neticists are always on the lookout for genetic diversity. Variety is said to be the spice of life. It is a staple necessity to geneticists. (This applies, of course, to Mendelian genetics proper. The great discoveries of the role of chromosomes in the development, and the relationships between DNA, RNA, and protein synthesis could conceivably have been made even if Mendel's laws remained unknown).

That every person differs from every other person is so obvious that this is taken usually for granted. What continues controversial is to what extent the human differences are due to genetic and in what measure to environmental variations. Though in a new guise, the old nature-nurture problem is still with us. Now, the individuality of flies is rather less evident than human individuality. I do not claim to recognize every *Drosophila* by her face. The drosophiline individuality is nevertheless easier to analyze, and this analysis helps to throw some needed light on human individuality.

The theory of genetic individuality is simple enough. It stems directly from Mendel's second law, the law of independent assortment. An individual heterozygous for *n* genes has the potentiality of producing 2*n* genetically different kinds of sex cells. Two parents, each heterozygous for the same *n* genes, can give rise to 3*n* genotypes among the progeny, and parents heterozygous each for *n* different genes may produce 4*n* genotypes. To be sure, not all of these genotypes are equally probable, because the linkage of genes in the same chromosome limits their independent assortment. Linkage disequilibrium delays but does not prevent eventual realization of the genetic variety. More important is the problem how large is *n*, that is, for how many genes an average individual is heterozygous, or how many genes are represented each by two or more variants in the populations of a species, such as man or a *Drosophila*.

The disagreement among geneticists on this point is rife. Those who espouse the classical theory of population structure believe that most genes are uniform, not only in all individuals of a species but even in different species not too remote in the biological system. The unfixed genes are a minority, perhaps of the order of some tens. Moreover, among the unfixed genes one variant, one allele, is normal and adaptively superior, while others are inferior and are maintained in populations by recurrent mutation. Though adherents of the classical theory are reluctant to admit this, the theory is a product of typological thinking. Lurking behind the facade of the variability, they like to envisage the Platonic archetype of the Normal Man, homozygous for all good with no bad genes.

The balance theory of population structure would assume numbers of variable genes of the order of hundreds, perhaps even thousands. An appreciable part of this variety is main-

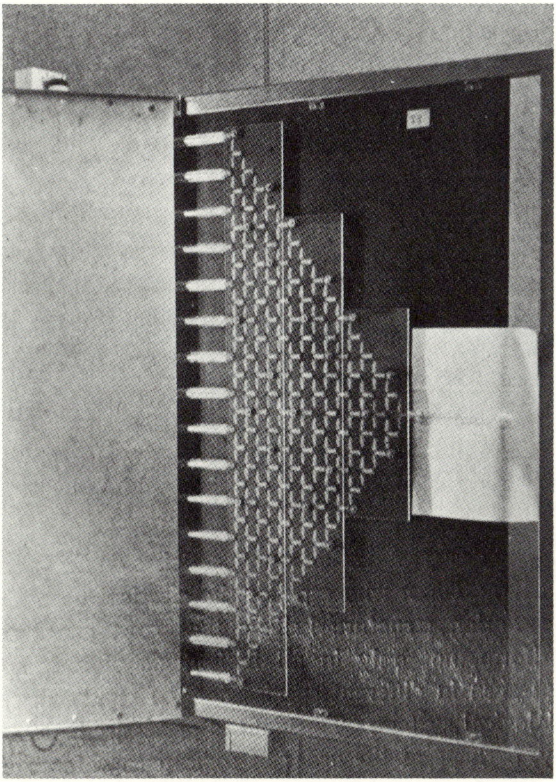

Fig. 1. The classification maze for selection of *Drosophila* flies for geotactic behavior. (The starting tube is on the right. The terminal tubes are on the left, No. 1 being the uppermost, and No. 16 the lowermost.)

tained in populations by several kinds of balancing natural selection. The kind most often discussed is the heterotic selection, operating because of hybrid vigor. There are, indeed, genetic variants which are adaptively favorable when heterozygous and unfavorable when homozygous. The gene which is homozygous condition causes sickle-cell anemia in man is a classical example; in heterozygous condition it confers a relative immunity to *falciparum* malaria. Perhaps even more important in evolution is diversifying natural selection. This can be explained most simply by pointing out that every living species faces not just one environment but a variety of environments. Human environments are certainly diverse, and moreover the diversity is growing. It is improbable that genes can be found to show optimal performance in all environments. More likely, different genes will be relatively more adaptive in different environments. Genetic variety is a method to cope with variety of environments.

Theoretical arguments cannot settle the questions for how many genes is an average individual heterozygous, and what proportion of the genes are represented by different alleles in different individuals of a species. Geneticists are busy working on these matters. I can cite here only the brilliant work of Lewontin and Hubby (1966), of the University of Chicago, as an example. Since the total number of genes is unknown, but is surely too large to have the whole set examined one by one, Lewontin and Hubby have decided to study what they believe is a random sample of genes. They chose a battery of enzymes that can be detected by electrophoresis in single individuals of the fly, *Drosophila pseudoobscura.* Some of these enzymes did and others did not show detectable genetic variations. The authors, after making a thorough examination of the possible biases and pitfalls, came to the conclusion that an individual fly was heterozygous for on the average between 10% and 15% of the genes in their sample. The numbers of kinds of genes in a sex cell can hardly be less than 10,000; an average fly may, then, be heterozygous for a number of genes of the order of 1,000.

Do these results have any bearing on man? Although man is not an overgrown *Drosophila,* he must have as many or more genes than *Drosophila* does. If the degree of heterozygosity in man is anything like it is in *Drosophila,* brothers or sisters are quite unlikely to inherit from their parents the same genes. The likelihood that any two unrelated persons are genetically identical is practically nil. Only identical twins may be genetically identical, since they arise by asexual division of a sexually produced fertilized egg. Even there the possibility of mutation and of cytoplasmic difference must be reckoned with. Human nature is, then, not unitary but multiform; the number of human natures is almost as great as the number of humans. Every person is unique, unprecedented, and unrepeatable.

The demonstration of the genetic uniqueness of individuals only opens, rather than solves, the problem as far as behavioral

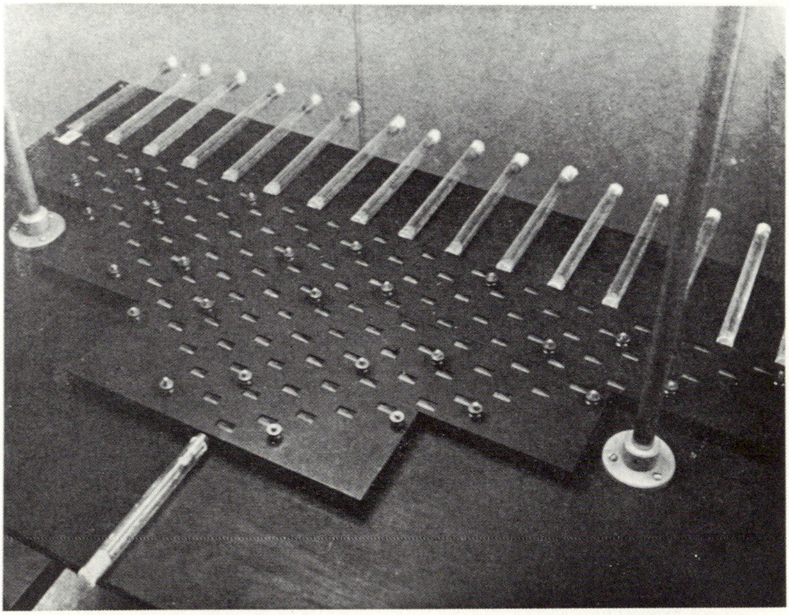

Fig. 2. The classification maze for selection of *Drosophila* flies for phototactic behavior. (The starting tube is on the lower left. The terminal tube No. 1, shown on the upper left, is reached by 15 dark passages. The terminal tube No. 16, the extreme right is reached by 15 light passages.)

and social sciences are concerned. There seems little point in belaboring the truism that behavior as such is not inherited. Only genes can be inherited, in the sense of being handed down from parents to offspring. Even so, I have mostly division products, true copies of the genes I have inherited from my parents, rather than these genes themselves. The skin color is not inherited either, because the skin pigment is not carried in the sex cells. However, I am yet to meet anybody who would contend that one's genes have nothing to do with one's skin color. Human, as well as animal, behavior is the outcome of a process of development in which the genes and the environment are components of a system of feedback relationships. The same statement can be made equally validly with respect to one's skin color, the shape of one's head, blood chemistry, and somatic, metabolic, and mental diseases.

There are some authors who go so far as to question the existence of problems of genetics of behavior, distinct from genetics of anything else. They are right only inasmuch as there is not likely to exist a special brand of DNA concerned with behavior, different from that in other kinds of genes; moreover there are no genes "for" behavior, as there are no genes "for" the shape of one's nose. The problem is more subtle. It is the problem, or rather problems, of the genetic architecture of behavioral differences. We want to know how many genes are usually involved in such differences, the magnitude of their effects, the nature of their interactions, the parts played by mutation pressure, hybrid vigor, environmental heterogeneities, and by all forms of natural selection in the formation, in maintenance, and in normal and pathological variations of behavior. In this sense, the genetics and the evolution of behavior may well be different from, let us say, the genetics and the evolution of blood chemistry, or of metabolism, or of chromosomal polymorphism, or of concealing colorations, or mimetic resemblances. And in this sense, which is the only meaningful sense, the genetics of behavior, especially the experimental genetics of behavior, is not yet even a fledgling field, although it has recently begun to chirp rather lively.

In this article I can discuss only one example of a study of genetics of behavior, that made by my colleague B. Spassky

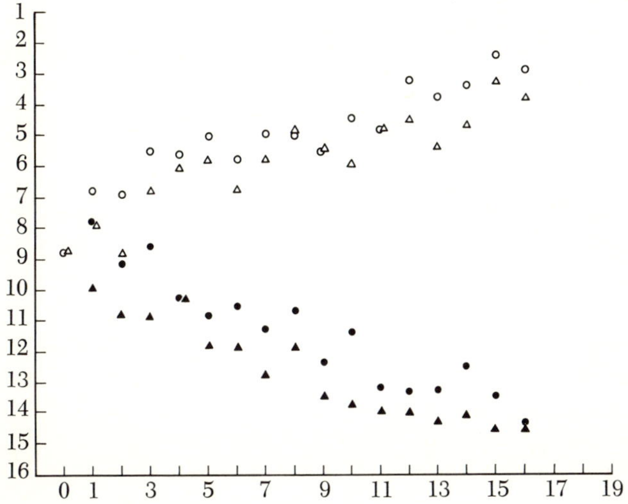

Fig. 3. Selection for negative (light symbols) and positive (dark symbols) phototaxis. (Circles—females, triangles—males. Abscissa—generations of selection, ordinate—the phototactic score.)

and myself on phototaxis and geotaxis in *Drosophila pseudoobscura*. Hirsch and his students (Erlenmeyer-Kimling, Hirsch, & Weiss, 1962; Hirsch, 1962; Hirsch & Erlenmeyer-Kimling, 1962) have constructed a classification maze (Figure 1), and selected populations of *Drosophila melanogaster* which were clearly positively and others negatively geotactic in their behavior. They showed furthermore that the genetic basis of this behavior was polygenic, the three large pairs of chromosomes all influencing the result. Hadler (1964a, 1964b) made a similar maze for selection for phototaxis (Figure 2), and succeeded in obtaining positively and negatively phototactic strains of *Drosophila melanogaster*. Dobzhansky and Spassky (1962), using Hirsch's maze, selected positively and negatively geotactic strains of *Drosophila pseudoobscura*. Their starting population was polymorphic for some inverted sections in the third chromosomes, and one of the variant chromosomes proved to favor negative geotaxis, while chromosomal heterozygosis favored positive geotaxis.

The results of newer experiments on selection for positive

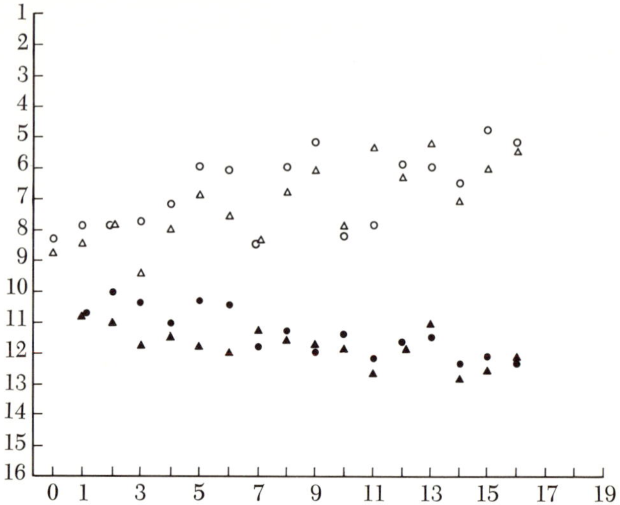

Fig. 4. Selection for negative (light symbols) and positive (dark symbols) geotaxis. (Circles—females, triangles—males. Abscissa—generations of selection, ordinate—the geotactic score.)

and negative phototaxis and geotaxis in *Drosophila pseudoobscura* are presented in Figures 3 and 4. The ordinates show the phototactic or the geotactic scores, i.e., the averages of the 16 terminal tubes of the mazes into which the flies distribute themselves. On the geotaxis maze the tube No. 1 is the uppermost and No. 16 the lowermost, on the phototaxis maze No. 1 is reached by 15 choices of dark passages and No. 16 by 15 choices of light passages. The selection is made by running through the maze 300 females or 300 males; the 25 most positive, or most negative, individuals of each sex are selected to be the parents of the next generation. The initial populations in our experiments were photo- and geotactically neutral on the average. Or, to be more precise, these initial populations had positive, neutral, and negative individuals in such proportions that the average scores were between 8 and 9 (an average of 8.5 is exact neutrality). After 15 generations of selection, the positively phototactic line had average scores 13.4 and 14.5 for females and males respectively, the negatively phototactic line 2.4 and 3.1, the positively geotactic line 12.1 and 12.5, and the negatively geotactic line 4.7 and 6.1. The frequency distributions overlap only slightly in the middle, i.e., only few flies of the selected strains end up in the terminal tubes Nos. 8 and 9.

Is it, then, the heredity which makes a *Drosophila* walk towards lights or darkness, climb up or descend? Even with flies, not only with men, the situation is more complex than that. From the effects of the selection in the first generation, the heritability of the photo- and geotactic responses can be calculated to lie between .15 and .20. This is somewhat oversimplifying the issue, but one can say that, as a first approximation, the genetic component of the behavior of the fly in our mazes is only 15% to 20%, while random chance and environment is responsible for 80% to 85%. Nor is this all. Taking the data for the 15 generations of selection as a whole, we can compute the so-called realized heritability, that is to say the efficiency of the response to the selection. This turns out to be very small, only about 9% for the phototaxis, and only about 3% for the geotaxis. In other words, a prediction of what the selection could accomplish in 15 generations, based on the initial heritability figure, would be a gross overestimate. There are several factors responsible for this situation, among which I shall single out just one, which seems most interesting.

In our first experiments (Dobzhansky & Spassky, 1962) we made selection in three populations of *Drosophila pseudoobscura* during 18 generations for positive and for negative geotaxis. After the positive and the negative populations have diverged about as much as the populations on Figure 4, the populations were split each into two. In one member of each pair the selection was reversed, i.e., a population formerly selected for the positive was now selected for a negative geotaxis, and vice versa. In another subpopulation the selection was relaxed, i.e., the subpopulation was propagated without selecting either the positive or the negative individuals. The selective gains obtained through 18 generations of the original selection were almost erased in 6 generations of the reverse selection. The simple relaxation of the selection resulted in a loss of about half of the selection gains.

A partial, or even complete, loss upon abandonment of selection of what had been gained by previous selection is a phenomenon well known to breeders of agricultural plants and animals. Lerner (1954) has called this the genetic homeostasis. Very simply, the average height, weight, speed of maturation, and many other characteristics of a population which are determined by cooperation of numerous polygenes, are held by natural selection at levels near optimal for the population in the environments in which that population usually lives. When a breeder selects toward higher or toward lower levels of certain characteristics, he does so for his benefit, not necessarily for the benefit of the animal or the plants in its original environments. In other words, the artificial selection is often pitted against natural selection. As the artificial selection progresses it becomes more and more frustrated by natural selection. When the artificial selection is stopped, natural selection is given an opportunity to undo what the artificial one had gained; and reverse selection is highly effective because the artificial and the natural selections then work in the same direction, in alliance rather than in opposition.

Biologically, adaptively, this is an excellent strategy for evolution to follow. It combines high adaptedness to the existing environmental conditions with high adaptability to environmental changes. This strategy is, however, not at all what the classical theory of genetic population structure envisages. If the environment were uniform, constant, and favoring phototactically and geotactically neutral *Drosophilae,* then the simplest solution of the adaptive problem would seem to be to make all members of the species homozygous for the genes favoring photo- and geotactic neutrality. "Normal" or "typical" flies would then be neutral, and positive and negative ones would be abnormal or atypical. But this is not what is observed. The populations, though neutral on the average, contain also positive and negative genetic variants.

The availability in the populations of this genetic variance confers upon them evolutionary plasticity. A change in the environment that favors a positive or a negative photo- or geotaxis makes the population respond rapidly by adaptive genetic changes. Such responses might occur also in a genetically uniform and homozygous population, but they would be much slower. They would have to wait for the occurrence of mutations. These mutations would have to produce genetic variants which were unfavorable in the old but adaptive in the new environments. The rapidity of the genetic adjustment is, however, not the whole story. A genetically polymorphic population not only responds adaptively to environmental challenges, but in so doing it does not, so to speak, burn the bridges for retreat. It is hedged against the contingency that the environmental change to which it is adapting may only be a temporary one. If it is indeed temporary, and the original environment returns, the population can readapt itself speedily, by returning to its former genetic composition.

And yet genetic homeostasis does not stand in the way of permanent, irreversible, progressive evolutionary changes. If a new environment or a new way of life endures, a new genetic system becomes stabilized. This genetic system will be buffered against the vagaries of the new environments, but no longer able to retrace its steps to the conditions of the bygone age. If these conditions returned, the species would probably become adapted to them in some new way. One of the most interesting lessons that evolutionary biology teaches us is that there may be many more than a single method to eke out a living from an environment. Major evolutionary changes are irreversible and unrepeatable.

This point is so central that it must be reiterated: Man is not just an overgrown *Drosophila*. We reject the belief that man is nothing but an animal. Yet he is, among other things, also an animal. Like *Drosophila*, he is a sexually reproducing, outbreeding species, and his populations are abundantly provided with genetic variability. The genetic diversity affects all kinds of traits—morphological, physiological, and behavioral. The discrete, clear-cut, and usually pathological genetic variations of behavior, such as the so-called Mongoloid idiocy or phenylketonuria, need not be considered in the present discussion. The genetic variations among healthy persons are no less interesting, though much harder to study. The same situation exists also in *Drosophila:* sharp, easily distinguishable, and poorly viable mutants of classical genetics, versus slight, quantitative, polygenic variations. The difficulty, in human as well as in *Drosophila* genetics, arises because in the phenotypic variance of the second kind of traits the genetic and the environmental influences are intermingled.

Neither in the most highly selected, nor in the unselected, photo- or geotactic lines of *Drosophila* is the behavior of an individual rigidly determined. We have seen that the heritability of these behavioral traits is rather low. Whether at a given point of the maze an individual climbs upwards or downwards, takes a light or a dark passage, is in part a matter of environment, or simply of chance. The evidence is nevertheless conclusive that the genotype does bias the choices. Some flies are inclined to walk more often upward and others downward. Are the behavioral traits in human populations also conditioned by genetic variations? I shall be among the first to insist that the evidence is incomplete, and that more data must be collected. Yet the existing evidence, for a variety of traits ranging from IQ measurements to smoking habits, indicates that at least some genetic conditioning is involved, of course relatively more for some traits and less for others.

It is no secret that the study of the genetic conditioning of human behavior is hampered by the emotional reactions which this issue elicits in many people. Some wish to give an aura of scientific respectability to their race and class biases. Differences in material well-being and in social position are represented as just and necessary outcomes of the genetic differences. Others cling obstinately to the old tabula rasa theory. Man is a product of his environment and social conditions, and his genes are simply irrelevant. I submit that, irrespectively of your preconceptions, modern biology makes it necessary to state the problems of genetic conditioning of behavior in terms rather different from the traditional ones. This is because one of the most significant changes in the biological theory in the recent decades has been a shift from typological to populational models and concepts. This conceptual reformation has been discussed with admirable clarity and discernment, particularly by Simpson (1961) and by Mayr (1963), making it possible to state what is essential for us here very briefly.

To a typologist, what is real and important is the species or the race to which an organism belongs. Differences among individuals of the same species and race are, of course, too obvious to be denied. A typologist regards them, however, as merely a kind of troublesome noise in the biological system. He tries, as it were, to recognize the melody obfuscated by the noise; he seeks to identify, classify, and name the species and the races. He hopes that once he can determine to which species and race an individual belongs, that individual is thereby adequately described.

A populationist, on the contrary, regards the individuals and their diversity as the prime observable reality. The biological validity of species and races is not thereby refuted (although some extremists, try to do just that, in my opinion ill advisedly). Species and races are, however, derivative from individuals, not the other way around. Species and races are Mendelian populations, reproductive communities of sexually reproducing organisms, forms of adaptive ordering of systems of individuals, evolved because they have made the evolutionary feedback processes between the organisms and their environments most efficient and successful.

Man in the street is a spontaneous typologist. To him, all things which have the same name are therefore alike. All men have the human nature, and an alleged wisdom has it that the human nature does not change. All Negroes are alike because of their negritude, and all Jews are alike because of their jewishness. Populationists affirm that there is no single human nature but as many human natures as there are individuals. Human nature does change. Race differences are compounded of the same ingredients as differences among individuals which compose a race. In fact, races differ in relative frequencies of genes more often than they differ qualitatively, one race being homozygous for a certain gene and the other lacking it entirely. The extremists who deny that races exist are disappointed typologists who have discovered for themselves the gene gradients between race populations. They fail to understand that such gradients elucidate the nature of race as a biological phenomenon; the facts warrant the conclusion that Platonic types of races do not exist, not that races do not exist.

The typological and populational operational approaches

are characteristically different. A race of typology is described in terms of means or averages of height, weight, cephalic index, intelligence, etc. Populationists regard variances at least as important as means. Genetic variance characterizes not only the status but also the evolutionary possibilities of a population. The *Drosophila* populations with which Hirsch and his colleagues as well as ourselves began our experiments were photo- and geotactically neutral on the average. Yet the experiments have shown that the average neutrality did not mean that all individuals were neutral. Selection has attested the presence in the populations of genetic elements for positive and negative photo- and geotaxis. This does not quite mean that the original populations contained individuals as sharply positive and negative as are individuals of the selected strains. Natural and artificial selection do not act as simple sieves which isolate genotypes which were there before selection. Selection creates novelty, because it compounds genotypes the origin of which without selection would be altogether improbable.

All human beings have certain universally recognized rights because they are members of the species Homo sapiens. Members of other species do not have the same rights. Cows are sacred to Hindus, but even in India cows are not treated exactly like humans. An imaginative French writer, Vercors, has given a thought-provoking discussion of legal and other problems that might arise if a hybrid of man and some anthropoid species were produced. Anyway, membership in a group, be that a species or a race, does not define all the characteristics of individuals. The notion that it does is implicit in race pride, exclusiveness, and bias.

Racists busy themselves attempting to scrape up any kind of evidence that Race X has a lower mean IQ, or smaller mean brain volume, or greater emotionality than Race Y. How large is the genetic component in such differences is questionable. The partitioning of the genetic and environmental variances obtained through studies on monozygotic and dizygotic twins cannot be used as a measure of the genetic and environmental components of the group differences. The basic assumption of the twin method is that the environments of the cotwins are uniform. This is obviously not true when different social classes, castes, and races are compared. Even if we had much more complete data on twins than are actually available, this would still leave the question of the magnitude of the genetic component in the group differences wide open. The argument that about one-half of the interracial variance in IQ must be genetic because this appears to be so among co-twins is a misinterpretation when it is not an intentional obfuscation.

To say that we do not know to what extent group differences in psychological traits are genetic is not the same as saying that the genetic components does not exist. It is a challenge to find out. If individuals within populations vary in some character, be that blood grouping, or stature, or intelligence, it is quite unlikely that the population means will be exactly the same. What matters is how great is the intrapopulational variance compared to the interpopulational variance. This is different for different characters. Skin pigmentation is individually variable in probably all races, but the interracial variance is evidently larger. Although precise data are not available, it is at least probable that the relation is reversed for psychological traits. In simplest terms, the brightest individuals in every class, caste, and race are undoubtedly brighter than the average in any other class, caste, or race. And vice versa—the dullest individuals in any of these groups are duller than the average of any group. There are sound biological reasons why this should be so. Very briefly, in the evolution of mankind the natural selection has worked, nearly always and everywhere, to increase and to maintain the behavioral plasticity and diversity, which are essential in all human cultures, primitive as well as advanced.

True enough, an individual taken from a population with a higher mean of some trait, say a higher intelligence, has a higher statistical probability to possess this trait more developed than an individual from a population with a lower mean. When we select *Drosophilae* for stronger or weaker photo- or geotaxis, we generally breed the high and the low selection lines separately. Spassky and myself have, however, some experiments in progress, in which pairs of populations exchange migrants in every generation. The migrants are selected for high or for low photo-, or geotaxis or for some other genetically conditioned trait. This may be considered to represent to some extent an experimental simulation of social mobility in human populations. The preliminary results of these experiments are at least to us, fascinating. Genetically selective social mobility seems to be a powerful evolutionary agent.

A day may conceivably arise when mankind will embark on some all-out eugenical breeding program. This day is not yet in sight, because mankind has not reached a level of wisdom when it could decide with anything approaching unanimity what combination of genetic qualities should the ideal man have. It is rather easier to agree what qualities he should not have. As for positive ideals, we can only recommend that a diversity of tastes, preferences, abilities, and temperaments should be preserved and perhaps even increased. Anyway, when we consider the social implications of the human genetic diversity we are not usually preoccupied with eugenical breeding programs. The genetic diversity is, for example, most relevant to educational problems. The students are, however, selected for study, not for stud.

Insofar as the genetic component is concerned, the intelligence, or temperament, or special abilities of the parents have little predictive value for these qualities in an individual child. This does not mean that such genetic components do not exist, as some authors have overhastily concluded. It means two things. First, the heritability is fairly low, as it is low in the photo- and geotactic behavior of our flies. In other words, the environmental variance is high, and in man the parent-offspring similarities in behavioral traits may well be due more to the cultural than to the biological inheritance. Second, one cannot too often be reminded of the fact that we do not inherit the genotypes of our parents but only one half of their genes. The genes do not produce their effects in development each independently of the others. The genes interact; the genetic "nature" of an individual is an emergent product of the particular pattern or constellation of the genes he carries. This is often the reason why a child is sometimes so strikingly dissimilar to his parents in some traits, even if the environment is kept constant.

How can I summarize the contents of this article, which is itself a summary of thinking concerning a variety of issues? Perhaps the best way is to say that genetics bears out John Deweys' emphasis of "the infinite diversity of active tendencies and combinations of tendencies of which an individual [human] is capable."

References

Dobzhansky, T., & Spassky, B. Selection for geotaxis in monomorphic and polymorphic populations of *Drosophila pseudoobscura*. *Proceedings of the National Academy of Science*, 1962, *48*, 1704-1712.

Erlenmeyer-Kimling, L., Hirsch, J., & Weiss, J. M. Studies in experimental behavior genetics. III. Selection and hybridization analyses of individual differences in the sign of geotaxis. *Journal of Comparative and Physiological Psychology*, 1962, *55*, 722-731.

Hadler, N. Genetic influence on phototaxis in *Drosophila melanogaster*. *Biological Bulletin*, 1964, *126*, 264-273. (a)

Hadler, N. Heritability and phototaxis in *Drosophila melanogaster*. *Genetics*, 1964, *50*, 1269-1277. (b)

Hirsch, J. Individual differences in behavior and their genetic basis. In E. L. Bliss, *Roots of behavior*. New York: Harper, 1962. Pp. 3-23.

Hirsch, J., & Erlenmeyer-Kimling, L. Studies in experimental behavior genetics. IV Chromosome analyses for geotaxis. *Journal of Comparative and Physiological Psychology*, 1962, *55*, 732-739.

Lerner, I. M. *Genetic homeostasis*. Edinburgh & London: Oliver & Boyd, 1954.

Lewontin, R. C., & Hubby, J. L. A molecular approach to the study of genic heterozygosity in natural populations. *Genetics*. 1966, *54*, 595-609.

Mayr, E. *Animal species and evolution*. Cambridge: Harvard University Press, 1963.

Simpson, G. G. *Principles of animal taxonomy*. New York: Columbia University Press, 1961.

Index